Construction
Change Orders

Construction Change Orders

Impact, Avoidance, Documentation

James J. O'Brien, P.E.
O'Brien Kreitzberg
A Dames & Moore Group Company
Lawrenceville, New Jersey

McGraw-Hill
New York   San Francisco   Washington, D.C.   Auckland   Bogotá
Caracas   Lisbon   London   Madrid   Mexico City   Milan
Montreal   New Delhi   San Juan   Singapore
Sydney   Tokyo   Toronto

Library of Congress Cataloging-in-Publication Data

O'Brien, James Jerome, date.
    Construction change orders: impact, avoidance, documentation /
James J. O'Brien.
      p.   cm.
   ISBN 0-07-048234-9
   1. Construction industry—Planning. 2. Construction contracts.
3. Construction industry—Specifications. I. Title. II. Title:
Change orders.
TH438.027   1998
624' .068—dc21   98-11752
                                        CIP

McGraw-Hill

A Division of The McGraw·Hill Companies

  2 3 4 5 6 7 8 9 0 FGR/FGR 9 0 3 2 1 0 9 8

ISBN 0-07-048234-9

*The sponsoring editor for this book was Larry S. Hager, the editing supervisor
was Stephen M. Smith, and the production supervisor was Tina Cameron. It
was set in Century Schoolbook by McGraw-Hill's Hightstown, N.J.,
Professional Book Group composition unit.*

Printed and bound by Quebecor / Fairfield.

McGraw-Hill books are available at special quantity discounts to use as
premiums and sales promotions, or for use in corporate training programs. For
more information, please write to the Director of Special Sales, McGraw-Hill,
11 West 19th Street, New York, NY 10011. Or contact your local bookstore.

 This book is printed on recycled, acid-free paper containing a minimum of 50% recycled de-inked fiber.

Contents

Preface

Construction change orders are an unwanted, but inevitable, reality of construction. Causes of change orders include the following:

- Unforeseen conditions
- Plans and/or specifications
- Value engineering
- Acceleration
- Addition and/or enhancement required by owner
- Accident or damage
- Force majeure

This book provides frank, objective views from the various construction vantage points: owner, designer (architect/engineer), project manager/construction manager (PM/CM), general (or prime) contractor, and subcontractor. Of course, each stakeholder sees things through self-serving eyes. For example:

Owner
- Is annoyed at additional costs and/or delay
- Assumes architect at fault
- Feels contractor is "taking over"

Architect
- Tries to limit (or even deny) validity of the change in scope
- Will take measures to avoid liability that can have an impact on the project
- Is concerned that owner (or contractor) may claim the design is deficient

Contractor
- Tries to negotiate to gain profit advantage
- Is concerned that too many changes will cause delay and/or disruption
- Anticipates impacts beyond the specific change itself

Subcontractor
- Same concerns as the contractor
- Sees contractor in role of owner
- Sees owner as remote factor

A defining factor in the form of change order is the type of contract:

- Fixed price, competitive bid
- Unit price
- Negotiated fixed price
- Cost plus fixed fee
- Time and materials

The change order process involves paperwork, negotiations, and approvals, all of which take up time. Typical steps are as follows:

Identify as change
- Request for information
- Unforeseen condition
- Directive by CM or architect
- Value engineering change proposal

Potential change order issues
- Background package assembled
- Estimate by contractor
- Estimate by CM

CM review (agree/disageee)

Negotiations (agree/disagree)

Process paperwork

This book addresses preconstruction methods of avoiding and/or mitigating impacts of changes: exculpatory clauses (such as no-damages-for-delay clauses), notice requirements, potential consequential damages considered, and others limiting the contractor's ability to claim a change order for changes in scope and/or unforeseen conditions.

The impact of change orders is described, including baseline schedules, preparation of time impact evaluations, as-planned and as-built approaches, and time extension requests.

The cost of time is addressed, including general conditions, home office expense, consequential damages, and various inefficiencies. Inefficiencies discussed include crowding, stacking of trades, and overtime.

Theories for contractor's recovery of cost overruns, including constructive change orders, directed acceleration, cardinal change, total cost claims, modified total cost claims, and the "jury method" approach, are also covered.

The approach of this book is to explain the change process, including the various pressures and/or biases, without bias toward (or against) any of the players.

Acknowledgments

- To the American Institute of Architects, for permission to use portions of its changes clause and related documents.
- To the Camden County Municipal Utilities Authority, for the change order example.
- To the Construction Management Association of America, for permission to use portions of its changes clause.
- To McGraw-Hill, for providing the John Doe case study from *CPM in Construction Management* and *Preconstruction Estimating*.
- To O'Brien Kreitzberg, for the daily report examples from its TRACK 5.0.
- To Postner & Rubin, for permission to use a case description from its newsletter.
- To the Southeastern Pennsylvania Transportation Authority, for the RailWorks change management and claims procedures.
- To the U.S. Postal Service, for information from its Contract Administration Course.

James J. O'Brien, P.E.

Change Orders

"In a perfect construction world there would be *no change orders*" (Anonymous). Flash—there is no perfect construction world.

What is a change order? A *change order* is a formal change to the construction contract that usually includes a change in work scope (usually an increase; however, a decrease is also possible). With a change in work scope, there is usually an increase in cost (again, a decrease is possible). Also, with a change in scope, there can be a change in the time to perform the work.

The Right to Make Changes

In virtually all contracts, the owner retains the prerogative of making changes to the work by the use of clauses, such as:

> The Director may, at any time during the progress of the work, authorize additions, deductions, or deviations from the work described in the specifications as herein set forth; and the contract shall not be vitiated or the surety released thereby. . . . Additions, deductions, and deviations may be authorized as follows at the Director's option:
>
> 1. On the basis of unit prices specified.
> 2. On a lump sum basis.
> 3. On a time and material basis.

Further, the contractor must accept this additional work as it is within the context of the contract. Although the terms "additional work" and "extra work" are used synonymously, there is an important legal distinction. Extra work involves the requirement for performance of work entirely independent of the contract. In *Public Construction Contracts and the Law* (McGraw-Hill, 1961), Henry A. Cohen describes extra work as: "The performance of work and the furnishing of the required labor and materials outside and entirely independent of and not necessary to complete the contract or something done or furnished in excess of the requirements of the contract, not contemplated by the parties, and not controlled by the contract" (p. 121). (Citations: Kansas City Bridge Co. v. State, 250 N.W. 343, S.D., 1993; Blair v. United States, 66 F. Supp. 405, Ala., 1946.)

Additional work is the work that can be imposed within the contract documents. It is a change or alteration to the plans or specifications for a number of reasons implicit in the original agreement. These reasons could include, but are not limited to, omissions in the design documents, recognition of better methods or materials to achieve the required effect, resolutions of problems recognized, resolution of unforeseen conditions not anticipated, and similar adjustments within the intent of the original contract.

Within the context of additional work change orders, the owner may impose changes to update the equipment, to recognize different functional requirements, and to otherwise improve upon the design. Changes of this nature could potentially be identified as extra work rather than additional work. As work above and beyond the contract, the contractor may choose to refuse extra work change orders. However, generally speaking, the contractor will prefer to accept both additional work and extra work change orders if they do not obviously impede progress in the field, and if they can be negotiated at an equitable figure. In negotiating change orders, the principal focus is upon units of work involved and the cost per unit. Where change orders delete some contract work, replacing it with other work, the identification of the exact scope of the change order can be very difficult.

Origins of Change Orders

There will be occasions where the contract does not suffice to cover either work conditions or the desired scope of work. Identification of these change conditions may emanate from the architect/engineer, field engineer at the job site (unforeseen site conditions), and/or the contractor.

The question of whether a change order will result in a credit or payment to the contractor is a matter of interpretation of the plans and specifications. The field engineer should exercise great care to avoid casual comment or agreement with purported change items, and should research a change/requirement in the plans and specifications to determine that it is not included in the contractual scope—either partially or completely.

Change orders are initiated through various methods. An RFI (request for information) usually precedes a change order, or a bulletin can be issued requesting a price from the contractor for proposed revisions. Sketches often accompany this bulletin. Direction can also be given verbally to make changes in the field. In addition, revised drawings can be issued.

Specification for Changes

The basis (i.e., authority and process) for making changes to the contract is usually in the general conditions portion of the specifications. Many large organizations have standard general conditions clauses. These are developed over time and evolve as experience points out weaknesses or needs in the clauses.

One such organization, the U.S. Postal Service, has this to say about changes[*]:

> Since the Postal Service's authority under any given contract is defined by the contract's clauses, it is necessary that those clauses contain provisions that allow the flexibility to alter the contract after award. It is also necessary that there be provisions that require the parties to equitably alter the delivery schedule or the price to be paid in correspondence with other changes in the contract's terms. Consequently, there are contract provisions which give the Postal Service the authority to make changes. There are also provisions which allow the Contractor or the Postal Service "relief" if the other party does something not contemplated under the original agreement or fails to do something contemplated by the original agreement, as well as provisions for "equitable adjustment" of performance time or price when changes are made.

The changes clause (Clause 11-26; Fig. 1-1) is probably the most frequently used clause in a Postal Service construction contract. This clause gives the Postal Service the right to make changes necessary to ensure that a contract's end results will fully meet USPS needs.

Some USPS terminology regarding change orders is as follows:

- *Change order.* A written order, signed by the contracting officer, directing the contractor to make a change that the changes clause permits the contracting officer to order without the contractor's consent. Change orders are another type of unilateral modification. They can be issued for several reasons, including a change in the needs of the requesting office, defects or ambiguities in the specifications, and factors (such as weather conditions) beyond the control of either party. A change order is a contract modification in the making.

- *Contracting officer.* The individual delegated the authority by the Deputy Postmaster General for Facilities to represent the USPS in all aspects of a contract including change orders.

- *Contract modification.* Any written alteration in the specification, delivery point, rate of delivery, contract period, price, quantity, or other contract provision of an existing contract.

There are two major types of contract modifications:

Unilateral. A unilateral modification is a contract modification that is signed only by the contracting officer. A unilateral modification may be either a purely administrative one or a substantive change order authorized by the changes clause. It may also be a change authorized by specific clauses or contract provisions (such as the exercise of an option of a suspension of work).

Bilateral. A bilateral modification is a contract modification that is signed by the contracting officer and the contractor. It is thus a modification made

---

[*] The following two extracts are from USPS Contract Administration Course, August 1992.

Clause 11-26   Changes (Construction) (October 1987)

a. The contracting officer may at any time, without notice to any sureties, by written order designated or indicated to be a change order, make changes in the work within the general scope of the contract, including changes:

1. In the specifications (including drawings and designs);
2. In the method or manner of performance of the work;
3. In the Postal Service–furnished facilities, equipment, materials, services, or site; or
4. Directing acceleration in the performance of the work.

b. Any other written or oral order (which, as used in this paragraph *b*, includes direction, instruction, interpretation, or determination) from the contracting officer that causes a change will be treated as a change order under this clause, provided the contractor gives the contracting officer written notice stating (1) the date, circumstances, and source of the order and (2) that the contractor regards the order as a change order.

c. Except as provided in this clause, no order, statement, or conduct of the contracting officer may be treated as a change under this clause or entitle the contractor to an equitable adjustment.

d. If any change under this clause causes an increase or decrease in the contractor's cost of, or the time required for, the performance of any part of the work under this contract, whether or not changed by any order, the contracting officer will make an equitable adjustment and modify the contract in writing. However, except for claims based on defective specifications, no claim for any change under paragraph *b* above will be allowed for any costs incurred more than 20 days before the contractor gives written notice as required. In the case of defective specifications for which the Postal Service is responsible, the equitable adjustment will include any increased cost reasonably incurred by the contractor in attempting to comply with the defective specifications.

e. The contractor must assert any claim under this clause within 30 days after (1) receipt of a written change order under paragraph *a* above, or (2) the furnishing of a written notice under paragraph *b* above, by submitting to the contracting officer a written statement describing the general nature and amount of the claim, unless this period is extended by the Postal Service. The statement of claim may be included in the notice under paragraph *b* above.

f. No claim by the contractor for an equitable adjustment will be allowed if asserted after final payment under this contract.

Figure 1-1   Clause 11-26, Changes. (*USPS Procurement Manual, 1993.*)

by mutual agreement. Bilateral modifications are used to make negotiated equitable adjustments resulting from the issuance of a change order and to reflect other agreements of the parties modifying contract terms. (The term *supplemental agreement* is often used synonymously with the term *bilateral modification* to refer to a contract modification that is accomplished by mutual action of the Postal Service and the contractor. It is a separate agreement that supplements the original contract, utilizing the original generic and special provisions.)

Several key phrases are used in the first sentence of the changes clause:

"the contracting officer may"   The clause restricts the authority to make a change to the contracting officer. A contractor should be alerted to the fact

that not every Postal Service representative involved in the contract is authorized to direct changes. However, if an individual without authority does direct change and the contractor acts on that direction, the Postal Service may be held to the change under the doctrine of constructive changes.

"without notice to any sureties" A *surety* is an individual or corporation that is legally liable for the debt, default, or failure of the contractor to fulfill its contractual obligations. This phrase says that the Postal Service has the right to change the contract without notifying the sureties and without getting their written agreement. The surety's obligation therefore remains unchanged.

"by written order" This requirement is self-explanatory: a written change order must be issued. The boards and courts have sometimes recognized oral changes, however, reasoning that ratification of those changes was only fair to the contractor. Also, the clause goes on to state that any other written or oral order (including direction, instruction, interpretation, or determination) from the contracting officer that causes a change will be treated as a change order under this clause, provided that the contractor gives the contracting officer written notice stating: (1) the date, circumstances, and source of the order; and (2) that the contractor regards the order as a change order.

"within the general scope" Controversies over what is or is not within the general scope of a contract have generated many board and court decisions, none of which has conclusively defined the term. The Supreme Court has said that the scope of a contract is "what should be regarded as fairly and reasonably within the contemplation of the parties when the contract was agreed to." If work ordered by a change is within the scope of the contract, it is binding on the contractor; if it is not within the scope, however, it is extra work that the contractor can legally decline to perform. The term has been interpreted broadly to give the contracting officer flexibility. The contracting officer must determine what lies within the general scope on a case-by-case basis, often through negotiations with the contractor. The determination is doubly difficult because, besides the issue of whether the contractor must accept the change, there is also the issue of whether the contracting officer can order the change without justifying it as a sole-source purchase. For this purpose, the contract's "scope" is interpreted much more narrowly. The general counsel has consistently required that any economically viable unit of work be done through competitive purchase unless a sole source justification is prepared.

"including changes" The clause permits unilateral changes in drawings, designs, or specifications, as well as in method of shipment or packing, place of delivery or performance of services, delivery or performance schedule, and Postal Service–furnished property or facilities.

If the change ordered by the contracting officer increases or decreases the contract price or the time needed for performance, the contractor is due a fair adjustment of either the contract price or the performance schedule or both.

(Further explanation of how equitable adjustments are figured can be found later in this chapter.)

Even if the aspect of the work that is changed does not involve an increase or decrease in price or time needed, the contractor may be due an equitable adjustment if the change order has the effect of increasing or decreasing the price or time needed for another aspect of the work.

The contractor's claim for equitable adjustment must be asserted within 30 days of receiving a written change order. A later claim may be acted upon—but not after final payment under this contract—if the contracting officer decides that the facts justify such action.

The contracting officer is allowed to exercise discretion and may accept and act on claims submitted after the 30-day period established by the clause. The boards and the courts have offered mixed guidance on this point. Some decisions have allowed claims presented after 30 days, while others upheld the right to insist on 30 days. The best policy for Postal Service representatives to follow is to try to convince the contractor of the need to meet the 30-day requirement. This type of prompt action benefits both the Postal Service and the contractor.

Failure to agree to any adjustment is a dispute under the claims and disputes clause (B-9). Nothing in that clause excuses the contractor from proceeding with the contract as changed. Every change order situation—and every contract modification situation—can potentially result in a dispute subject to the disputes clause. Clause B-9 gives the contractor the duty to "proceed diligently" with the work as changed. If there is no agreement on price adjustment when the change order is issued, the contractor must go on working and try to negotiate an equitable adjustment later. If the contractor walks off the job, the contractor risks the possibility of a termination of the contract for default.

Standard Forms

A number of organizations and associations have prepared their own standard contract forms. These include the American Institute of Architects (AIA), the Association of General Contractors (AGC), the Construction Management Association of America (CMAA), and the Engineers Joint Contract Documents Committee (EJCDC).

Standard forms offer a convenient way to acquire the tested (often court-tested) experience of others. Those forms that are widely used provide the parties with mutually understood "rules of engagement."

AIA

The AIA changes clause is in its general conditions documents (either A201 or A201/CM). Figure 1-2 is the changes clause from A201/CM, June 1980 edition. The clause is slightly larger than one double-column, single-spaced page.

Figure 1-2   Changes clause from AIA A201/CM, June 1980 edition. (*AIA, used with permission.*)

Article 12
Changes in the Work

12.1   Change Orders

12.1.1   A Change Order is a written order to the Contractor signed to show the recommendation of the Construction Manager, the approval of the Architect and the authorization of the Owner, issued after execution of the Contract, authorizing a change in the Work or an adjustment in the Contract Sum or the Contract Time. The Contract Sum and the Contract Time may be changed only by Change Order. A Change Order signed by the Contractor indicates the Contractor's agreement therewith, including the adjustment in the Contract Sum or the Contract Time.

12.1.2   The Owner, without invalidating the Contract, may order changes in the Work within the general scope of the Contract consisting of additions, deletions or other revisions, the Contract Sum and the Contract Time being adjusted accordingly. All such changes in the Work shall be authorized by Change Order, and shall be performed under the applicable conditions of the Contract Documents.

12.1.3   The cost or credit to the Owner resulting from a change in the Work shall be determined in one or more of the following ways:

   .1   by mutual acceptance of a lump sum properly itemized and supported by sufficient substantiating data to permit evaluation;

   .2   by unit prices stated in the Contract Documents or subsequently agreed upon;

   .3   by cost to be determined in a manner agreed upon by the parties and a mutually acceptable fixed or percentage fee; or

   .4   by the method provided in Subparagraph 12.1.4.

12.1.4   If none of the methods set forth in Clauses 12.1.3.1, 12.1.3.2 or 12.1.3.3 is agreed upon, the Contractor, provided a written order signed by the Owner is received, shall promptly proceed with the Work involved. The cost of such Work shall then be determined by the Architect, after consultation with the Construction Manager, on the basis of the reasonable expenditures and savings of those performing the Work attributable to the change, including, in the case of an increase in the Contract Sum, a reasonable allowance for overhead and profit. In such case, and also under Clauses 12.1.3.3 and 12.1.3.4 above, the Contractor shall keep and present, in such form as the Owner, the Architect or the Construction Manager may prescribe, an itemized accounting together with appropriate supporting data for inclusion in a Change Order. Unless otherwise provided in the Contract Documents, cost shall be limited to the following: cost of materials, including sales tax and cost of delivery; cost of labor, including social security, old age and unemployment insurance, and fringe benefits required by agreement or custom; workers' or workmen's compensation insurance; bond premiums; rental value of equipment and machinery; and the additional costs of supervision and field office personnel directly attributable to the change. Pending final determination of cost to the Owner, payments on account shall be made on the Architect's approval of a Project Certificate for Payment. The amount of credit to be allowed by the Contractor to the Owner for any deletion or change which results in a net decrease in the Contract Sum will be the amount of the actual net cost as confirmed by the Architect after consultation with the

---

(Continued)

Figure 1-2 *(Continued)*

<table>
<tr><td></td><td colspan="2">Construction Manager. When both additions and credits covering related Work or substitutions are involved in any one change, the allowance for overhead and profit shall be figured on the basis of the net increase, if any, with respect to that change.</td></tr>
<tr><td></td><td>12.1.5</td><td>If unit prices are stated in the Contract Documents or subsequently agreed upon, and if the quantities originally contemplated are so changed in a proposed Change Order that application of the agreed unit prices to the quantities of Work proposed will cause substantial inequity to the Owner or the Contractor, the applicable unit prices shall be equitably adjusted.</td></tr>
<tr><td>12.2</td><td colspan="2">Concealed Conditions</td></tr>
<tr><td></td><td>12.2.1</td><td>Should concealed conditions encountered in the performance of the Work below the surface of the ground or should concealed or unknown conditions in an existing structure be at variance with the conditions indicated by the Contract Documents, or should unknown physical conditions below the surface of the ground or should concealed or unknown conditions in an existing structure of an unusual nature, differing materially from those ordinarily encountered and generally recognized as inherent in work of the character provided for in this Contract, be encountered, the Contract Sum shall be equitably adjusted by Change Order upon claim by either party made within twenty days after the first observance of the conditions.</td></tr>
<tr><td>12.3</td><td colspan="2">Claims for Additional Cost</td></tr>
<tr><td></td><td>12.3.1</td><td>If the Contractor wishes to make a claim for an increase in the Contract Sum, the Contractor shall give the Architect and the Construction Manager written notice thereof within twenty days after the occurrence of the event giving rise to such claim. This notice shall be given by the Contractor before proceeding to execute the Work, except in an emergency endangering life or property in which case the Contractor shall proceed in accordance with Paragraph 10.3. No such claim shall be valid unless so made. If the Owner and the Contractor cannot agree on the amount of the adjustment in the Contract Sum, it shall be determined by the Architect after consultation with the Construction Manager. Any change in the Contract Sum resulting from such claim shall be authorized by Change Order.</td></tr>
<tr><td></td><td>12.3.2</td><td>If the Contractor claims that additional cost is involved because of, but not limited to, (1) any written interpretation pursuant to Subparagraph 2.3.11, (2) any order by the Owner to stop the Work pursuant to Paragraph 3.3 where the Contractor was not at fault, or any such order by the Construction Manager as the Owner's agent, (3) any written order for a minor change in the Work issued pursuant to Paragraph 12.4, or (4) failure of payment by the Owner pursuant to Paragraph 9.7, the Contractor shall make such claim as provided in Subparagraph 12.3.1.</td></tr>
<tr><td>12.4</td><td colspan="2">Minor Changes in the Work</td></tr>
<tr><td></td><td>12.4.1</td><td>The Architect will have authority to order minor changes in the Work not involving an adjustment in the Contract Sum or extension of the Contract Time and not inconsistent with the intent of the Contract Documents. Such changes shall be effected by written order issued through the Construction Manager, and shall be binding on the Owner and the Contractor. The Contractor shall carry out such written orders promptly.</td></tr>
</table>

The clause (Article 12, Changes in the Work) has four sections. The first (12.1) starts with a definition of what a change order is, followed by the right of the owner to order changes in "the Work."

The third subsection of 12.1 lists the usual three methods of determining the cost of the change: agreement on a lump sum, by unit prices previously agreed upon, or cost (time and material) plus an agreed percentage or fee. In addition, the subsection lists a fourth method, which comes into play if the parties cannot agree on one of the usual three. In the fourth method, the architect ("after consultation with the Construction Manager") determines the cost of the changed work. The subsection (12.1.4) goes on to list those components to be included in the cost:

- Materials (including sales tax and delivery charges)
- Labor (including fringes and taxes)
- Worker's compensation and bond cost
- Rental value of equipment and machinery
- Additional supervision and/or field office personnel

The final subsection of Section 12.1 deals with unit prices, where the quantities are so changed that application of the original unit price would be inequitable to either owner or contractor. In this case, "the applicable unit prices shall be equitably adjusted." (In other unit price specifications, it is usual to set limits—such as plus-or-minus 20 percent—that define the points at which unit prices are negotiated.)

The first section is the heart of the changes clause. It constitutes 60 percent of the document. The remaining three sections address special circumstances: concealed conditions, claims for additional cost, and minor changes in the work.

Section 12.2, Concealed Conditions, covers the following:

- Concealed conditions encountered below ground and at variance with the contract documents or of an unusual nature.
- Concealed or unknown conditions in an existing structure at variance with the contract documents or of an unusual nature.

In this case, an equitable change order is made upon claim by either party; the claim must be made within 20 days after first recognition of the concealed conditions.

Section 12.3, Claims for Additional Cost, requires the contractor to give written notice of intent to make a claim within 20 days after occurrence of the situation. Failure to comply with this notice requirement will invalidate any such claim.

Section 12.4, Minor Changes in the Work, gives the architect the authority to order minor changes without compensation in money or time.

Figure 1-3 shows the change order form from AIA Document G701/CM, June 1980 edition. Figure 1-4 shows the form for application and certification for

CHANGE ORDER
CONSTRUCTION MANAGEMENT EDITION
AIA DOCUMENT G701/CM

Distribution to:
OWNER ☐
ARCHITECT ☐
CONSTRUCTION MANAGER ☐
CONTRACTOR ☐
FIELD ☐
OTHER ☐

PROJECT:
(name, address)

TO (Contractor):

CHANGE ORDER NUMBER:

INITIATION DATE:

ARCHITECT'S PROJECT NO:

CONSTRUCTION MANAGER'S
PROJECT NO:

CONTRACT FOR:

CONTRACT DATE:

You are directed to make the following changes in this Contract:

Not valid until signed by the Owner, the Architect and the Construction Manager.
Signature of the Contractor indicates agreement herewith, including any adjustment in the Contract Sum or the Contract Time.

The original (Contract Sum) (Guaranteed Maximum Cost) was .... $.
Net change by previously authorized Change Orders .... $.
The (Contract Sum) (Guaranteed Maximum Cost) prior to this Change Order was .... $.
The (Contract Sum) (Guaranteed Maximum Cost) will be (increased) (decreased) (unchanged)
   by this Change Order .... $.
The new (Contract Sum) (Guaranteed Maximum Cost) including this Change Order will be ... $.
The Contract Time will be (increased) (decreased) (unchanged) by   (   ) Days.
The Date of Substantial Completion as of the date of this Change Order therefore is   .

Recommended:

CONSTRUCTION MANAGER

ADDRESS

BY   DATE

Agreed To:

CONTRACTOR

ADDRESS

BY   DATE

Approved:

ARCHITECT

ADDRESS

BY   DATE

Authorized:

OWNER

ADDRESS

BY   DATE

Figure 1-3 Change order form from AIA G701/CM. (*AIA, used with permission.*)

APPLICATION AND CERTIFICATE FOR PAYMENT   AIA DOCUMENT G702

TO (Owner):

PROJECT:

ATTENTION:

CONTRACT FOR:

APPLICATION NO:

PERIOD FROM:
TO:

ARCHITECT'S
PROJECT NO:

CONTRACT DATE:

PAGE ONE OF   PAGES

Distribution to:
☐ OWNER
☐ ARCHITECT
☐ CONTRACTOR
☐
☐

CONTRACTOR'S APPLICATION FOR PAYMENT

CHANGE ORDER SUMMARY	ADDITIONS	DEDUCTIONS
Change Orders approved in previous months by Owner TOTAL		
Approved this Month		
Number	Date Approved	
TOTALS		
Net change by Change Orders		

The undersigned Contractor certifies that to the best of his knowledge, information and belief the Work covered by this Application for Payment has been completed in accordance with the Contract Documents, that all amounts have been paid by him for Work for which previous Certificates for Payment were issued and payments received from the Owner, and that current payment shown herein is now due.

CONTRACTOR:

By: Date:

Application is made for Payment, as shown below, in connection with the Contract.
Continuation Sheet, AIA Document G703, is attached.

The present status of the account for this Contract is as follows:

ORIGINAL CONTRACT SUM .... $

Net change by Change Orders .... $

CONTRACT SUM TO DATE .... $

TOTAL COMPLETED & STORED TO DATE .... $
(Column G on G703)

RETAINAGE _____%
or total in Column I on G703 .... $

TOTAL EARNED LESS RETAINAGE .... $

LESS PREVIOUS CERTIFICATES FOR PAYMENT ...... $

CURRENT PAYMENT DUE .... $

State of: County of:
Subscribed and sworn to before me this   day of   , 19
Notary Public:
My Commission expires:

ARCHITECT'S CERTIFICATE FOR PAYMENT

In accordance with the Contract Documents, based on on-site observations and the data comprising the above application, the Architect certifies to the Owner that the Work has progressed to the point indicated; that to the best of his knowledge, information and belief, the quality of the Work is in accordance with the Contract Documents; and that the Contractor is entitled to payment of the AMOUNT CERTIFIED.

AMOUNT CERTIFIED .... $
(Attach explanation if amount certified differs from the amount applied for.)
ARCHITECT:

By: Date:

This Certificate is not negotiable. The AMOUNT CERTIFIED is payable only to the Contractor named herein. Issuance, payment and acceptance of payment are without prejudice to any rights of the Owner or Contractor under this Contract.

AIA DOCUMENT G702 • APPLICATION AND CERTIFICATE FOR PAYMENT • APRIL 1978 EDITION • AIA® • © 1978
THE AMERICAN INSTITUTE OF ARCHITECTS, 1735 NEW YORK AVENUE, N.W. WASHINGTON, D.C. 20006

G702 — 1978

Figure 1-4   Application for payment from AIA G702. (*AIA, used with permission.*)

payment from AIA Document G702, April 1978 edition. As change orders are approved, they are added to the line "Net Change by Change Orders."

CMAA

The CMAA changes clause is Article 11, Changes, part of CMAA Document No. A-3 (1988 edition). Figure 1-5 shows the CMAA changes clause.

Figure 1-5   Article 11, Changes, from CMAA Document No. A-3. (*CMAA, used with permission.*)

Article 11   Changes

11.1   Authorized Variations in Work

11.1.1   The Construction Manager may authorize minor variations in the Work from the requirements of the Contract Documents that do not involve an adjustment in the contract price or the contract time and that are consistent with the overall intent of the Contract Documents. These shall be communicated by the Construction Manager to the Contractor who shall perform the Work involved promptly.

11.2   Changes in the Work

11.2.1   Without invalidating this Contract and without notice to any surety, the Owner may at any time or from time to time order additions, deletions, deductions or revisions in the Work. These shall be authorized by a change order. Upon receipt of such document, the Contractor shall promptly proceed with the Work involved which shall be performed under the applicable conditions of the Contract Documents.

11.2.2   The Contractor shall not be entitled to an increase in the contract price or an extension of the contract time with respect to any Work performed that is not required by the Contract Documents as amended, modified and supplemented, except as specifically provided herein.

11.2.3   If notice of any change affecting the general scope of the Work or the provisions of the Contract Documents including, but not limited to, contract price or contract time is required by any surety providing a bond on behalf of the Contractor, the giving of any such notice shall be the Contractor's responsibility and the amount of each applicable bond shall be adjusted accordingly.

11.3   Change of the Contract Price

11.3.1   The contract price constitutes the total compensation payable to the Contractor for performing the Work. All duties, responsibilities and obligations assigned to or undertaken by the Contractor in performing the Work shall be at the Contractor's expense without change in the contract price.

11.3.2   The Owner shall initiate the change order procedure by issuing a request for proposal to the Contractor, accompanied by technical drawings and specifications. The Contractor shall, within the time period stated in the request for proposal, submit to the Construction Manager for evaluation detailed information concerning the costs and time adjustments, if any, as may be necessary to perform the proposed change order work. When approved by the owner, change orders shall be incorporated into the Contractor's Construction Schedule.

Figure 1-5 *(Continued)*

11.3.2.1 The contract price may only be changed by a change order. Any request by the Contractor for an increase or decrease in the contract price shall be based on written notice stating the general nature of the request delivered by the Contractor to the Construction Manager within ten (10) days after the beginning of the occurrence of the event giving rise to the request. The proposed cost of the request, with supporting data, shall be delivered within thirty (30) days after the end of such occurrence and shall be accompanied by a written statement that the amount requested includes all known amounts, direct, indirect and consequential, incurred as a result of the occurrence of the event. No request for an adjustment in the contract price shall be valid if not submitted in accordance with this paragraph.

11.3.3 The value of any work included in a Change Order or in any request for an increase or decrease in the contract price shall be determined in one of the following ways:

11.3.3.1 By application of unit prices to the quantities of the items involved, subject to the provisions of paragraphs 11.3.8 through 11.3.10, inclusive;

11.3.3.2 By mutual acceptance of a lump sum that may include an allowance for overhead and profit; or

11.3.3.3 On the basis of the cost of the work, determined as provided in paragraphs 11.3.4 and 11.3.5, plus a Contractor's Fee for overhead and profit, determined as provided in paragraphs 11.3.6 and 11.3.7.

11.3.4 The term "cost of the work" means the sum of all costs necessarily incurred and paid by the Contractor in the proper performance of the Work. Such costs shall be in amounts no higher than those prevailing in the locality of the project and shall include only the following items:

11.3.4.1 Actual payroll costs for employees in the direct employ of the Contractor in the performance of the work. Payroll costs for employees not employed full time on the Work shall be apportioned on the basis of their time spent on the Work. Payroll costs shall include, but not be limited to, the audited cost of salaries and wages, plus the cost of fringe benefits that shall include social security contributions, unemployment, excise and payroll taxes, Workers' or Workmen's Compensation, health and retirement benefits, bonuses, sick leave, vacation and holiday pay applicable thereto. The Contractor's employees shall include superintendents and foremen at the site. The expenses of performing work after regular working hours, on Saturday, Sunday or legal holiday shall be included in the above only to the extent authorized in writing by the Construction Manager;

11.3.4.2 Cost of all materials and equipment furnished and incorporated in the Work, including costs of transportation and storage thereof, and suppliers' field services required in connection therewith. All cash discounts shall accrue to the Contractor unless the Owner deposits funds with the Contractor with which to make payments and in which case the cash discounts shall

Figure 1-5 *(Continued)*

accrue to the Owner. All trade discounts, rebates and refunds and all returns from sale of surplus materials and equipment shall accrue to the Owner and the Contractor shall make provisions such that the monies may be obtained;

11.3.4.3 Payments made by the Contractor to the subcontractors for work performed. If required by the Construction Manager, the Contractor shall obtain competitive bids from subcontractors acceptable to the Construction Manager and shall deliver such bids to the Construction Manager who shall then determine which bids shall be accepted. If a subcontract provides that the subcontractors be paid on the basis of cost of the work plus a fee, the subcontractor's cost of the work shall be determined in the same manner as the Contractor's cost of the work. All subcontracts shall be subject to the other provisions of the Contract Documents;

11.3.4.4 Costs of special consultants including, but not limited to, engineers, architects, testing laboratories, surveyors, attorneys and accountants, employed for services specifically related to the work; or

11.3.4.5 Other costs including the following:

11.3.4.5.1 The proportion of necessary transportation, travel and subsistence expenses of Contractor's employees incurred in discharge of duties connected with the Work;

11.3.4.5.2 The cost, including transportation and maintenance, of all materials, supplies, equipment, machinery, appliances, office and temporary facilities at the site and hand tools not owned by the workers that are consumed in the performance of the Work and the cost, less market value, of the items used, but not consumed, that remain the property of the Contractor;

11.3.4.5.3 Rentals of all construction equipment and machinery and the parts thereof whether rented from the Contractor or others in accordance with rental agreements approved by the Construction Manager and the costs of transportation, loading, unloading, installation, dismantling and removal thereof, all in accordance with terms of the rental agreements. The rental of any such equipment, machinery and parts shall cease when the use thereof is no longer necessary for the Work;

11.3.4.5.4 Sales, consumer, use or similar taxes related to the Work and for which the Contractor is liable, imposed by laws and regulations;

11.3.4.5.5 Deposits lost for causes other than acts of the Contractor, any subcontractor or anyone directly or indirectly employed by any of them or for whose acts any of them may be liable and royalty payments and fees for permits and licenses;

11.3.4.5.6 Losses, damages and related expenses not compensated by insurance and sustained by the Contractor in connection with the performance and

Figure 1-5 *(Continued)*

furnishing of the Work, provided they have resulted from causes other than the acts of the Contractor, any subcontractor or anyone directly or indirectly employed by any of them or for whose acts any of them may be liable. Such losses shall include settlements made with the written consent and approval of the owner. No such losses, damages or expenses shall be included in the cost of the Work for the purpose of determining the Contractor's fee. If, however, any such loss or damage required reconstruction and the Contractor is placed in charge thereof, the Contractor shall be paid for services a fee proportionate to that stated in paragraph 11.3.6.2;

11.3.4.5.7 The cost of utilities, fuel and sanitary facilities at the site;

11.3.4.5.8 Incidental expenses such as telegrams, long distance telephone calls, telephone service at the site, express packages and similar items in connection with the Work; and

11.3.4.5.9 Cost of premiums for additional bonds and insurance required because of changes in the work.

11.3.5 The term "cost of the work" shall not include any of the following:

11.3.5.1 Payroll costs and other compensation of the Contractor's officers, executives, principals of partnerships and sole proprietorships, general managers, engineers, architects, estimators, attorneys, auditors, accountants, purchasing and contracting agents, expediters, timekeepers, clerks and other personnel employed by the Contractor whether at the site or in the Contractor's principal or a branch office for general administration of the Work and not referred to in paragraph 11.3.4.1 or specifically covered by paragraph 11.3.4.4, all of which are to be considered administrative costs covered by the Contractor's fee;

11.3.5.2 Expenses of the Contractor's principal and branch offices other than the Contractor's office at the site;

11.3.5.3 Any part of the Contractor's capital expenses, including interest on the Contractor's capital employed for the Work and charges against the Contractor for delinquent payments;

11.3.5.4 Costs due to the negligence of the Contractor, any subcontractor or anyone directly or indirectly employed by any of them or for whose acts any of them may be liable including, but not limited to, the correction of defective Work, disposal of materials or equipment wrongly supplied and making good any damage to property; or

11.3.5.5 Other overhead or general expense costs of any kind and the costs of any item not specifically and expressly included in paragraph 11.3.4.

11.3.6 The Contractor's fee allowed to the Contractor for overhead and profit shall be determined as follows:

11.3.6.1 A mutually acceptable fixed fee; or, if none can be agreed upon,

11.3.6.2 A fee based on the following percentages of the various portions of the cost of the Work;

(Continued)

Figure 1-5 *(Continued)*

> 11.3.6.2.1 For costs incurred pursuant to paragraphs 11.3.4.1 and 11.3.4.2, the Contractor's fee shall be fifteen (15) percent;
>
> 11.3.6.2.2 For costs incurred pursuant to paragraph 11.3.4.3, the Contractor's fee shall be five (5) percent. If the subcontract is on the basis of cost of the Work plus a fee, the maximum allowable to the Contractor on account of overhead and profit of all subcontractors shall be ten (10) percent;
>
> 11.3.6.2.3 No fee shall be payable on the basis of costs itemized under paragraphs 11.3.4.4, 11.3.4.5 and 11.3.5;
>
> 11.3.6.2.4 The amount of credit to be allowed by the Contractor to the owner for any such change which results in a net decrease in cost shall be the amount of the actual net decrease, plus an amount equal to ten (10) percent of the net decrease; and
>
> 11.3.6.2.5 When both additions and credits are involved in any one change, the adjustment in the Contractor's fee shall be computed on the basis of the net change in accordance with paragraphs 11.3.6.2.1 through 11.3.6.2.4, inclusive.
>
> 11.3.7 Whenever the cost of any work is to be determined pursuant to paragraph 11.3.4 or 11.3.5, the Contractor shall submit in form acceptable to the Construction Manager the itemized cost, together with such supporting data as may be deemed necessary by the Construction Manager.
>
> 11.3.8 When the Contract Documents provide that all or part of the Work be unit price work, initially the contract price shall be deemed to include, for all unit price work, an amount equal to the sum of the established unit prices for each separately identified item of unit price work times the estimated quantity of each item as indicated in the contract document. The estimated quantities of items of unit price work are not guaranteed and are solely for the purpose of comparison of bids and determining an initial contract price. Determinations of the actual quantities and classifications of unit price Work performed by the Contractor shall be made by the Construction Manager in accordance with paragraph 8.3.5.
>
> 11.3.9 Each unit price shall be deemed to include an amount considered by the Contractor to be adequate to cover its overhead and profit for each separately identified item.
>
> 11.3.10 When the quantity of any item of unit price work performed by the Contractor differs more than twenty-five (25) percent from the estimated quantity of such item indicated in this Contract and there is no corresponding adjustment with respect to any other item of Work, the Construction Manager and Contractor shall determine a mutually acceptable price for the changed item.
>
> 11.3.11 If the Contractor and owner are unable to arrive at an agreement as to the change in the contract price or performance time, the Contractor shall nevertheless proceed with the change if so ordered in writing by the Construction Manager and the value of the work included in the change order shall be determined in paragraphs 11.3.3 through 11.3.10.

The CMAA clause is three and a half pages (double column, space and a half). There are three sections to the clause:

11.1   Authorized Variations in Work

11.2   Change in the Work

11.3   Change of the Contract Price

The first section (11.1) is only one paragraph. It gives the construction manager authority to authorize minor variations that do not change the contract price or time for performance. (This is similar to Section 12.4, Minor Changes in the Work, of AIA A201/CM.)

The second section (11.2) is three paragraphs:

- Owner may at any time order "additions, deletions, deductions or revisions."

- Contractor is not entitled to an increase in contract price or an extension of time for work not required by the contract documents as amended except as specified in Article 1.1.

- If notice of any change is required by any surety, the contractor is responsible.

Section 11.3, Change of the Contract Price, has 11 subsections. It takes up 85 percent of the clause:

11.3.1 Unless changed, the contract price represents the total amount payable to the contractor.

11.3.2 The owner initiates the change order procedure by issuing a request for proposal, accompanied by drawings and specifications. The contractor submits, in a timely manner, detailed information to the CM. When approved, change orders are incorporated into the construction schedule.

    11.3.2.1 Contract price may be changed only by change order. Any contractor request for change shall be submitted within 10 days of the start of an occurrence, and the proposed cost shall be submitted within 30 days of the conclusion.

11.3.3 The value of work included in a change order shall be determined in one of the following ways:

    11.3.3.1 Unit prices.

    11.3.3.2 Mutual acceptance of lump sum.

    11.3.3.3 Cost of the work plus a fee for overhead and profit.

11.3.4 "Cost of the work" is the sum of all costs necessarily incurred and paid, at prevailing rates, by the contractor, including the following:

    11.3.4.1 Actual payroll costs for those in direct employ of the contractor including audited salaries and wages plus taxes, insurance, and fringes.

    11.3.4.2 Cost of materials and equipment furnished and incorporated in the work.

    11.3.4.3 Payments made by the contractor to subcontractors for work performed.

11.3.4.4 Costs of special consultants (such as engineers, architects, testing labs, attorneys, etc.) for services specific to the work.

11.3.4.5 Other costs including:

11.3.4.5.1 Transportation and subsistence expenses incurred by employees in relation to the work.

11.3.4.5.2 Cost of office equipment, office supplies, and hand tools consumed at the site.

11.3.4.5.3 Construction equipment rentals.

11.3.4.5.4 Sales or use taxes.

11.3.4.5.5 Deposits lost.

11.3.4.5.6 Losses, damages, and related expenses not covered by insurance

11.3.4.5.7 The cost of utilities at the site.

11.3.4.5.8 Incidental expenses such as telegrams, long-distance telephone calls, express packages, etc., at the site.

11.3.4.5.9 Cost of premiums for additional bonds and insurance.

11.3.5 The "cost of the work" shall not include:

11.3.5.1 Administrative salaries considered covered by the contractor's overhead fee.

11.3.5.2 Expenses for the contractor's offices other than site.

11.3.5.3 Contractor's capital expenses.

11.3.5.4 Costs due to negligence.

11.3.5.5 Other overhead or general expense costs not included in 11.3.4.

11.3.6 The contractor's fee for overhead and profit is determined as follows:

11.3.6.1 Mutually agreed fixed fee.

11.3.6.2 Fee based upon the following percentages of various portions of the work:

11.3.6.2.1 Costs (labor and materials) incurred times 15 percent.

11.3.6.2.2 Costs (subcontract) times 5 percent.

11.3.6.2.3 No fee on special consultants, "other costs," and costs not allowed.

11.3.6.2.4 Credit shall be cost plus 10 percent.

11.3.6.2.5 When additions and credits are involved, fee is computed on basis of net change.

11.3.7 Cost of work shall be submitted in a form acceptable to the CM, including itemized costs.

11.3.8 The estimated quantities for unit price work are not guaranteed.

11.3.9 Each unit price is deemed to be adequate to cover overhead and profit as well as the item.

11.3.10 When the quantity of any unit price item differs more than 25 percent from the estimated quantity, the CM and contractor shall determine a mutually acceptable price.

11.3.11 If the contractor and owner are unable to arrive at agreement to change in price and/or performance time, the contractor shall nevertheless proceed with the change (if so ordered by the CM). Value of the change order is then determined by processes in 11.3.3 through 11.3.10.

Points of View on Changes

The significance of changes in construction, as in most things, lies in the eyes of the beholder.

The Owner's Right to Make Changes

The owner's right to make changes, which was once absolute, has evolved over time. Much of the law in United States claims cases flows from the Supreme Court review of United States v. Rice (317 U.S. 61, 63 S. Ct. 120; 1943). This case involved an interpretation of paragraphs 3 and 4 of the standard construction contract form used by the U.S. government. In paragraph 3, Changes, the government reserved the right "at any time, by a written order, and without notice to the sureties, [to] make changes in the drawings and/or specifications of this contract and within the general scope thereof."

The clause provides that "an equitable adjustment shall be made" for any increase or decrease in the amount due.

In paragraph 4, Change Conditions, a similar prerogative to make changes was reserved in the event of "conditions at the site materially different from those shown on the drawings or indicated in the specifications."

The Supreme Court was asked to decide whether damages for delay could be recovered under either paragraph. A unanimous court found that the damages for delay could not be recovered, principally on the basis that the paragraphs called for an alteration in "the amount due." The court held that had an alteration in time been contemplated, then that dimension would specifically have been mentioned.

While enforcing the so-called Rice Doctrine, courts have generally recognized that an independent action could be entered by the contractor for breach of contract. This was not initially true. In Great Lakes Construction Co. v. United States (95 Ct. Cl. U.S. 479; 1942), some 109 changes were made to the contract. The contractor was required to prepare a bid despite certain omissions in the plans and specifications. A corrected set of plans and specifications was delivered, shortly after the contract was signed, that indicated some 700 additions and corrections. The parties disputed the value of changes, but work continued. The contractor's claim for damages due to delay was rejected by the court on the basis of paragraph 3 of the standard government contract. Further, a claim for delay due to breach of contract was not allowed by the court.

In George A. Fuller Co. v. United States (69 F. Supp. 409, Ct. Cl.; 1947) the court found as follows:

> [The contractor] would have no right to complain if the [owner's] exercise of its reserve right to make changes set its work schedule awry.... At least this would be so if [the owner] has acted with due alacrity. . . .
>
> [The contractor] had contracted to do the work in accordance with the specifications [the owner] had prepared, and [the owner] was, therefore, under a duty not to render the project more expensive than it would have been if the contractor could have complied with the plans. . . . If faulty specifications prevent or delay completion of the contract, the contractor is entitled to recover damages for the [owner's] breach of its implied warranty.
>
> This exculpatory rule [of Rice] is not applicable to a situation in which unreasonable delays were the result of the defendant's failure to promulgate properly drawn specifications.

Similarly, in Luria Bros. v. United States (369 F. 2d 701), the court held: "Where the change is necessitated by defective plans and specifications [the owner] must pay the entire resulting damage without any deduction for time to make changes, as would be the case if the redesign were necessitated by a changed condition or the like."

In Continental Illinois Bank & Trust Co. v. United States (101 F. Supp. 755; 1952), the government decided to redesign a building and halted work. Work was delayed 175 days. An evaluation of a reasonable time for design was estimated at 40 calendar days. The court awarded the contractor damages for delay for 135 days, saying: "We think that the government's taking of 175 days for the redesign of a boiler house was inconsiderate of the harm which was being caused to the contractor, and was a breach of contract. The right reserved in the contract to make changes in the work does not mean that the government can take as much time as it pleases to consider such changes regardless of the consequences to the other party."

Typical Causes of Change Orders

Unforeseen conditions. Conditions in the field don't match the contract documents (i.e., plans and specifications). This most often occurs in regard to under ground conditions, such as uncharted utilities, uncharted existing foundations, rock or other strata at higher elevation than expected, high groundwater, and so on.

Plans and/or specifications. During implementation it is discovered that, if done per contract plans and/or specifications, the work cannot be installed, or if installed, will not operate. This most often is discovered through the preparation of shop drawings by the subcontractor and/or coordination drawings by the various trade contractors.

Scope change (additional or enhancement) by owner.   Changes of choice by the owner. Examples include the following:

- Technological changes require owner to upgrade installation—hospitals, process industries, high-technology installations
- Funding availability (through unexpected sources and/or favorable bids)—owner decides to expand scope
- Change in projected requirements as a result of change in projections and/or demographics

Value engineering.   An example of this cause would be a value engineering change proposal (VECP) that is proposed by the contractor with savings shared (usually 50 percent to 50 percent) with the owner.

Force majeure.   Change order is for time delay due to forces beyond the control of the contractor. These include accident, fire, flood, strikes, severe weather (beyond average), and so forth. Almost all contracts allow the contractors time, but not the cost of extended conditions.

Acceleration.   Change order initiated by the owner to either expedite an earlier completion date or regain all or part of lost time. If an owner wants to hold an end date unreasonably (i.e., refuses to allow reasonable time extensions), the result can be "constructive acceleration," and the contractors can recover costs.

Typical Reactions/Positions to Change Orders

The owner

The owner's position rarely gets any stronger than it is at the start of the construction contract: is still on budget, has a design, and has established a schedule based upon needs. If properly structured, the contract is a very strong one: it attempts to give the owner a maximum of prerogatives, while holding the owner harmless from as many actions (or lack of action) as is possible.

During the construction stages, inevitable—and often unforeseen—problems arise. The schedule tends to become expanded, while the contract tends to increase because of the additional costs of unforeseen conditions, delays, and out-of-scope work.

The realistic, experienced owner expects some delay and to pay more than the base contract price. Cost is generally better controlled than time, with an acceptable cost increase range on the order of 5 percent. (In many public situations, the executive in charge, such as the commissioner of public works, may authorize changes up to an aggregate of 5 percent without additional approval.)

Time overruns tend to be much higher in terms of percentage of the time specified. This causes hardship to all parties. The contractor's overhead increases, while the owner's costs for lack of availability of required space can range from loss of rental to loss of contracts because of lack of production facilities.

When change orders come up, the owner's reactions may be some (or all) of the following:

- Is annoyed at additional costs and/or delay
- Assumes the architect (or architect/engineer) is at fault
- Assumes contractor is "taking over"

On the other hand, if the change is an owner-ordered enhancement, the reaction might be:

- "I want what I want."
- "The architect should have thought of this."
- "The cost is minimal."
- "This shouldn't take any more time."

The contractor

The contractor starts most contracts in the worst position possible, agreeing to, as a meeting of the minds, a contract that has been unilaterally drafted by the owner and the owner's representatives—particularly the architect/engineer. More than that, the contractor is almost inevitably the low bidder and is counting and recounting the money "left on the table."

Throughout the project, the contractor negotiates, operates, and maneuvers to make the contract work to the greatest extent possible. In almost all instances, the contract does undergo an evolution: some of the more preemptive rights are forfeited, certain clauses through performance (or lack thereof) are waived, some of the controls that the owner retained (or thought he or she retained) are muted.

When change orders come up, the contractor's reactions may be some or all of the following:

- "How can we negotiate to gain profit advantage?"
- "Too many changes will cause delay and/or disruption of work."
- "How else will these change impact the job?"
- "These change orders could cause cash flow problems."
- "Can we recover soft costs (estimating, negotiating, shop drawings, administration, purchasing, etc.)?"
- "Can we get a time extension?"

Subcontractors

Subcontractors have the same concerns as the general (prime) contractor but are one level removed from the ability to negotiate and/or control their own situation.

The architect (or architect/engineer)

For the architect, a change order can have a number of ramifications—none of them good. When it comes to change orders, the architect can only tie or lose; there is no win. This leads the architect to "play it safe," applying one or more of the following approaches:

- Claim that there is no change, that the contract documents include this scope
- Claim that even if there is a change, applicable code or regulation requires the contractor to include the change—or that custom-of-the-trade requires the contractor to include the change
- Claim that the general (exculpatory) specification language requires the contractor to provide an operating system that would incorporate the change
- Extend negotiating time with refusal to accept or admit the scope change

The reason for all of the above is twofold: first, the architect is concerned that the owner will claim the design effort was not adequate (i.e., negligence); or second, the contractor will claim breach of contract due to deficient design documents.

The project manager (PM or CM)

When change orders build up, the project manager may come to believe that perhaps the contractor "bought" the contract with a low bid, hoping to "make up" the difference through change orders.

To the project manager, change orders mean the following:

- More work (to negotiate, track, analyze, and control) is involved.
- Impact on the budget must be monitored.
- Schedule impact(s) of the changes must be monitored.
- Responsibility for each change must be assessed.

Change versus Contract Type

Figure 3-1 lists the following nine types of construction contracts. They range from firm fixed price (lowest owner risk) to time and materials, or T&M (highest owner risk):

1. Firm fixed price (lump sum)

2. Indefinite quantity (unit price)

3. Fixed price with economic price adjustment

4. Fixed price with incentive (guaranteed maximum price—GMP)

5. Design-build

6. Cost plus incentive fee

7. Cost plus incentive fee with performance incentives

8. Cost plus fixed fee

9. Time and materials and labor hours

From the contractor's viewpoint, the processing of change orders ranges from most difficult (firm fixed price) to easiest (time and materials).

Firm Fixed Price (Lump Sum)

With the firm fixed price contract, the project is taken through the complete design cycle, then through the bidding process, with the contract going to the lowest responsible bidder. This approach gives the owner the most competitive price. The time frame from concept to completion is the longest because full design has to be completed before the bidding process. In effect, the design, bidding, and construction is in series. Because of this time component, the firm fixed price contract makes best sense for projects that are readily designed, such as schools, office buildings, and commercial buildings.

Contract Type	Risk/Decision Factors
1. Firm fixed price (lump sum)	Cost risk entirely to the contractor Lowest degree of owner administration
2. Indefinite quantity (unit price)	Final quantities indeterminate upon receipt of offer Verification of installed quantities required
3. Fixed price with economic price adjustment	Potential for labor/price changes over course of project Inflationary economy
4. Fixed price with incentive (guaranteed max.)	Uncertainties in cost/manpower required or an accurate estmate impossible Desire to avoid large contingencies added to lump sum price
5. Design-build	Uncertainties in cost/manpower required Design and construction responsibilities assumed by design-build contractor
6. Cost plus incentive fee	Contractor unable or unwilling to agree on GMP ceiling price
7. Cost plus incentive fee with performance incentives	Contract scope not well-defined in terms of quality
8. Cost plus fixed fee	High degree of uncertainty with target cost
9. Time and materials and labor hours	Scope of work undetermined Fee undetermined Urgent need to start work

Figure 3-1 Nine types of construction contracts.

The contractor has great incentive to control costs in order to earn the maximum profit. That incentive is in play when the contractor is negotiating the price of a change order. One of the early submittals by the contractor is a cost breakdown by category (usually by CSI division and subdivisions; see Fig. 3-2). This cost breakdown is used for progress payments, which are usually monthly. The cost breakdown is useful to the owner in evaluating purposed change order costs.

When a change is identified, the specifications set forth the procedures for preparing the change order. The choice is usually a negotiated lump sum. When agreement cannot be reached, the contract usually allows the owner to direct the contractor to proceed on a time/material basis. In federal contracts, the contracting officer can issue a *unilateral change order*. In a unilateral change order, the contracting officer sets the initial price (subject to later negotiation or claim).

Indefinite Quantity (Unit Price)

Unit price contracts are used on heavy construction work that involves repetitive operations such as excavating, backfilling, compacting, paving,

CSI Format

The Construction Specifications Institute developed a method of organizing construction information according to a system of classification and numbering that is widely accepted as a standard in the industry. Architects follow this format to develop specifications. Contractors follow this same format to assemble bid estimates and to check that all segments of the proposed project have been included in their bids.

Following is a brief description of each of the major divisions:

Division 1   *General requirements*: Overhead and profit; mobilization; temporary construction and utilities; rental equipment; non-manual-labor distributables; scaffolding.

Division 2   *Sitework*: Demolition; excavation; pilings and caissons; underground utilities; electrical distribution; pavings and walks; irrigation and landscaping.

Division 3   *Concrete*: Formwork; reinforcing steel; concrete in place, precast concrete; tilt-up construction; cement decking.

Division 4   *Masonry*: Brick and stone masonry; concrete unit masonry; fireplaces.

Division 5   *Metals*: Structural steel; preengineered steel buildings; metal decking; metal stairs; metal fabrications.

Division 6   *Carpentry*: Rough carpentry; wood framing and decking; glu-lam construction; millwork; prefinished wood paneling; wood staircases.

Division 7   *Thermal and moisture protection*: Waterproofing, insulation; roofing; composite building panels; sheet metal and flashing; rooftop accessories.

Division 8   *Doors and windows*: Metal and wood doors and hardware; specialty doors; glass and glazing; curtain wall and storefront systems.

Division 9   *Finishes*: Lath and plaster; metal studs; partition walls; gypsum wallboard; floor, wall, and ceiling finishes; painting.

Division 10   *Specialties*: Chalkboards and bulletin boards; toilet partitions and accessories; folding and demountable partitions; louvers and access flooring.

Division 11   *Equipment*: Specialized equipment for banks, churches, schools, food services, libraries, prisons, etc.

Division 12   *Furnishings*: Furniture; blinds and shades; office workstations; multiple seating.

Division 13   *Special construction*: Air-supported structures; athletic rooms; clean rooms; cold storage rooms; preengineered structures, pools, and ice rinks.

Division 14   *Conveyances*: Elevators; escalators; moving sidewalks; pneumatic tube systems.

Division 15   *Mechanical*: Plumbing equipment; pipes and valves; heating, ventilating, and air-conditioning equipment; ductwork; piping and insulation; fire protection systems.

Division 16   *Electrical*: Electrical service and distribution; lighting and power; motor control center, communication, and alarm systems; lightning protection; electric heat systems.

Division 17   May be used for project-specific requirements. For example, corrections projects often include the following:

               Security/communication: Electronic security/communication systems—door locking, closed-circuit television, radio, intercom and paging, and fire alarm.

Figure 3-2 CSI divisions and subdivisions.

and pipelines. The work has to be capable of measurement by units such as square feet, square yards, cubic yards, gallons, tons, square meters, kilograms, liters, and so on. This approach permits competitive bidding by furnishing a list of basic materials (i.e., bill of materials). The bid is developed to a single-figure bottom line by extending each category quantity by the bidder's price and then totaling.

The indefinite quantity keeps the competitive advantage for the owner. Since quantities are paid for on a measured basis, the design does not have to be as precise as in a fixed price contract; similarly, the bids are easier to prepare. The bidders focus on categories that have large quantities and/or high unit prices. They will do the best estimate possible on these. For instance, if the bid sheet has the following:

Common excavation	50,000 cubic yards (c.y.)
Rock excavation	10,000 c.y.
Total	60,000 c.y.

A contractor might prepare the following bid:

Common excavation	50,000 c.y. @ $8.00 = $400,000
Rock excavation	10,000 c.y. @ $30.00 = $300,000
Total	60,000 c.y. $700,000

But the contractor, after studying the geotechnical information, believes that the rock quantity has been underestimated by 100 percent. Based on this, the contractor could unbalance the bid as follows:

Common excavation	50,000 c.y. @ $4.00 = $200,000
Rock excavation	10,000 c.y. @ $50.00 = $500,000
Total	60,000 c.y. $700,000

This keeps the contractor at the same level competitively. Of course, the contractor is hoping for the following result:

Common excavation	40,000 c.y. @ $4.00 = $160,000
Rock excavation	20,000 c.y. @ $50.00 = $1,000,000
Total	60,000 c.y. $1,160,000

A variation on the unbalanced bid is the *penny bid*. On unit price pipelines, when support piling is called for, contractors make their best guess as to the linear feet of piling. They incorporate the piling cost into the pipeline cost per foot and bid a penny a foot for piling.

Change orders are very easy in unit price categories. No change order is necessary until the base category amount is exceeded. In addition, unit price specifications usually permit negotiation of the unit price (up or down) once the base quantity is either exceeded or underrun by 20 percent.

In Foundation Co. v. State of New York (233 N.Y. 177), the state attempted to require the contractor to perform pumping operations on a unit cost basis as contemplated in the contract. However, the contractor had anticipated that he could pump from one location when developing this unit price. The situation at the site required seven pumping operations, substantially increasing the cost. The court agreed that the price included in the contract as part of the bid was inapplicable and that the contractor could recover on a quantum meruit basis for the reasonable value of the work performed.

Similarly, in Depot Construction Co. v. State of New York (41 Misc. 2d 764, 246 N.Y.S. 2d 527), Depot was required to perform work in a manner that had not been contemplated by the contractor. The state attempted to hold the contractor to unit prices per the contract. The court of claims held that these unit prices no longer applied. One reason for this holding was that the state had caused the conditions that required additional work be accomplished. Since it was outside any reasonable contemplation at the time of the contract, the court held that the contractor was entitled to compensation on a quantum meruit basis.

The court made the important distinction that where there is a qualitative change in the nature of the work described by a unit price in the contract, the contractor is thereby not bound to the unit prices in the contract.

Fixed Price with Economic Price Adjustment

This approach is used in an environment of high inflation and/or very long duration. The price adjustment is based on the consumer price index (CPI) and applied to the labor. Change orders are processed in the same manner as the firm fixed price contract, except that the CPI is applied to the labor in the change order work.

Fixed Price with Incentive (Guaranteed Maximum Price)

This approach is used on fast-track construction where the design and construction overlap (i.e., construction starts before design is complete). The GMP has to include substantial contingency money. The incentive is a sharing formula for any underrun in cost, usually 50 percent owner–50 percent contractor. Change orders are given only for changes in scope. The contractor will try to characterize normal design evolution or development as change in scope.

This approach is principally used in the private sector by developers. The contract may be competitively bid or negotiated.

Design-Build

With the design-build method, a single legal entity holds direct responsibility (and liability) for project design and construction. The owner deals with one entity rather than two entities (i.e., the architect to prepare documents and the contractor to build from these documents). In other words, one organization accepts the requirements and then meets those requirements.

Design-build organizations can be formed in many ways: an owner may form an in-house design-build team, a designer may hire a contractor, a designer may form a joint venture, or a contractor may employ a design professional, thereby having in-house design-build capabilities.

Because the owner wants either a fixed price or (more usually) a guaranteed maximum price (GMP), the contractor controls the venture because the contractor is the risk taker. The design-build entity usually employs the fast-track approach in which the designer prepares the up-front portion of the building package and then turns it over to the construction division to perform. While that portion of the work is under way, the designer continues with the next portion, and so forth. Owners, especially private owners, want to have a fixed/guaranteed cost going into a project.

One-time owner-builders are usually disenchanted with the results of this "guarantee." The designer-builder is taking a higher risk in offering the GMP because it is based upon less-than-complete design information. Accordingly, the designer-builder includes a contingency factor in the GMP.

The main reason for fast-tracking design-build projects is to compress the time frame and to give the owner a building before the same building would be completed under the conventional method of delivery. However, employing the fast-track system on a sophisticated building can be financially disastrous because the design professional must have a complete understanding of the sophisticated building before starting to put out construction packages. Taking the time to do extensive research that incorporates all of the complicated functions required in a sophisticated building can nullify the time compression potential. See Fig. 3-3.

In design-build, the owner states criteria that the project is to provide. The designer-builder provides either a fixed price or, more usually, a guaranteed maximum price. Both are based upon a scope. If the owner increases the scope, then the cost increases.

Change orders in design-build situations are similar to those in GMP (i.e., the contractor will try to characterize normal design development as scope change). In design-build, the owner should keep in mind that the contractor team has design responsibility. The contractor will try to pass design costs on to the owner as change orders.

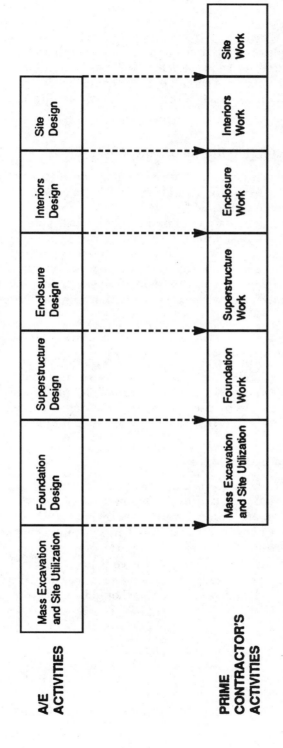

FAST TRACK

| A/E ACTIVITIES | Mass Excavation and Site Utilization | Foundation Design | Superstructure Design | Enclosure Design | Interiors Design | Site Design |

| PRIME CONTRACTOR'S ACTIVITIES | Mass Excavation and Site Utilization | Foundation Work | Superstructure Work | Enclosure Work | Interiors Work | Site Work |

EACH ACTIVITY IS PERFORMED BY THE SAME PRIME CONTRACTOR. THE DESIGN PROFESSIONAL IS REQUIRED TO PROVIDE THE CONTRACTOR WITH THE DOCUMENTS FOR THE NEXT ACTIVITY BEFORE THE PREVIOUS ACTIVITY IS COMPLETE (ALLOW TIME FOR ESTIMATING AND PURCHASING). LONG LEAD ITEMS ARE CONSIDERED UP-FRONT TO MAINTAIN CONTINUITY OF SEQUENCE.

Figure 3-3 Fast-track design-build contract.

33

Cost plus Incentive Fee

The cost plus incentive fee approach is similar to GMP; however, instead of a GMP, a target price is set. The fee is based on a formula related to the target price. When changes come up, the contractor will try to adjust the target price upward. This contract is often used with architect/engineers and other consultants.

Cost plus Incentive Fee with Performance Incentives

This approach is used where time is of the essence but the work scope is too uncertain to use GMP. It is also used for projects where the work is R&D, complex, or original (i.e., no "go-bys").

The contract sets a minimum fee and cost-plus. Provision is made for an "awarded" fee based upon a subjective review after completion. The exposure for the owner is similar to cost plus fixed fee. The approach is favorable to the contractor (usually an architect/engineer or consultant) only if the contractor has special expertise in this type of contract. Since there is no GMP or target price, change orders are cost-plus without impact (or meaning).

Cost plus Fixed Fee

With this method, the contractor is reimbursed for all costs, including overheads. The fixed fee is based on pre-award negotiations. Since the contractor takes no risk, the fees are in the range of 2 percent of the costs.

A variation is to use reducing percentages of cost as the cost increases. Change orders that increase the scope usually increase the fixed fee.

Time and Materials and Labor Hours

In the time and materials approach, the contractor is reimbursed for all materials at actual cost. (A fee of 2 to 3 percent is sometimes allowed.) Direct labor is paid at an agreed upon rate (usually hourly) that varies with labor classification. The hourly rate includes fringes, contractor overhead, and profit (at 5 to 10 percent). The contract usually allows the owner the right to audit project and home office records of the contractor.

In this contract mode, there is no need for change orders. An as-built record of events, and changes in events, should be kept.

4

Using the Contract
to Control Risk

Over time, the authors of construction contracts have developed a variety of approaches to protect owners (and themselves). The goal in each case is to shift any unexpected risk to the contractors without necessitating a change order.

Exculpatory Clauses

A standard designer's ploy is the inclusion of exculpatory clauses in the general conditions (i.e., the "boilerplate" section). These clauses seek to excuse errors, omissions, or other problems that may evolve from the plans and specifications.
Here are some typical exculpatory clauses:

> The survey, and in particular the elevations and contours, are furnished as information to the bidders. Bidders shall verify survey information, and the owner assumes no responsibility for the accuracy of the survey as furnished.

> Data logs shown in the plans are for information only, and the contractors are warned that reliance on this information shall be at their own risk and that neither the owner nor the engineer shall be liable for errors.

> Contractor shall be responsible for interpretation of subsurface soil conditions as described in the engineer's boring data. Further, all soil information, including boring data, is for information only and shall not be considered part of the contract documents.

> The contractor shall visit the site and shall become satisfied as to actual site conditions. Contractor shall verify all existing dimensions, and owner assumes no responsibility for variation in said existing dimensions.

> Whenever provisions of any section of the plans and specifications conflict with any union, trade association, or agency that regulates work described, the contractor shall make necessary arrangements to reconcile any such conflict without recourse to the owner and without delaying the progress of the work.

It is the intent of the plans and specifications to produce first-class-quality work. The contractor shall request an interpretation in any situation that would prevent the placing of first-class work. Should such a conflict occur in or between the drawings and/or specifications, the contractor is deemed to have estimated on the more expensive way of doing the work.

It is the intent of the contract document to provide for complete installation of all portions of the work. All items, materials, and equipment are to be furnished and installed complete and ready for operation or use. The contractor will be deemed to have based the bid on a complete installation.

Where additional or supplemental details or instructions are required to complete an item, the contractor shall request information from the architect. No work that depends upon this information shall be installed or fabricated without the written approval of the architect. Further, the furnishing of the data required shall not be the grounds for a claim for extra work. The contractor shall be deemed to have made an allowance in the bid for completing such work consistent with adjoining or similar details and/or the best accepted practices of the trade, whichever is more expensive.

Where the scope of the work of this contract in the plans and specifications requires supports, connections, or installation of any group of items furnished by other contractors, the omission of a given item from the drawings of this contract shall not relieve the contractor from the responsibility for installing, connecting, or supporting such item at no increase in contract cost.

Wherever additional materials or work not shown or specified is required to complete the work in accordance with the obvious intent thereof, the contractor shall provide these materials or work at no additional cost to the owner.

Approach Specified

One of the classic cases in the interpretation of exculpatory language in specifications is MacKnight Flintic Stone Company v. City of New York (160 N.Y. 72, 54 N.E. 661; 1899). The plans and specifications had been prepared by the city, and inspection work was by the city. As part of its work for the completion of a district courthouse and prison for the City of New York, the contractor was to furnish "all materials and labor for the purpose to make tight the boiler room, coal room of the courthouse and prison." The design documents contemplated a water problem, since the floor of the boiler room was 26 feet below curb level. The specifications described in detail the manner in which waterproofing was to be accomplished. Further, the specifications required that the facility was to be turned over to the city "in perfect order and guaranteed absolutely waterproof and damp proof for five years."

Leaks appeared, and the city refused to pay for the work. The contractor brought suit on the basis that he had followed the specifications exactly, and that nevertheless leaks had developed. The court agreed with the contractor. It noted that the construction contract was an agreement not simply to do a particular thing (i.e., waterproof the basement), but to do it in a particular way and to use specified materials in accordance with the design. Accordingly, in the failure of the specific design, the contractor could not be held accountable for the more general requirement to waterproof the basement.

Codes and Regulations

Implicit in every building contract is the legally constituted agreement that the contract shall neither request nor allow any illegal activity. Often a general conditions clause is included to the effect that all actions under the contract shall comply with applicable rules and regulations of governmental bodies and agencies. This approach is reasonable; the contractor is expected to obey the laws of the locality, county, state, and federal government. Specifically, in construction, the laws of the federal government include the regulations promulgated by the U.S. Department of Labor under the Occupational Safety and Health Act (OSHA).

A contractor's failure to be cognizant of these generally distributed rules and regulations is not an acceptable defense. In fact, the expectation that the contractor shall meet the rules and regulations of the government building codes goes a step further. On occasion, the contract general conditions include language similar to the following:

> The latest requirements of the National Electric Code shall govern the work specified under this section. In case of conflict between the specifications, drawings, and/or the National Electric Code, the more stringent requirement shall prevail.

> The contractor shall be required to comply with applicable regulations, rules, and requirements of the New York City Building Code and the requirements of all other authorities having jurisdiction. The latest editions or revisions will govern, and where conflicts occur, the more restrictive requirement shall be deemed to govern.

While on the face of it an experienced contractor would be expected to be familiar with the governing codes, the courts have found it unreasonable to expect the contractor to undertake a detailed comparison of the design plans and specifications with the applicable codes in the limited time available for bidding. To the extent that the governing codes and the plans and specifications conflict, the courts find the situation to be one of ambiguity and rule in favor of the contractor.

In Blount v. United States (346 F. 2d 962), the court recognized that contractors are businesspeople. Further, it specifically recognized that contractors are usually pressed for time during the bidding phase. The court defined the contractor's obligation as follows: "They are obligated to bring to . . .attention major discrepancies or errors which they detect in the specifications or drawings. . .but they are not expected to exercise clairvoyance in spotting hidden ambiguities in the bid documents."

In the case of Green v. City of New York (283 App. Div. 485, 123 N.Y.S. 3d 715; 1945) the case involved a contract where the specifications permitted the use of plasterboard partitions in bathrooms. However, in the general conditions, there was a statement that the contractor would have to comply with the rules and regulations of the state. Further, it provided that "this rule shall take precedence over any requirements of those specifications where a conflict occurs." The Department of Housing and Building of the State of New York ruled that plaster over lath was required by the Multiple Dwelling Law in partitions enclosing bathrooms.

In the suit, the court ruled against the city on the following basis:

> It could not be expected that the bidders would examine the various laws and building codes as to each item specified to see if the codes required some different method of construction. Such procedure would make the bidder's interpretation of the law, rather than the specifications, controlling. It would make the specifications so indefinite and uncertain as to destroy the validity of any contract awarded pursuant thereto.

In the Town of Poughkeepsie v. Hopper Plumbing & Heating Corp. (260 N.Y.S. 2d 901), a somewhat different situation arose. Two school districts had awarded contracts to Hopper Plumbing & Heating for plumbing work in new elementary schools. The town entered an action to prevent performance of that work until the plans had been submitted for a plumbing permit. This would be in keeping with the town plumbing code and ordinance requiring that all new plumbing work be inspected.

The commonsense issue would be one of public interest, specifically, public safety. However, the legal evaluation requires interpretation of the status of the two school districts. The court found that under the Education Law of New York State, which required approval of the plans and specifications by the Commissioner of Education, the New York State legislature had preempted the area of review from the local jurisdiction and vested it exclusively with the state. Similar exclusions occur in federal and state work for projects that occur within local jurisdictions. In essence, the local site is generally considered to become a federal or state reservation. (There are often specific exclusions to the exclusion itself, and each case must be examined on its own merits.)

The guiding principle continues to be that the contractor has the right to rely upon the specific information in the plans and specifications and is not called upon to identify those areas of overlap or ambiguity with existing regulation. Where such ambiguity or overlap occurs, the owner is responsible if the resolution imposes additional expense.

Once again, the contractor must be alert to maintain its rights and prerogatives. In Reetz v. Stackler (24 Misc. 2d 291, 201 N.Y.S. 2d 54; 1960), a contractor was to provide an acoustical ceiling in a bowling alley. Three of the general conditions invoked exculpatory language in terms of building codes and laws. The town building department advised the contractor that a two-hour fire rating would be required for the ceiling. A change to the contract was executed to that effect. After completing the work in conformance with the building department instructions (which was made a change), the contractor submitted a claim for extra work, citing the fact that the installed work was more expensive than that described in the contractual plans and specifications. However, the court found that the contractor had assumed the specific obligation of erecting the more expensive ceiling by agreeing to the form of the contract rider as it was presented.

No-Damages-for-Delay Clause

This usual clause in construction contracts states that, in the event of delay, the contractor will be compensated only by an extension of the contract end date. Such extension automatically extends the period of time before imposition of liquidated damages. In some cases, the clause refers to unforeseen delays rather than the broader term of delay.

In Kaplen & Son Ltd. v. Housing Authority (126 at 2d 13, N.J.; 1956), the plaintiff was the contractor for the building of public housing. Before construction could start, occupants of existing buildings had to be moved. A delay occurred, and the contractor sought to recover delay damages resulting from five months' delay. The contract included a no-damages-for-delay clause. The clause included language to the effect that such problems would include "hindrances or delays from any cause in the progress of the work, whether such hindrances or delays be avoidable or unavoidable, and the contractor agrees that he will make no claim for compensation, damages, or litigation of liquidated damages for any such delays, and will accept in full satisfaction for such delays said extension of time."

The court ruled against the contractor's claim because the clause was found to be binding. The court went further, finding that such a clause was "conceived in the public interest in protecting public agencies. . .against vexatious litigation based on claims, real or fancied, that the agency has been responsible for unreasonable delays."

In Psaty & Fuhrman v. Housing Authority (68 at 2d 32, R.I.; 1949), the contract included a no-damages-for-delay clause that applied "whether such delays be avoidable or unavoidable." The contractor made a claim on the basis that the clause excused reasonable delay only. The court held that the clause was valid and binding unless malicious intent, fraud, or bad faith could be proven.

However, in Cauldwell-Wingate Company v. State of New York (276 N.Y. 365, 12 N.E. 443; 1938), there was a no-damages-for-delay clause. Nevertheless, the court held that the clause did not apply to the delays incurred because of direct interference and misrepresentation of the state.

Similarly, in DeRiso Brothers v. State (161 Misc. 934, N.Y.; 1937), the no-damages-for-delay clause was not a defense to a claim for delay caused by failure of the governmental owner to provide temporary heat for interior finishing work, as specifically guaranteed in the contract. The court saw this omission as active interference, and a condition that could not be foreseen by the contractor.

In another case, an owner and his architect used the premise that the acts of the owner's architect were an unforeseeable cause of delay. Accordingly, a precedent set by Henry Shenk v. Erie Company (319 Pa. 100; 1935) would apply. In Shenk the contract included a no-damages-for-delay clause and related this to delays due to unforeseeable causes of delay. In the instant case, the contractor's counsel successfully argued that "acts of one's agents are patently not beyond one's control. . ." Further, he cited George A. Fuller Company v. United States (69 F. Supp. 409, Ct. Cl.; 1947) in which the court said the following:

> Nor has it ever been thought that the provisions. . .providing for an extension of time for completion of the work on account of delays due to unforeseen causes, including those caused by the government would serve to relieve defendant of liability for damages for such delays. . . .This provision for an extension of time related to the assessment of liquidated damages against the contractor and had no reference to the recovery of damages by the contractor for delays caused by the government.

Counsel also cited J. D. Hedin Construction Co. v. United States (347 F. 2d 235, Ct. Cl.; 1965) and Hall Construction Co. v. United States (379 F. 2d 559, Ct. Cl.; 1966).

In an arbitration between a contractor and subcontractor, the subcontractor cited Gasparini Excavating Co. v. Pennsylvania Turnpike Commission (409 Pa. 465; 1936), claiming that the general contractor had not made the promised work site available. While the facts did not support the subcontractor's claim, the attorney for the general contractor also challenged the applicability of the law in Gasparini, because he held that the subcontractor had not demonstrated a material interference.

Another case cited by the subcontractor was Sheehan v. Pittsburgh (213 Pa. 113; 1905). The court refused to apply the no-damages-for-delay clause on the grounds that the delay in the minds of the parties at the time of contract was not of a nature that could subject the contractor to a large loss. The parties were referring to "loss of damage arising out of the nature of the work," while the true cause for loss was the failure by the city to procure right-of-way. Further, the fact that the city had not condemned the required right-of-way was a fact unknown to the city, and therefore, unforeseen.

The courts, in finding that a no-damages-for-delay clause should not be enforced, look to both the unforeseeability of the cause and the nature of the interference on the part of the owner. In Shenk, the court defined "unforeseeability" as follows:

> Where a party under a delay and time extension provision on entering a contract foresees or should foresee that the work might be delayed by the failure of the owner or another contractor to perform the remedy therefor. . .[is an] extension of time on the part of those who perform the work, and the presumption arises that this was intended to measure the rights of the contractor thereunder.

Early cases often combined the right of the owner to make changes and the no-damages-for-delay clause to refuse legitimate damage claims within the contract. Thus, in Wells Bros. Co. v. United States (254 U.S. 83; 1920), the contractor was refused a claim for delay damages. The contract was for a federal post office/courthouse building. Immediately after the contract was signed, the government "ordered and directed" the contractor not to order the limestone specified in the contract because a change was anticipated. The contractor agreed to wait 2 weeks for the change, but in fact, was not given the new specifications for about 10 months, at which time an appropriation had been made by Congress. The change substituted marble for limestone, increased the contract by $210,500, and made an extension of time necessary.

During this 10-month period, the contractor had completed excavation, foundation, and the structural steel erection. The government, anticipating a major change in the scope of work with congressional approval of parcel post, ordered a delay that shut the job down for six months. The contractor sued for damages because of both delay periods.

The court interpreted the contract provisions that permitted the government to extend all or part of the work "without expense" but with an additional day-for-day time extension and also a "no-damages-for-delay" clause. The court refused the contractor's claim on the basis that the contract clauses were plain and comprehensive. The court, however, went on to make a philosophical statement that included a major presumption with regard to the manner in which contractors prepare their bids:

> Men who take million dollar contracts for government buildings are neither unsophisticated nor careless. Inexperience and inattention are more likely to be found in other parties to such contracts than the contractors, and the presumption is obvious and strong that the men signing such a contract as we have here protected themselves against such delays. . .by the higher price exacted for the work.

The federal courts continued to find in favor of the government position in the enforcement of the "no fault to the government" interpretation of the change and delay clauses. A 1946 case (United States v. H. P. Foley Co., 329 U.S. 64) involved a lawsuit for damages for delay. The delay occurred at the beginning of the project when the government was unable to make a site available to Foley, an electrical contractor who had bid a fixed fee for the installation of a field lighting system at National Airport.

The lighting system involved the runways and taxiways that were under construction. Before the areas could be stabilized enough for the electrical work to go in, hydraulic dredging and subsequent earth work had to be completed. The dredging required more time than the engineers had anticipated, due to instability of some of the dredged soil.

The court found that the government and its engineers had been diligent and stated: "The government cannot be held liable unless the contract can be interpreted to imply an unqualified warranty to make the runways promptly available." Finding that no such warranty was expressed in the contract, the court disallowed the claim. In its opinion, the court relied upon H. E. Crook Co. v. United States (270 U.S. 4; 1926) and United States v. Rice (317 U.S. 61, 63 S. Ct. 120; 1943). The court stated:

The contract reserved a governmental right to make changes in the work which might cause interruption and delay, required (the contractor) to coordinate his work with the other work being done on the site, and clearly contemplated that he would take up his work on the runway sections as they were intermittently completed and paved. . .This contract. . . set out a procedure to govern both parties in case of [contractor's] delay in completion. . . If delay were caused by [the contractor], the government could terminate the contract, take over the work, and hold the [contractor] and its sureties liable. Or. . .the government could collect liquidated damages. If, on the other hand, delay were due to "unforeseen causes" procedure was outlined for extending the time in which respondant was required to complete his contract, and relieving him from the penalties of contract termination or liquidated damages.

In the Crook and Rice cases, we held that the government could not be held liable for delay in making its work available to contractors, unless the terms of the contract imposed such liability.

A 1955 case signaled a turning point in the federal court interpretations of delay damages. In Ozark Dam Constructors v. United States (120 Ct. Cl. U.S. 354; 1955), a joint venture contracted for the construction of a dam on the White River in Arkansas for the government. The government was to furnish cement at Cotter, Arkansas, and the contractor would unload and transport it at his expense. The general conditions to the contract included the standard Article 9 and damages for delay, and also noted that for causes beyond the control of the contractor, including strikes, the contracting officer would extend the contract time on a day-for-day basis.

The contract promised that upon 30 days' notice, the government would ship cement by railroad to Cotter, Arkansas. The contractor requested 12,000 barrels of cement for September delivery and 41,000 for October. A railroad strike prevented the deliveries, and the suit by the joint venture was for damages for nondelivery. Evidence established that both the contractor and the government had anticipated the possibility of a railroad strike. However, the government did nothing to seek out possible alternate methods of making delivery. Further, evidence indicated that delivery could have been made by truck.

The contract also included a clause that stated: "The government will not be liable for any expense or delay caused the contractor by delayed deliveries except as provided in Article 9 of the contract." After administrative reviews by the contracting officer and the secretary of the army, suit was entered. The opinion of the court stated: "A contract for immunity from the harmful consequences of one's own negligence always presents a serious question of public policy." In regard to the specific failure to deliver the cement, the court stated:

The possible consequences were so serious, and the actions necessary to prevent those consequences were so slight, that the neglect was almost willful. It showed a complete lack of consideration for the interest of the [contractor]. . . . Our conclusion is that the nonliability provision in the contract, when fairly interpreted in the light of public policy, and of the rational intention of the parties, did not provide for immunity from liability in circumstances such as are recited in the [contractor's] petition.

This court, viewing construction in a broader context, felt that such immunity clauses, particularly when poorly administered in the field, could only serve to require all contractors bidding for government work to include a prohibitive contingency in their price. In an earlier finding, the court held that the government had the responsibility to be reasonable in its administration of contracts. In Severin v. United States (102 Ct. Cl. U.S. 74; 1943) a hospital was to be constructed within 500 calendar days. While the work was in progress, the government stopped all work above the ninth floor pending certain changes. This delay continued for five months. At that time, the contractor was ordered to proceed with changed work.

After some of this work had been completed in its new configuration, the government issued an order canceling the changes and reverting to the original plans. A change order was allowed, increasing the price and establishing a three-month extension. However, the court found that the change orders were simply the actual cost of the actual work and had nothing to do with the delay. Despite the standard federal phrases in the general conditions, the court allowed damages because the contractor "was delayed by the non-use of his equipment, the idleness of his supervisory employees, rental cost of equipment, and extra costs of operating his sand and gravel pit, and the extra costs of the subcontractors. . . Due to the [government's] procrastination and its inability to decide definitely what it proposed to do . . .the contractor's extra costs were all in addition to the work which was performed under the change orders."

The findings here did not actually run against findings in cases such as United States v. Rice but saw the unwarranted interference with the contractor's operations as being beyond the contract and therefore not limited by the clauses of the contract.

The Rice doctrine, over the years, has given license to the owner (and the owner's agents) to impose unfair advantage upon the contractors in the form of delays that are not compensable because of the bar against claims for the damages due to the delay. In the case of Ross v. United States (115 F. Supp. 187, 192, Ct. Cl.; 1953), the court found that delay was a breach of contract. Although a change order had been given for work involved, the court held that the amount included in the change order did not compensate for the actual cost of the delay but only for the work, stating: "In this case, [the contractor] accepted the $4000 without prejudice or any claim for increased costs resulting from delays in performance; but, independent of this reservation, we think [the contractor] is entitled to recover, because this was a breach of contract."

This approach was reaffirmed in F. H. McGraw & Co. v. United States (103 F. Supp. 394; 1955), where it took more than 1000 days to construct an addition to a veteran's hospital as compared to the contract time of 400 days. The contracting officer had found that the contractor was not responsible and had issued extensions of time. At least 159 days of the delay was due to an ambiguous partial stop order that was finally rescinded. The court held that the government had delayed the contractor by taking an unreasonably long period of time to issue the change orders, stating: "It is settled that the [government] is allowed under the contract only a reasonable time within which to make permitted changes in the specifications, and that the [government] is liable for breach of its contract if it unreasonably delays or disrupts the contractor's work."

In addition to material breach, the courts will find that a prospective breach can be proper cause for discontinuance of activity. Thus, in J. D. Hedin Construction Co. v. United States (347 F. 2d 235, Ct. Cl.; 1965), when the government sought to require the contractor to do exploratory work on the nature of the subsurface before the government would revise plans for footings, the contractor refused to do so, and ultimately sued for delay damages. The government defended on the grounds that the delay was caused not by its failure to review the plans, but rather by the contractor's failure to perform the exploratory work that had been ordered (but which was clearly beyond the scope of the contract). The court rejected the government argument summarily: "The short answer to this. . . is that the [government] could not impose on [the contractor] such a duty without first authorizing a change order or a proceed order assuring [the contractor] compensation for this work which was beyond the scope of the contract."

The finding of a breach of contract can open the door to recovery on the part of the contractor of all losses or costs (as the case may be) from the point of breach forward. Once the contract has been breached, the method contained in the contract for making adjustments in compensation is inapplicable. This was the finding in Merritt-Chapman & Scott Corp. v. United States (429 F. 2d 431, Ct. Cl.; 1971): "A true breach of contract claim which, by definition, is outside of the scope of the contract, is subject to neither equitable adjustment under the contract nor to administrative review or resolution. Jefferson Construction Co. v. Untied States, 392 F. 2d 1006, Ct. Cl.; 1968."

In George A. Fuller Co. v. United States (108 Ct. Cl. U.S. 70; 1947) the decision discussed breach due to interference by the owner:

> One who, while preventing the other party from carrying out the contract, nevertheless hinders or delays him in doing so, breaches the contract, and is liable for the damage which the injured party has sustained thereby. The Supreme Court so held in United States v. Smith, 94 U.S. 214. In this case, it was said "under such circumstances, the law implies that the work should be done within a reasonable time, and that the United States would not unnecessarily interfere to prevent this."

This case also introduces the concept of implication of performance in a reasonable time and avoidance of unnecessary interference.

In the nonfederal environment, an earlier case discussed breach of contract as a means of providing an equitable solution to the no-damages-for-delay clause in a contract. The case was Selden Breck Construction Co. v. Regents of University of Michigan (274 F. 982, Mich.; 1921). The project was started late and finished late because the university was unable to deliver the site at the time intended. Under the no-damages-for-delay clause, the university sought to avoid responsibility for the delay damages. However, the court held that the contractor's willingness to complete the project after the delayed delivery did not preclude his rights to sue for damages under breach of contract. The decision indicated:

> The correct rule is that upon breach of a building contract by failure of the owner to perform his obligations under such contract, which delays the contractor in completing his work thereunder, the latter is not obligated to abandon such work, but may elect to continue therewith after such breach and, upon performance of the contract on his part, is entitled to recover the damages sustained by him as a result of the delay caused by such owner.

In a 1970s case (Kalish Jarcho v. City of New York, 58 N.Y. 2d 377) the court found in favor of the contractor and awarded damages for delay. On appeal, the verdict was overturned. The appeal court decision (448 N.E. 2d 413, 461 N.Y.S. 2d 746; 1983) found that damages for delay can be awarded only if the cause of delay could not have been contemplated at the time the bid was prepared.

Delays due to Change Orders

Most contracts start with a very limited time frame for the accomplishment of a substantial amount of work. The majority of contracts, for a variety of reasons, exceed the initial completion estimate. Accomplishment of the basic contract work is usually difficult enough in a timely fashion. However, the owner typically reserves the right to change the contract. This right, combined with the no-damages-for-delay clause, operates to require the contractor to take on additional work, even when meeting the base contract schedule is difficult. In the negotiation of each change order, there should be an evaluation of the impact upon the overall time frame, and additional days should be added to the base contract.

The ability of the government (owner) to impose additional work on a contract, even when behind schedule, was affirmed in the case of Crook Co. v. United States (270 U.S. 4; 1925). The court stated:

> The contractor. . . must satisfy the government of his having the capital, experience, and ability to do the work. Much care is taken, therefore, to keep him up to the mark. Liquidated damages are fixed for his delays. But the only reference to delays on the government's side is the agreement that if caused by its acts, they will be regarded as unavoidable, though probably inserted primarily for the contractor's benefit as a ground for extension of time, is not without bearing on what the contract bound the government to do. . . The [contractor] agreed to accept in full satisfaction for all work done. . . the contract price. . . Nothing more is allowed for changes, as to which the government is master. . .The [contractor's] time was extended, and it was paid the full contract price. In our opinion, it is entitled to nothing more.

In Frazier-David Construction Co. v. United States (97 Ct. Cl. U.S. 1; 1942) the contractor filed a claim for additional costs due to subsurface conditions different from those shown in the plans and specifications. A change order had been issued for the basic increase in costs, and an extension of time was granted. However, the claim was for additional costs due to the delay, which the court dismissed.

There is obviously a point at which the number and value of change orders will materially change the nature of the contract either by increasing the contract value or by requiring the contractor to employ a different level of resources, funding, equipment, and supervision than had been contemplated. Based on the authority of officials in many jurisdictions to approve change orders up to 5 percent of the contract value without going back to higher authority for approval, it is the general custom of the construction trade that change orders in this magnitude are to be expected. (The actual number of change orders has little meaning, since small changes can be packaged into individual larger change orders.)

In Magoba Construction Co. v. United States (99 Ct. Cl. U.S. 662; 1943) the court held that the issuance of 62 changes during the progress of work did not constitute a change in the basic scope of the contract. The contract was for approximately $2 million for post office reconstruction.

An even more extreme example occurred in Great Lakes Construction Co. v. United States (95 Ct. Cl. U.S. 479; 1942) where, during the construction of a federal penitentiary, 109 change orders were issued. However, this understated the real situation. The basic bid was made on an incomplete set of plans and specifications. Shortly after the contract was made, a new set of plans and specifications was delivered, which included some 700 additions, revisions, and corrections. The court of claims found that this number of changes did not violate or breach the contact, and a claim for damages due to delay was disallowed.

An important case that counteracted the premise that the owner could impose virtually unlimited change upon a contract is Ross v. United States (115 F. Supp. 187, 192, Ct. Cl.; 1953). In this case, the contractor was granted a number of time extensions for changes, but the change orders compensated him only for the cost of the changes and not for the loss due to delay. The court found that the contract was "entitled to recover because this was a breach of contract."

Notice of Delay

Contracts usually include in their general conditions a requirement that the contractor shall notify the owner of delay, usually within 10 days from the beginning of the delay. The following clause is taken from the general conditions form used by the state of New Jersey: "The contractor shall, within ten days from the beginning of such delay, unless the Director shall grant a further period of time prior to the date of final settlement of the contract, notify the State, in writing, of the causes of the delay." The general conditions of the New

York State Dormitory Authority contracts include a similar clause: "The contractor shall, within ten days from the beginning of such delay, notify the Owner, in writing, of the causes of the delay."

The giving of notice in writing is a part of normal project administration. The form need not be argumentative. Some contractors are reluctant to continually give notice of apparent delay because of the overtones of litigation that are implied. For instance, each day of weather considered to be in the category of force majeure must be the subject of formal notification. To meet this requirement, the contractor should set up criteria for determination that weather is excessive in comparison to an established prior record from the U.S. Weather Service. Failure to give due notice of delay as required by the contract can be overcome, but not in all cases. Therefore, prudence indicates that a documentation approach should be established.

One of the early cases supporting the requirement for notice was Louisville and N.R. Co. v. Hollerback (5 N.E. 28, Ind.; 1886). In this case, the defendant delayed the plaintiff's work by failing to prepare adequate foundations. The plaintiff sent a notice indicating the cause of his inability to complete. The court affirmed recovery of delay damages noting: "The contractor may not acquiesce in the suspension in silence. . . but if notice be given of his readiness and willingness to prosecute the work to completion within the time agree upon, and that its suspension will involve him in loss, we can discover no principle upon which it can be held upon the contractor."

Peter Kiewit Sons Co. v. United States (60—1 BCA 2580; 1960) was heard before the Federal Board of Contract Appeals. Despite contractual requirements for an on-site source for subbase material, this material had to be obtained elsewhere at greater cost. Kiewit sued for damages due to the extra work and delay involved. The board noted:

> The government's defense is in part based upon the fact that it did not receive notice of this claim until. . .after completion of the construction work called for by the contractor, but prior to final payment. The government was, however, completely familiar with the difficulties appellant was experiencing. . .Under these circumstances, a formal written notice of changed conditions or a formal notice that a claim would be filed was unnecessary.

However, in the two preceding cases, the owner had either interfered with the contractor or there were changed conditions that created a situation beyond the scope of the contract. Other citations that permit recovery without notice also involved changes in the contract beyond contract's scope. These include Nat Harrison Associates Inc. v. Gulf States Utilities Co. (491 F. 2d 578, 5th Cir.; 1974), Mullen Construction Inc. v. United States (72—1 BCA 9227; 1972), and Eisen-Magers Construction Co. v. United States (59—1 BCA 2234; 1959).

It is clear that failure to fulfill the requirement for notice to the letter and the spirit of the contract can jeopardize a contractor's proper claim to damages for delay.

Written Approval for Change in Scope

In Citizen's National Bank of Meridian v. Glasscock (243 So. 2d 67; 1967), the contractor sued Citizen's National Bank for work beyond the scope of the contract. This work had been ordered by the owner's engineer. The trial court awarded the contractor $8902 on a quantum meruit basis. However, the reviewing authority reversed the decision, since the contract clearly set forth that "no extra work change shall be made unless in pursuance of a written order from the owner." On this basis, the reviewing authority held that the owner had to pay for extra work only if it was authorized in writing prior to its execution. This court interpreted that the contract between the parties spoke directly to this point, leaving no leeway for an award on a quantum meruit basis.

The requirement for a written change order does not necessarily preclude a finding for quantum meruit. In some jurisdictions, including federal and New Jersey, there can be a waiver for a requirement for a written change order. In New York, similarly, there can be a waiver, but not if the contract expressly provides that the agreement cannot be changed orally.

In Lord Construction Co. v. United States (28 F. 2d 340, 2d Cir.; 1928) the court stated:

> The right of the plaintiff to recover for extras not covered by the contract and for which the plaintiff produced no written orders signed by the engineer. . . [wherein] the contract provided that the plaintiff should not be entitled to receive payment for any extra work as extra work unless such bill for extras be accompanied by an order in writing from the engineer, . . .the court admitted proof which tended to show that at the regular meetings . . . the engineer and the plaintiff fully discussed and considered such extra items and work, and the plaintiff was then directed to proceed with them . . . That a contract requirement such as here provided may be subsequently waived by the parties is established by the authorities.

The general conditions usually include a statement that changes to the work will not be authorized unless they are approved in writing by the owner. Also, the contractor generally includes specific language limiting the authority of the architect or engineer to make authorized changes. In Reid Co. v. Find (139 S.E. 2d 829), deviations from the plans and specifications by the contractor were found to be unauthorized, even though the architect and engineers for the owner had approved them. The court held:

> Whatever may have been said or done by the architect or engineer to lead the [contractor] to believe they had approved the changes, their actions were in direct conflict with the provisions of the contract which required that written approval should be had before major change could be made. . . These contractual requirements were well known to the (contractor) and equally as binding upon it, as they were to the [owner], architect and engineer.

While there is precedent for ratification of oral approval of changes in the work, the contractor who accepts such an informal direction is in great potential jeopardy. Similarly, the contractor should be alert to promptly notify an owner of the intention to protest or appeal a change order, when that notice is required by the contract.

There are occasions when the owner waives the right to require notice. One such instance was the Arunel Corp. v. United States (96 Ct. Cl. U.S. 77; 1942) in a government contract for the construction of a lock and dam on the Savannah River. The contractor had a proper request for claim on a particular change but had failed to give notice. The contracting officer, however, considered the claim on its merits without reporting or noting that the claim had been filed late. The court rejected a defense on the basis of failure to make due notice of claim, deciding that the contracting officer's action had shown that he did not rely on the failure to file the protest and thereby waived the prerogatives for that particular notice under the contract.

The general conditions usually include a requirement to the effect that "any claim arising from such delay must be made in writing to the engineer immediately upon occurrence of the delay." Or, "if the contractor anticipates any claim arising from delay during construction, notice of such claim must be made in writing to the owner or authorized representative immediately upon occurrence of the delay." While sometimes difficult to implement, since the actual impact of a delay may require some substantial time to evaluate, nevertheless, it is relatively easy to notify the owner that delay has occurred and that its impact is being evaluated. In fact, a form letter can be a mutually convenient means of so stating.

In Roanoke Hospital Association v. Doyle & Russell (Cir. Ct. Roanoke, Va.; review by S.C. of Va. #740212; 1975), the Doyle & Russell general superintendent had testified, in effect, that his organization had not been regularly making written notice of delay because it did not want to irritate the owner. Constant notices of delay and claims for impact from the contractor can create an image of a contractor who is preparing a major claim. Generally, it is the project management representatives of the owner who are annoyed, but any attempt to modify or waive this requirement would be frowned upon by the owner's legal counsel.

The courts have acknowledged that notice provisions are salutory in construction contracts. In United States v. Cunningham (125 F. 2d 28, D.C. Cir.; 1941), one of the contract provisions read: "The contractor shall within ten days from the beginning of any such delay notify the contracting officer in writing." There was some delay in commencing work, but the contractor did not give any written notice. Instead, he gave oral notice. The court held that the oral notice was not in compliance with the contract, stating: "Obviously, the intent of this provision is to inform the government of the cause of delay and afford an opportunity to remove it, and likewise to warn the government of the intention of the contractor to insist upon it as a means of prolonging the stipulated time for completion of the work."

In Scott Township School District Authority v. Branna Construction Corp. (409 Pa. 136; 1962) the Pennsylvania Supreme Court held that the contractor's claim for extra work was barred because the contractor had not complied with the contract provision that required communications between the parties to the contract to be in writing. The court held: "Where a public contract states the procedure in regard to work changes and extras, claims for extras will not be allowed unless these provisions have been strictly followed."

The filing of appropriate notice of both delay and the probable impact can be of mutual benefit to contractor and owner when the evaluation of the claim is made. Such notice will serve both parties best if it is factual and to the point. Argumentative, self-serving discourses are inappropriate as part of a notice of delay and an indication of intent to claim for damages.

Authority to Approve Change

In a recent Virginia case (Trustees of Asbury United Methodist Church v. Taylor and Parish, Inc., No. 940162, Va.; Jan. 13, 1995), the Virginia Supreme Court strictly construed the term "contract documents." Taylor and Parish had a guaranteed maximum price contract with the Asbury United Methodist Church to construct a new church facility. During construction, two of the church's trustees entered into a change order, which added about $300,000 in cost. The church refused to pay the change order, holding that the two trustees did not have the authority to bind the church.

The contractor invoked an arbitration clause, compelling arbitration of "claims, disputes, and other matters arising out of, or relating to, the construction documents or the breach thereof." The contractor sued the owner under theories of both breach of contract and quantum meruit (unjust enrichment). The arbitration proceeded and, in the course thereof, the arbitrator found that, indeed, the two trustees could not bind or commit the church, and that therefore the church was not liable for the additional costs of the invalid change order. The arbitrator did award the contractor over $260,000 as quantum meruit relief. The circuit court confirmed that award.

The Virginia Supreme Court agreed that the two trustees could not bind the church. This court explained that the contract documents were limited to the standard form of agreement, specific documents listed, and "all modifications issued after execution of the contract," including any "change order," which was defined as "a written order to the contractor signed by the owner and the architect. . . authorizing. . . an adjustment in the contract sum." Because the two trustees could not properly execute a change order, in the eyes of the court, the change order did not exist. Accordingly, under the arbitration clause, the contractor did not have the right to demand arbitration regarding the change clause.

Further, since a quantum meruit claim does not arise out of the contract documents but is based on equitable resolution, the Virginia Supreme Court found that the quantum meruit claim was not valid. The language of the arbitration clause clearly limited all claims and disputes to the confines of "the contract documents."

Ambiguous Plans and Specifications

Although the law, as stated in Rosenman v. United States (U.S. Ct. Cl., #211-65), is generally in favor of the contractor—or at least the contractor's choice of interpretation if two reasonable interpretations (or more) can be drawn from a particular set of plans and specifications—the contractor should not be complacent in this position. In Meyers v. Housing Authority of Stanislaus County (50 Cal. Reptr. 856), the court found that the drawings were definitely ambiguous. Nevertheless, the trial court and the appeals court ruled against the contractor. The contractor had bid on two housing projects. Part of the sewer and drainage lines shown on the drawing (and necessary to connect the dwelling units to the existing utility mains) extended beyond limits that were shown variously as contract limits, property line, and project limits shown on the plans. The contractor contended that the drawings thus limited their responsibilities to the area within the designated boundaries.

While the court found that the drawings were definitely ambiguous, they also noted that the drawings constituted only one document in an integrated contract. Since all documents that are part of the contract must be construed together, the drawings were interpreted by the court in light of the specifications. In the specifications, several systems were delineated that were to be done by others at no expense to the contractor. These exempt systems did not include the connection to utility mains. The court interpreted failure to include "off-site" plumbing in this itemization as meaning that it was included in the work.

Clearly, the specifications in this case were clear enough, while the drawings were definitely ambiguous. The owner won by a slim margin. During reviews of the plans and specifications, the design team and the owner's representative should review the drawings from the viewpoint of the contractors who are going to bid on them. This is particularly important in a project that is subdivided into a number of contracts. Clear delineation must be made of those areas that are "not in contract," or "by others." Further, in reutilizing the same sheets for different contracts, it is preferable to renumber the sheets by using reproducible master copies to avoid any confusion. Failure to do so can produce a small economy in drawing preparation at a great expense to the owner and/or the designer.

In New England Foundation Company v. Commonwealth (327 Mass. 587, 100 N.E. 2d 6; 1951), a more classical ambiguous situation was involved. The contractor was required to install more than 4000 cast-in-place concrete piles with a safe working load capacity of 20 tons. The specification described a method of testing to validate the safe capacity of a typical pile. The specification also gave a pile-driving formula (similar to the *Engineering News-Record* formula).

Each pile was driven according to the formula. A load test was conducted after 4325 piles had been driven; unfortunately, this test indicated that the piles would not hold a 20-ton load within the criteria. The owner insisted on the driving of an additional number of piles to make up the deficit in load capacity. The contractor sued for the cost of the extra piles, claiming that they were extra time and extra work. The court agreed with the contractor that, given two methods of determining the pile capacity, the contractor had the prerogative of assuming that the formula would meet the specifications.

It is a recognized legal principle that an ambiguity will be resolved against the party who drafted the specification. If there are two reasonable interpretations, the contractor's interpretation will prevail over the designer's interpretation, with the exception of design-build situations. Since the designer is the owner's representative, this, in effect, means that the contractor's position will prevail over the owner's position.

Citations and quotations to this effect were listed in the Egan v. City of New York (17 N.Y. 29 90, 18 A.D. 2d 357) brief as follows:

- In Camarco Contractors, Inc. v. State of New York (40 Misc. 2d 4S6; 491, 243 N.Y.S. 2d 240), the Court said: "The court is bound to resolve an equivocal provision of the contract or the interpretation thereof against the one who drew it. That is fundamental law."

- In Heating Maintenance Corp. v. State of New York (206 Misc. 605, 134 N.Y.S. 2d 71), the Court of Claims said: "The State prepared the entire contract on its own forms and all reasonable doubts as to the meaning thereof are to be resolved against it, as it is responsible for the language used and the uncertainty thereby created."

- In Frye v. State of New York (192 Misc. 260, 78 N.Y.S. 2d 243) the Court of Claims repeated the rule of law that, where there is an ambiguity in the contract, the same must be resolved against the State that prepared it. Specifically:

    In construing the contract in question, the same rules of construction are applicable as between individuals. . .If the language used is capable of more than one construction, the Court must resolve all doubts against the person who uses the language and most beneficially to the promisee.

- In Evelyn Building Corp. v. City of New York (257 N.Y. 499, 513), the Court of Appeals said: "The contract in question was prepared by the City, and, in case of doubt or ambiguity must be construed most strongly against it."

Quality of Plans and Specifications

The architect or architect/engineer developing a design has the problem of transposing all of his or her thoughts into the plans and specifications. The goal is to transmit, in a clear and concise fashion, the intent of the design. It is quite common that the effort will fall short in some cases, causing confusion and leaving the owner open to a claim for added scope or extra work on the basis of quantum meruit. There are cases, too, where the designer says too much and there is overlap. In these cases, ambiguity develops and imposes a different set of problems.

In the situation of ambiguity, the designer often believes that she or he has the prerogative to direct the contractor to utilize the interpretation that the designer had always intended. The courts do not agree. In Rosenman v. United States (U.S. Ct. Cl. #211-65), the court stated: "Although the specifications and drawings may have been clear as a bell in the mind of [the] architect, it is not the subjective intent that is the legal determinant. . . rather, it is the representations of the specifications and drawings themselves." The court went on to note a well-defined principle of law: "When the government draws specifications which are fairly susceptible of a certain construction and the contractor actually and reasonably so construes them, justice and equity require that the construction be adopted."

This legal hazard, implicit in any situation where the plans and specifications may be variously construed, is often ignored or unrecognized. Further, the general conditions section often contains exculpatory clauses or language through which the designer hopes to be able to impose intent, even in the absence of definitive plans and specifications.

If the quality of the plans and specifications is such that an area of ambiguity does exist, the question of legal responsibility can become quite difficult. In Reid v. Fine (139 S.E. 2d 829), the contractor sued because the owner refused to make final payment for an air-conditioning and heating system. The court found that the contractor had performed an unauthorized deviation from the plans and specifications.

However, the contractor presented evidence that showed that the plans were deficient in a number of areas. The ducts from floor to floor could not be installed as described on the drawings because of an interference by existing electrical conduits. Further, the room that had been designated to contain the HVAC equipment was not large enough for the equipment specified. The contractor advised the designer that he would be unable to install the work.

Based upon informal concurrence by the designer, the contractor made changes in both the equipment and the methods of routing same to remedy the situation. However, the resulting HVAC system was approximately 50 tons lower in capacity than that required, the primary air handling was below capacity by almost 8000 cfm, and a number of other instances were cited where either capacity or quality was less than that required.

The contractor lost his suit, principally because he had not proceeded properly upon determination of the problems. The owner had been left out of the dialogue regarding the problems and was awarded a suitable adjustment in contract costs in compensation for the performance of the contractor.

Warranty of Design

In Luria v. United States, the court stated that the owner implies a warranty of design that, in effect, says: "If the specifications are complied with, satisfactory performance will result." Basically, the warranty states that a contractor who bids on the basis of plans and specifications has the right to depend on those documents to accurately indicate the conditions of the job.

Further, if actual conditions are otherwise than those described, then the contractor is entitled to proper compensation for any additional costs.

One of the leading cases cited is Hollerbach v. United States (233 U.S. 165; 1914). In this case, the contractor was to repair a dam. The specifications qualified that the quantities given were approximate and that the bidder should visit the site to determine the nature of the work. However, the specifications then went on to specifically describe the material that would be found in the dam. These materials were generally described as broken stone, sawdust, and sediment. When the contractor proceeded with the work, he found that the backing was made up principally of sound wooden cribbing of logs filled with stone. The cribbing was so sound that removal was much more expensive than the contractor had anticipated.

The government held that the specifications had been approximate (per statement), and that the contractor was supposed to visit the job site. However, that court held—and many have held since—that a contractor site visit as a prerequisite is not a bar to a claim for unforeseen conditions. Further, even if a contractor does visit a site, it is not contemplated that the contractor will be required to take borings or make unusually extensive explorations that are more in the province of the owner and the designer.

Another long-established decision, Christie v. United States (234 U.S. 234; 1915), discusses the question of soil borings. The specifications for construction of locks and dams provided that "the material to be excavated, as far as known, is shown by borings, drawings of which may be seen at this office, but bidders must inform and satisfy themselves as to the nature of the material." The borings showed gravel, sand, and clay, but no other materials.

During excavation, the contractor encountered large stumps below the surface of the earth, as well as buried logs that were cemented together with sand and gravel. The penetration of this material during the boring phase had been difficult, and it turned out that the government engineers had continually moved to areas where the rig would drill through. The court found that even though there was no ulterior motive by the government engineers in boring in the easy locations, this did not shift the responsibility for the difficult foundations work over to the contractor.

The owner has two basic choices. To provide information in the plans and specifications, the owner must stand by that information. If the actual situation proves to be more difficult, then legal precedent indicates that the owner will have to pay for the additional expense, whether it be because of delay, additional equipment, or additional work.

On the other hand, the owner choosing to be cautious (or devious) may choose to provide little or no information, shifting the entire responsibility to the contractor. In such a case, the owner who discloses less than is known because it may be inaccurate will at least pay whatever risk factor is introduced into the contractor's bid. In this situation, the courts would find a contract enforceable, but the prudent contractor will have put in a substantial amount to cover the uncertainties. Many of the bids that far exceed the owner's estimates for major projects, such as dams and highways, are a direct result of timidity on the part of the owner in refusing to limit the risk to the contractor.

The owner who knows that a subsurface situation is difficult and deliberately withholds the information may well be held responsible for deliberate or constructive fraud. Clearly, owners should be willing to tell all they know or think they know about the site. The courts have found that the owner and designer, having much more time to examine the situation, enjoy a position of superior knowledge. Accordingly, the courts inevitably favor the contractor, who is required to bid upon the basis of plans and specifications prepared by others.

In cases of misrepresentation or constructive fraud, the general factors that must be present, include the following:

- The plans and specifications must make a misrepresentation of facts.

- The fact misrepresented must be a material one.

- The misrepresentation must have been made knowledgeably. (If the misrepresentation is not willful, it will be construed to be willful if the background, information, and experience of the maker would have normally led to an opposite presentation of the facts.)

- The owner or agent must have induced the contractor to rely upon the misrepresentation.

- The contractor must have relied upon the information as given.

- The contractor's belief of the information must have been justified (i.e., it cannot fly in the face of his or her experience and knowledge).

- The misrepresentation must have caused actual damage.

The contractor who relies upon plans and specifications and then discovers a misrepresentation before completion of the contract may choose one of two paths: continue the project and then enter suit to recover additional costs, or refuse to continue to work on the basis that the misrepresentation constitutes a breach (United States v. Spearin, 248 U.S. 132; McConnell v. Corona City Water Company, 85 P. 929, Calif.; Lentilhon v. New York, 185 N.Y. 548; Carroll v. O'Connor, 35 N.E. 1006).

Evidence of an actual fraud or misrepresentation is often difficult to produce. Ironically, constructive misrepresentation or fraud, whether intentional or unintentional, is often much easier to identify—for instance, if an owner had borings taken that would have disclosed a difficult situation but inadvertently failed to provide the information to the contractor. Soil borings are often inaccurate because of failure to promptly locate the boring hold and other factors. For this reason, it is not unusual for an owner to be uneasy about furnishing the information, fearing that it might be inaccurate.

5

The Change Order Process

Change orders originate in a variety of ways. At a point in the process, the potential change order (PCO; also preliminary change order) is recognized by the project manager [PM; also construction manager (CM)] and given a PCO number (chronological).

Upon identification as a PCO, the PCO enters a process. This process varies greatly with the owner's organization. Most governmental agencies have a rather cumbersome approval process, often mandated by charter, by law, or by regulation. Quasi-public agencies (transportation authorities, etc.) are similarly encumbered by lengthy approval processes.

From PCO identification to the start of the final approval process, the scope of the change order (CO) must be described by the architect (A/E), reviewed separately by the contractor and the PM/CM, negotiated (contractor and PM/CM), and then submitted to the approval process.

Request for Information (RFI)

This document is generated by the prime/general contractor. It may pass through RFIs from subcontractors or other primes. The issuance of RFIs usually peaks during the shop drawing phase of the construction process. The process is a combination of legitimate questions, failure to comprehend the contract documents, and setting up potential claims for defective design/contract documents. See Fig. 5-1.

It is important to respond to RFIs in a timely manner. The response time and the total number of RFIs can be factors in a contractor claim of shoddy management and/or deficient design. The RFI can also be used as initial notice to the owner that a problem exists. See Fig. 5-2.

RFIs played a major role when Bechtel National, Inc. (BNI) submitted a $4,155,176 claim against NASA (90-1 BCA 22549, NASA BCA No. 1186-7). On December 23, 1983, NASA awarded BNI a fixed price contract in the amount of $13,075,376. The contract required BNI to perform structural, mechanical,

XYZ Construction Company
Request for Information
John Doe Project

RFI # _____   Specification Section _____

Date _____   Drawing _____

Subject:

Please provide the following information:

Figure 5-1   Request for information (RFI) form.

TO: XYZ Construction Company

RE: John Doe Project

FM: Construction Manager (OK)

RFI# _____   Response Date _____

Category   Clarification
                                  Start Change Order Process (PCO)
                                  Submit Schedule Impact Analysis
                                  Submit Cost Estimate
                                  Coordination, within Scope
                                  Other _____

Response:

Figure 5-2   Response to RFI form.

and electrical work to reconfigure mobile launch platform No. 3 (MLP-3), constructed as part of the Apollo program to accommodate the space shuttle. The contract incorporated detailed plans and specifications for the work. MLP-3 was intended for use in the launch of STS-44, then scheduled for November 1986. MLP-3 was needed by January 1986 for testing and validation work to be done by other contractors.

The contract provided a substantial completion milestone and a final completion date based on the notice to proceed. NASA issued the notice to proceed on January 3, 1984. The substantial completion milestone date, June 26, 1985, was 550 days from receipt of the notice to proceed. The contract completion date, October 24, 1985, was 660 days from the receipt. The contract specified that, in order to meet the schedule, more than one shift and more than a five-day work week might be required.

The system used by the parties to resolve problems encountered during the performance of the contract involved the exchange of RFI forms on which Bechtel referenced contract requirements and wrote a statement of the problem on the top half of the page, and NASA wrote its directions for correction of the problem on the bottom half of the page. The contract required prompt reporting of "all construction problems and design deficiencies encountered during construction so that solutions or workarounds may be readily provided by the Government."

BNI had not expected many changes or design errors because the MLP-3 was the third of a kind, and the designer had developed the contract plans and specifications based on plans and specifications used in the earlier configuration of MLP-1 and MLP-2. BNI presented evidence of inaccuracies found in the plans and specifications.

The main purpose of the RFIs was to define and correct discrepancies in the specifications and drawings, but they were also used for certifications, deviation waivers, and general information. During the terms of the contract, 1209 RFIs were written. In preparing the claim, BNI summarized the RFIs by giving each one a descriptive title, the dates issued and answered, a classification as to the discipline involved (structural, mechanical/piping, or electrical), a judgment of causation, and a judgment of impact. This summary, "NASA Impact RFIs," is Exhibit 4 to the claim.

The conclusion in Exhibit 4 is that 848 of the 1209 RFIs resulted from NASA-caused problems and had impact ("NASA RFIs"). BNI classified the type of impact as "major" if the problem was discovered while the work was in progress, "minor" if discovered before the work began, and "cumulative" if only engineering time was involved to address the problem. There was no impact if the RFI was unnecessary or classified as a Bechtel responsibility.

The classification of an RFI as "major" signified to BNI that it caused "a definite disruption to our crews in the field." BNI did not summarize how many NASA-caused RFIs had "major" impact or when they were issued, and did not acknowledge that problems discovered while work was in progress might not cause substantial disruption. According to BNI, the extent of the impact on an individual basis from particular RFIs could not be determined.

A majority, but not all, of the matters dealt with through the RFI system arose from inaccuracies or deficiencies in drawings and specifications. Some RFIs were prepared for BNI's own benefit to obtain a deviation waiver for cost savings, suggest an improvement, correct mistakes that had been made, or for other reasons. Some RFIs asked unnecessary questions.

One RFI about a vendor's telephone number was considered frivolous. Other RFIs resulted from problems that Federal Steel or other material suppliers had caused. BNI acknowledged these sources of RFIs and did not maintain that all 1209 RFIs were attributable to factors for which NASA was responsible.

The period of heaviest impact from the NASA RFIs was claimed for March–August 1985, with multiple RFIs classified as among the 848 NASA RFIs written on the majority of workdays. The highest number of NASA RFIs written in a single day was 12, and almost half of the NASA RFIs were written during the March–August period.

A structural design engineer (Foster) for the MLP-3 employed by the designer (Reynolds Smith & Hill) prepared an analysis throughout the project that showed his assessment as to the cause or source of each of the RFIs. He was responsible at the Merritt Island office of RS & H for receiving all of the RFIs and coordinating the responses. His final report on all 1209 RFIs attributed 660 RFIs, or 55 percent of the total number of NASA causes, which he described in his chart as follows:

Existing conditions found to be different from those anticipated.

Drawings or specs not clearly defined or in error—RS & H drawings.

Drawings not clearly defined or in error—NASA drawings.

NASA/new requirements.

Foster attributed the remaining 549 RFIs, 45 percent of the total, to the following BNI causes:

BNI proposal to fabricate or install different (equal to) from design drawings.

Clarification, information, or certification.

Misfabrication.

BNI suggestion for improvements.

The breakdown between the government and BNI causes was to show that the RFI emanated either from government action or inaction, or from some BNI action or problem. The charts were not circulated or discussed within RS & H or NASA as they were prepared.

BNI's theory of loss of productivity resulting from the RFIs was presented in an exhibit, "Impact of Major RFIs on Productivity." BNI witnesses explained at the hearing how the chart shows that productivity is affected when work is in progress and a problem arises. Various activities occur to resolve and work around the problem. Different rates of productivity ranging from 25 to 90 percent were assigned to the various activities.

NASA's expert witness, Thomas J. Driscoll, Senior Vice President at O'Brien-Kreitzberg & Associates, challenged the chart as not reflecting the actual conditions of the job and not using accurate productivity figures. In his opinion, there would be no crew standby time because the Bechtel foremen would reassign the crews to other available work. Driscoll gave the chart no credibility. The chart was prepared "to explain the manner in which cumulative impact costs are incurred as a result of RFIs." The exhibit was "only a representative example of what happened when there was a major impact on the job—that is, an impact that affected a work operation that was in progress." BNI agreed that not every RFI caused all the impacts depicted but considered the chart a reasonable representation of impact caused by major RFIs.

There is no significant discussion of the RFIs in the contemporaneous project records kept by the contractor and the government in the record of the appeal. BNI did not establish the extent of the disruptive effects of the NASA RFIs that are claimed with documentary evidence. BNI project personnel did not keep notes or maintain any documentation of disruptive impacts caused by the RFIs. There was no contract requirement, and BNI did not have a company policy or requirement that project personnel maintain logs or diaries. Therefore, BNI did not have such logs or diaries to document job occurrences or to use in the preparation of the claim. The logs and diaries of BNI personnel that are in the record were submitted to show that only one person maintained a daily record on the job that is available as evidence of what occurred.

Weekly status meetings were held involving representatives of BNI, NASA, and RS & H to discuss work activities of the previous week, what was planned for the next week, and problems that were presented, but there was no regular practice of reviewing the status of RFIs at these meetings.

BNI's notice, dated June 20, 1985, of its cumulative impact claim referred to previous discussions of the claim with NASA. BNI stated its intent to claim compensation for "the loss of productivity caused by the cumulative impact of the many disruptive changes and design deficiencies recorded in the RFIs, change orders, and other project documents." BNI also stated that it was preparing analyses that would demonstrate "occurrence and effects" of the disruption and exclude increased costs not attributable to NASA. NASA did not respond and did not change any of its contract administration practices upon receipt and distribution of the notice. BNI also did not change any of its practices to maintain a record of the impact or the costs that would be claimed.

The NASA board agreed that Bechtel was entitled to recovery for loss of productivity caused by NASA RFIs, but only to the extent that the cumulative impact was not the result of delay and disruption from BNI steel subcontractors. The latter delay was the responsibility of BNI and not NASA. The BNI productivity analysis was not consistent with the facts and therefore did not provide a measure of entitlement. In short, Bechtel convinced the board they had entitlement, but their records were so insufficient that the board could not measure the damages.

For the project files, the CM should prepare a preliminary entitlement evaluation for each RFP response. This should be a confidential file, not shared with either contractor or architect/engineer. (Even though "confidential," always be aware that in a litigation, any party can request, and receive, copies of this confidential file.) The file is used in risk analysis by the CM for information/advice to the owner in the event of a claim or litigation. The preliminary entitlement evaluation can set other actions in motion, such as cost estimate and schedule impact study.

The following examples are from the John Doe Project (App. A):

Figure 5-3 is RFI 001, which describes an unexpected below-ground interference. In Fig. 5-4, the response to RFI 001, the owner arranges for the electric utility to resolve the conflict; therefore, no change order is required.

Figure 5-5 is RFI 002, which describes the discovery of a sinkhole. Figure 5-6 is the response to RFI 002 and indicates that a change order will be forthcoming.

Figure 5-7 is RFI 003. Figure 5-8, the response to RFI 003, says that the contractor has submitted insufficient information to get an answer on a substitution. Further, the response directs the contractor to submit detailed information through the regular submittal process.

Figure 5-9 is RFI 004. Figure 5-10, the response to RFI 004, says that a change order bulletin is forthcoming.

XYZ Construction Company
Request for Information
John Doe Project

RFI # _____ 001 _____   Specification Section _____ 16375 _____

Date _____ 7-31-98 _____   Drawing _____ E 002 _____

Subject: 13.2-kV entry duct interference with excavation at Plant area.

Please provide the following information:

Unidentified electric duct bank encountered. It is not on the as-built utility drawings furnished as part of the bid set.

Figure 5-3   RFI 001.

TO: XYZ Construction Company

RE: John Doe Project

FM: Construction Manager (OK)

RFI# _____ 001 _____   Response Date _____ 8-12-98 _____

Category   Clarification
                                  Start Change Order Process (PCO)
                                  Submit Schedule Impact Analysis
                                  Submit Cost Estimate
                                  Coordination, within Scope
                          X   Other _____ Owner will correct _____

Response:

          The Electric Company has identified the duct bank and will relocate
          it by 9-15-98.

Figure 5-4 Response to RFI 001.

XYZ Construction Company
Request for Information
John Doe Project

RFI # _____ 002 _____   Specification Section _____ 02220 _____

Date _____ 8-21-98 _____   Drawing _____ S 001 _____

Subject: Sinkhole at col. lines Ax11

Please provide the following information:

A sinkhole has been discovered at the plant corner location column lines Ax11.
Please advise action to take.

Figure 5-5 RFI 002.

TO: XYZ Construction Company

RE: John Doe Project

FM: Construction Manager (OK)

RFI#ーーーーー002ーーーーー   Response Date ーーーーーーーーー8-27-98ーーーーーーーーー

Category   Clarification
                    X   Start Change Order Process (PCO)
                               Submit Schedule Impact Analysis
                    X   Submit Cost Estimate
                               Coordination, within Scope
                               Other ーーーーーーーーーーーーーーーー

Response:

ーーーーーーーーー Structural engineer is preparing a change order to correct the
                 situation on an expedited basis. Bulletin due 9-11-98.

Figure 5-6 Response to RFI 002.

XYZ Construction Company
Request for Information
John Doe Project

RFI #ーーーーーーー003ーーーーーーーー   Specification Section ーーーー16300ーーーー

Date ーーーーー9-09-98ーーーーーーーー   Drawingーーーー E 032 ーーーーーーーー

Subject: Request information on 100-HP motor starter

Please provide the following information:

The electrical specification for the 100-HP starter (MCC Class I, Type B combined
MCP FVNR) gives Acme Motors model 100G as an acceptable unit. We believe
that Electric Equipment model 100Q will be suitable.

Please confirm

Figure 5-7 RFI 003.

TO: XYZ Construction Company

RE: John Doe Project

FM: Construction Manager (OK)

RFI#   003   Response Date   9-18-98

Category   X   Clarification
                                Start Change Order Process (PCO)
                                Submit Schedule Impact Analysis
                                Submit Cost Estimate
                                Coordination, within Scope
                                Other _____

Response:

          The Electric Equipment model number is insufficient for a
          decision. Suggest you submit detailed information on the Electric
          Equipment model 100Q starter through the regular submittal
          process.

Figure 5-8 Response to RFI 003.

XYZ Construction Company
Request for Information
John Doe Project

RFI #_____004_____   Specification Section _____15060_____

Date _____9-15-98_____   Drawing_____M 018_____

Subject: Hot Water Piping - HVAC

Please provide the following information:

The heat supply lines show no return system. Will the water go to sewer, or
should a return system be furnished?

Please advise.

Figure 5-9 RFI 004.

```
┌──────────────────────────────────────────────────────────────────┐
│                                                                    │
│   TO:      XYZ Construction Company                                │
│                                                                    │
│   RE:      John Doe Project                                        │
│                                                                    │
│   FM:      Construction Manager (OK)                               │
│                                                                    │
│   RFI# _____004_____    Response Date _____9-30-98_____│
│                                                                    │
│   Category                          Clarification                 │
│                                X    Start Change Order Process (PCO)│
│                                     Submit Schedule Impact Analysis│
│                                X    Submit Cost Estimate           │
│                                     Coordination, within Scope     │
│                                     Other _____ │
│                                                                    │
│   Response:                                                        │
│                                                                    │
│              A return system is required.  The engineer will furnish a change│
│              bulletin by October 16, 1998.                         │
│                                                                    │
└──────────────────────────────────────────────────────────────────┘
```

Figure 5-10 Response to RFI 004.

Request for Information Log

An RFI log is the heart of the clarification, change order, and dispute process. The log tracks each RFI for initiation date, identification number, description, date sent to reviewer, reviewer's name, date sent from reviewer to CM, and dates returned to contractor; it also allows for resolution comments. The RFI procedure is used to document all requests for information/clarifications received from the contractor and the related response by the architect/engineer (or others). It is also the precursor to potential change orders or disputes and is therefore central to the control system.

If the contractor either discovers conflicts, omissions, or errors in the contract documents or has any questions concerning interpretations or clarifications of the contract documents, then, before proceeding with the work affected, the contractor must immediately notify the construction manager in writing on the request for information form. RFIs are numbered consecutively. The following information is logged:

Field	Description
Attention To	The person accountable for responding to this RFI.
Drawing #	The drawing that is applicable to this RFI, if any.
Specification #	The specification section that covers this RFI.

Field	Description
Requests	A memo field available for entering the request. It will print under "Information Needed" on the request for information form.
Description	A summary/title describing the request.
From Contractor	The date the RFI was received from the contractor.
Requested By	The person requesting the information.
Response Required By	The response date.
To Review	The date the RFI went to review.
Reviewer	The person doing the review.
From Reviewer	The date the RFI is returned to the CM.
To Contractor	The date the RFI is returned to the contractor.
Reply	A memo field available for recording the information requested. This information will print under "Reply" on the request for information form.
Remarks	Available for comments or notes concerning the RFI.
Preliminary Evaluation	Classification of RFI by PM/CM.

Figure 5-11 summarizes information on RFIs 001 to 004.

Other Ways to a Change Order

The contractor can take several other paths to a change order. First, the contractor, believing that a change order is required, can write a letter to the project/construction manager stating that opinion. Second, the contractor can submit a value engineering change proposal (VECP). The owner can also initiate a change order by requesting changes, usually through the architect/engineer. Similarly, a change may be suggested by the architect/engineer or project/construction manager and approved by the owner.

Value engineering change proposal

The contractor can submit a VECP, usually in letter format. The VECP does not have to be fully developed. It is reasonable for the contractor to provide the following:

- Scope of the proposed change
- Estimated design cost
- Range of potential savings
- Estimated schedule impact

RFI #	Description	Drawing #	Specification #	Sent to Review	Reviewed	Classification (Confidential)	Next Action
001	13.2-kV Entry Duct Interference	E002	16375	7-31-98	8-12-98	OA	9-15-98
002	Sinkhole at Ax11 (column lines)	S001	02220	8-21-98	8-27-98	PCO	9-11-98
003	Request for Approval of Substitute 100-HP Motor Starter	E032	16300	9-09-98	9-18-98	CL	--
004	Hot Water Piping Return	M018	15060	9-15-98	9-30-98	PCO	10-16-98

Classifications:

PCO	Potential Change Order
CL	Clarification
DD	Design Deficiency
OA	Owner Action

Figure 5-11 RFI log.

   This information should be sufficient for the PM/CM to proceed (or not) with the formal submittal of the VECP. Figure 5-12 is a sample VECP by letter.

Change bulletin—owner and PM/CM

   The owner, through the PM/CM, can issue a change bulletin when deemed necessary. The change bulletin should direct the contractor to proceed with those actions possible. If revised drawings and/or specifications are not issued, contractor actions will not start until they are. Figure 5-13 shows a sample bulletin issued by the PM/CM.

Potential Change Orders (PCOs)

   As potential change orders are identified, a potential change order log is developed. The potential change order log follows the change order process from the request for a price quote from the contractor, through negotiations, and finally to the approved change order. It is cross-referenced to the request for information, the change estimate, and the change order logs. PCOs are numbered consecutively. The following information is logged:

Field	Description
Specification #	Applicable to this potential change order.
Drawing #	The drawing relevant to this PCO, if applicable.
Initiated By	A pick-list with the following choices: ■ Contractor ■ Construction manager ■ Owner ■ Other
Reason for Change	A pick-list with the following changes: ■ Owner change ■ Differing site conditions ■ Clarification of decision ■ Other
Justification for Change	A memo field for describing why the change is being made. It will print in Section 2 on the justification for contract modification form.
Requesting Official	The individual requesting the change.
Request Date	The date the individual requested the change.
RFQ	A request for quote. Fields for the following information are included: ■ Number ■ Date

To: Project Manager (OK)   Date: 9-03-98

Fm: XYZ Construction Company

Re: VECP #1
      John Doe Project
      WHATCO

Sirs:

We propose to proceed with development of the following VECP, which is as
follows:

Scope

Roof framing for the plant area was originally as shown in Attachment A. By
Addendum #3, a lighter roof girder system described by Attachment B was
made part of the bid package. [Attachments A and B are located in App. B of
this book.]

XYZ Company proposes the use of long span joists in lieu of the girders. The
joists will be the same height as the girders (56"), so that no change is required
in columns, building walls, etc. The Vulcraft designation is DLH 13 @ 33#/foot.

The joist spans are to be spaced at 10'-0" on center, so that the specified
$1\frac{1}{2}$" roof deck can be used. However, at this 10'-0" O-C, the roof deck does not
require secondary joists.

Design Cost

The cost of designing the joists is included in the cost per ton of the long span
joists. There will be a small charge for redesigning connections (typical) at the
longitudinal support beam between columns.

Range of Potential Savings

Based on preliminary calculations, savings of over $200,000 is anticipated.

Schedule Impact

Assuming an O.K. to proceed with development of the VETC, one week to
prepare it, and approval/disapproval in one week, no schedule impact is
anticipated.

Yours very truly,

XYZ

Figure 5-12   VECP letter request.

Field	Description
	■ Time extension (in days as quoted)
	■ By (company or individual making the quote)
	■ Cost (amount quoted)
C/E	Change estimate. Fields for the following information are included:

To: XYZ Construction Company   Date: 9-03-98

Fm: Project Manager (OK)

Re: Change Bulletin #1
         John Doe Project
         WHATCO

You are directed to make the following change to Specification 12.000:

   Change chiller 30 ton 1200 CFM

   to 3 chillers 15 ton 600 CFM

and associated piping and electrical changes as described on drawing package to
be issued within one calendar week.

Yours very truly,

Project Manager (OK)

Figure 5-13   Change bulletin issued for owner.

Field	Description
	▪ Number
	▪ Date
	▪ Time extension (in days estimated)
	▪ By (individual making the estimate)
	▪ Cost (amount estimated)
CO	Change order. Fields for the following information are included:
	▪ Number
	▪ Date
	▪ Time extension (in days as negotiated)
	▪ Cost (amount negotiated)
Status (values)	The following fields are included:
	▪ Pending (a PCO in process; unresolved)
	▪ Approved (a PCO approved for payment and/or time extension becomes a pending change order)
	▪ Canceled (a PCO that has been resolved without further payment or time extension)
Justification of Extension	A memo field for explaining why contract days need to be extended.
Resolution	A field for comments or notes dealing with the resolution of this potential change order.

Figure 5-14 is an example of a PCO log for the John Doe Project.

The U.S. Postal Service *Construction Administration Handbook*, June 1992, describes the following responsibilities of the project manager and contracting, support, and field personnel as they relate to potential changes (PCs):

The Project Manager, RE or designee maintains a Proposed Change (PC) log for each prime contract. PC logs are to be current at all times and available for review by the Contracting Officer.

The Project Manager assigns a PC number to each proposed change. (This is done for tracking purposes, even though the PC may be abandoned before an RFP is sent to the Contractor.)

The Project Manager assigns a sequential PC number starting with PC #1 and enters the PC number into the PC log, along with a brief description of the change.

All correspondence and documents related to the change will reference the PC number. This includes letters issued by USPS or by the Contractor.

The Project Manager instructs the A-E to reference the PC number on all revised or new drawings, sketches, or specifications. The A-E references each drawing revision number to the appropriate PC number. This is done near the sheet title block.

When a PC number is assigned, the Project Manager opens a separate file to hold all correspondence related to the proposed change.

The Project Manager promptly reviews each proposed change to determine the following:

a) Is it actually a change in contract scope, i.e., not part of the original contract or any previous modification?

b) What is the reason for the change?

c) Why is the change necessary or desirable?

d) Has the A-E concurred with the proposed change from a design standpoint?

e) What revisions to contract drawings and specifications are needed to effect the change?

f) If the PC has not been caused by design errors or omissions, what is the preliminary A-E cost estimate to prepare revised drawings or specifications (only if needed) to define the PC scope of work?

g) How long will it take the A-E to prepare revised drawings and specifications (only if needed) for issuance to the Contractor?

h) What is the preliminary construction cost estimate?

i) What is the preliminary construction schedule impact?

j) Are uncommitted funds available?

k) Is the work within the general purview of the contract?

l) Is it practical to obtain competitive proposals for the additional work?

The USPS provides a form, shown as Fig. 5-15, to record the results of the review. Figure 5-16 shows a sample USPS PC log format.

PCO #	RFI #	Description	Drawing #	Specification #	Initiated by	Reason		Status
001	002	Sinkhole at A-11	S 001	02220	CT	DSC	Pending	Pending
002	004	Hot Water Piping	M 018	15060	CT	DD	Pending	Pending
003	--	Long Span Joists - Plant	S 005	05120	CT	VE	Pending	Pending
004	--	Change from one chiller to three smaller	M 010	15650	OW	OW	Pending	Pending

Initiated
OW   Owner
CM   Construction Manager
CT   Contractor
OT   Other

Reason
OW   Owner Change
DSC Differing Site Conditions
CL   Clarification
DD   Design Deficiency
VE   VECP

Figure 5-14 PCO log.

PRELIMINARY PC REVIEW
Page 1 of 2

Project Name _____	Date _____
A-E Name:_____	Contractor Name:_____
A-E Contract #:_____	Contract #:_____
A-E PC #:_____	Construction PC #:_____

To Contracting Officer:

BRIEF PC DESCRIPTION:_____

_____

_____

Project Manager's review of the Proposed Change (PC) indicates the following:

A.   The PC work is outside of the current contract scope of work: Yes___ No___

B.   This is a design-phase PC due to USPS changes in design criteria: Yes___ No___

C.   This is a construction-phase PC: Yes___ No___

If *YES*, this PC is caused by:

☐   Design Errors and Omissions   ☐   Value Engineering

☐   Differing Site Conditions   ☐   Functional Design Changes

☐   Time Extension due to excusable delays

(DESCRIBE:_____)

☐   Contractor Failure to Provide Specified Materials/Workmanship

☐   Assessment of Liquidated Damages

☐   Other:_____)

D.   The PC originates in correspondence, requests, or discussions with or from:

☐   Contractor   ☐   A-E   ☐   USPS D&C   ☐   USPS Other:_____

☐   USPS Occupying Org.   ☐   Local Agency   ☐   Utility Company

E.   The PC is in the interest of USPS because it will:

☐   Lower Construction Costs   ☐   Result in Earlier Completion

☐   Increase Operational Efficiency ☐   Improve Construction Quality

☐   Delay Construction If Not Implemented

☐   Detrimentally Affect Use Of The Facility If Not Implemented

☐   Result In An Unsafe Or Code-Violating Condition If Not Implemented

☐   Other:_____

Figure 5-15   PC review form. (*USPS, used with permission.*)

Estimates

When a PCO is identified, the PM/CM orders its staff to prepare an estimate of the work scope. The contractor, in turn, is requested to prepare an estimate as a precedent to negotiations. Contractor estimates are usually made up of quotes from subcontractors. For change order review, these quotes must include estimated trade man-hours and material costs times markup. Markups must be within contract limits—usually 20 percent of overhead

PRELIMINARY PC REVIEW
Page 2 of 2

Project Name :	Date:
A-E PC #:	Construction PC #:

F.   The A-E has concurred with the PC from a *design* standpoint: Yes__ No__ N/A__

G.   Are Revised drawings/specs needed to define the PC scope of work?

    ☐   YES

        The A-E estimates revised drawings/specs can be ready in __ calendar days.

        ☐   PC IS *NOT* CAUSED BY DESIGN ERRORS/OMISSIONS

            The preliminary A-E fee estimate is $_____.

            The PC to the A-E contract should be handled as an: RFP__ RFP/NTP__

        ☐   PC *IS* CAUSED BY DESIGN ERRORS/OMISSIONS

            A letter will be sent to the A-E directing the A-E to prepare revised drawings/specs and a definitive construction cost estimate by _____, for no change in fee amount.

    ☐   NO

        ☐   PC *IS* CAUSED BY ERRORS/OMISSIONS

            A letter will be sent to the A-E directing the A-E to prepare and submit a definitive construction cost estimate by _____.

H.   The preliminary construction cost estimate is:

        ADD_____   DEDUCT_____   AMOUNT:_____

I.   The preliminary schedule impact is:

        ADD_____   DEDUCT_____   AMOUNT:_____

J.   Uncommitted funds are available to pay for this PC: Yes__ No__

K.   The PC falls within the general purvue of the contract: Yes__ No__

    Maybe ____

    If *NO* or *MAYBE*, the Legal Department should review this issue: Yes__ No__

L.   It is practical to get competitive proposals for the PC: Yes__ No__

    Maybe ____

    If *YES* or *MAYBE*, explain_____

_____

M.   The construction PC should be handled as an: RFP____ RFP/NTP____

N.   Additional Remarks:_____

_____

_____

Submitted By:	Contracting Officer Action:	
Project Mgr._____ Date:____	Approved:_____	Date____
	Disapproved:_____	Date____

Figure 5-15 *(Continued)*

and profit of the subcontractors, and a 5 percent markup by the general contractor.

In assembling their cost proposal, the prime/general contractors usually use the subcontractor proposals "as is." Therefore, it is necessary for the PM/CM to be prepared to analyze the cost proposal in detail. The basis for this analysis is a reasonably detailed estimate.

United States
Postal Service

PROPOSED CHANGE/PC LOG

		Project Name and Location:						A-E or Contractor Name:						Contract No.:			Page:
PC #	MOD #	BRIEF PC DESCRIPTION	DATE PPCR APP	REV. DOC'S REQ.	PC CLAUSE	PC JUST.	PC ORIG	USPS ROM EST.	USPS DETAIL EST.	RFP DATE	DATE PROP REC.	CONTRACT PROP AMT.	NTP DATE	NTP AMT.	NEG. AMT.	REMARKS/STATUS

ABBREVIATIONS:

NTL = Not less than
ROM = Rough Order of Magnitude
E&O = Design Errors and Omissions
NTP = Not to Proceed
NTE = Not to Exceed
ORIG = PC Originator
PPCR = Preliminary PC Review Form
JUST = PC Justification
APP = Approved
REV = Revised
RFP = Request for Proposal
AMT = Amount

PC CLAUSES:

1. E&O
2. Differing Site Conditions
3. Value Engineering
4. "Functional Design" Changes
5. Time extensions due to excusable delays
6. Contractor non-compliance with specifications
7. Assessment of liquidated damages
8. Other

PC JUSTIFICATIONS:

1. Lower construction cost.
2. Increase operational efficiency.
3. Delay construction if not implemented.
4. Detrimentally affect facility if not implemented.
5. Unsafe condition or code-violation if not implemented.
6. Result in earlier completion.
7. Improve construction quality.
8. Other.

PC ORIGINATORS:

1. Contractor
2. A-E
3. USPS D&C
4. USPS Occupying Organization
5. Utility Company
6. Local Government Agency
7. Other

Figure 5-16 PCO log format. (*USPS, used with permission.*)

Figure 5-17 shows a sample letter format requesting the contractor to submit a detailed cost proposal. Figure 5-18 shows an estimate of the cost to resolve PCO #1 (sinkhole). This cost estimate is prepared by the PM/CM. The cost per cubic yard (in the ground) is $4590 ÷ 125 c.y. = $36.72/c.y. Factored by the 20 percent expansion when measured by truckload, each cubic yard is $30.60. The importance of the unit price is that the sink-hole might well be larger. It should be agreed that any increase in size be compensated.

Figure 5-19 shows the cost estimate for the 2-inch steel pipe hot-water return (uninsulated) in the plant and warehouse. (The office hot-water system was already a closed loop.) Again, this is an estimate by the PM/CM.

Negotiation

Through the process, a PCO has been identified and an estimate prepared by staff (or consultant) is in hand. It's time to negotiate the cost of the change.

As a society, we want to be treated fairly and equitably. When purchasing or buying something, we expect to be asked for, and to pay, a fair price. A responsible purchaser should have a fair price in mind so that the asking price can be evaluated.

If a fair price is not intuitively obvious, the purchaser can comparison shop from various sources. Even then, if a "fair" asking price is over the purchaser's budget, sale opportunities can be sought (if time is available). If time is not available, then the purchaser must pay the list or asking price, which doubtless includes a premium cost.

In some areas, such as health care, the prudent purchaser can't comparison shop. The price schedule is in the hands of the health insurance companies. We expect them to get us a fair price for health services, and we rely on government regulation and oversight to monitor this.

In at least two areas, cars and houses, we expect the seller to build a premium into the asking price. In those cases, we know we have to negotiate to reach an acceptable price. We know that if we can't get an acceptable price, we can walk away. In a construction change situation, on the other hand, the owner usually can't walk away. This contractor is the one the owner has to negotiate with, and the contractor knows it.

There may be special circumstances, however, particularly near the end of a project, where the owner can walk away. To do this, the owner deletes the work to be changed, deferring it to be accomplished by in-house forces or another contractor. Doing this is unusual, but it is sometimes necessary, for instance, if a contractor is in financial difficulty and can't take on added work.

When deleting any work and negotiating for a credit, there are two downsides:

1. The owner will never get full credit for deletions (i.e., overhead and profit stays with the contractor).

2. The owner will probably lose warranty on any systems relating to the deleted work.

SAMPLE

RE-14
Procedure
370.70

**11.5 Request for Cost Proposal for a Proposed Change
(after contract award)**

UNITED STATES POSTAL SERVICE

Contractor's Name
Contractor's Address

Dear _____:

Subject: [Project Name and Location]
Contract No:_____
RFP #:_____

Please submit to this office your detailed cost proposal for all labor, material, and equipment necessary to accomplish the following change. Please remember that in your contract requirements (Clause FB-271), overhead, profit, and commission <u>are not</u> allowed on FICA/FUTA at <u>any</u> tier. All further correspondence regarding this work shall be identified to the above referenced RFP until incorporated into this contract by written modification.

Your proposal should be submitted for the following work (narrative--if insufficient space, provide attachments):

In addition to your cost proposal, please indicate any changes to the construction schedule as a result of the proposed change.

This letter is a request for a price proposal only and is not an authority to proceed. This authority is being withheld pending receipt and review of your detailed cost proposal. Please submit your cost proposal no later than [date] to [name of Project Manager].

Very truly yours,

_____
Contracting Officer

Figure 5-17 Sample letter format requesting contractor to submit detailed cost proposal. (*USPS, used with permission.*)

Figure 5-18 Cost estimate for PCO #1—sinkhole.

PCO #1: Sinkhole

Scope

Probes of the sinkhole established an area of 15' × 15' × 15' deep or (15 × 15 × 15)/27 or a volume of 125 cubic yards.

Both parties agree on this as a base amount. The PM/CM ran some tests that show the material measured in place expands 20 percent when dumped into a truck. Accordingly, the truck measurements will be 150 cubic yards.

The excavated material cannot be used as backfill. It can be spread on a low place on site.

The unit prices on the quantities are taken from the cost estimate [see App. B of this book.]

The cost estimate is below.

(Continued)

Negotiating techniques

In negotiating, you (the PM) may be subjected to a wide variety of ploys or techniques. Here are some possible scenarios:

Good guy–bad guy. The contractor's negotiator is an obvious bad guy (rude, loud, and unreasonable). After a no-progress session, when you have been put on the defensive, one of the contractor's staff contacts you and apologizes for the bad guy's rude and unreasonable demands. This good guy says, "Look, let me see what I can do." Later, through the good guy, a better offer/proposal is made, perhaps with the proviso, "the jerk is out of town; but I can only keep this open till Friday." Better proposal, no jerk to deal with—watch out.

The poker player. You (the PM) and the contractor have reached positions that are substantially apart. Time passes and pressure is on you because the submitted change may start to delay the project. The contractor continues to delay. Pressure builds (apparently only on you). The contractor may be under similar pressure but is a good poker player (i.e., bluffer). On the other hand, the contractor may know that there are other problems that will delay the project and wants this change to cause the delay. The contractor would appear to have the better cards (i.e., two possibilities vs. one). The PM could raise the stakes with a threat to either direct a time/material change or issue a unilateral change order if the contractor fails to negotiate an acceptable agreement.

Bait and switch. The contractor offers a proposal too good to be true, but when you are ready to agree, the contractor links the "great deal" to one or more not-so-great deals for a proposed package settlement.

The kitchen sink. The contractor's proposal for trade labor and materials compares well with your estimate, but the proposal total is well above yours. By

<table>
<tr><td colspan="2">

O'BRIEN-KREITZBERG & ASSOCIATES, INC.

</td><td>

Project
_____John Doe_____

Location
_____WHATCO_____

Architect Engineer
_____PCO #1_____

</td></tr>
</table>

CONSTRUCTION COST ESTIMATE			DATE PREPARED 8-31-98				SHEET 1 OF 1	
DRAWING NO. 5001			ESTIMATOR			CHECKED BY:		
	QUANTITY			LABOR		MATERIAL		TOTAL
SUMMARY	NO. UNITS	UNIT MEAS.	PER UNIT	TOTAL	PER UNIT	TOTAL		COST
Sinkhole @ C.L. A/11								
Excavation	125	CY	$5.36		CSI	02220		$670.00
Spread Spoil	150	CY	$5.22		CSI	02220		783.00
Borrow	150	CY	$3.10		CSI	02220		465.00
Deliver Borrow	150	CY	$9.34		CSI	02220		1,401.00
Fill	150	CY	$2.64		CSI	02220		396.00
Compact	125	CY	$0.88		CSI	02220		110.00
								3,825.00
				20% Overhead	and	Profit		765.00
						TOTAL		$4,590.00

Figure 5-18   *(Continued)*

examining the contractor's breakdown, you determine that the specific overhead items assigned contain many (if not all) items that you believe are covered by the flat overhead percentage applied to the change labor and materials.

Phantom costs.   At first, the contractor's breakdown looks good. But on closer inspection, you notice a painter has substantial costs in for scaffolding. Because you are familiar with that painter on that job, you know that his painters reach long and high with rollers on 10- and 15-foot extension poles. You can therefore hold scaffolding to a minimum. The lesson here: be sure to have a hands-on familiarity with the project.

High-cost factors. The contractor's unit time proposal for various pieces of equipment is reasonable, but the cost proposal is extended at daily rates.

Project
___John Doe___

O'BRIEN-KREITZBERG & ASSOCIATES, INC.

Location
___WHATCO___

Architect Engineer
___PCO #2___

CONSTRUCTION COST ESTIMATE			DATE PREPARED 9-22-98				SHEET 1 OF 1	
DRAWING NO. M018			**ESTIMATOR**			**CHECKED BY:**		
SUMMARY	QUANTITY		LABOR		MATERIAL		TOTAL	
	NO. UNITS	UNIT MEAS.	PER UNIT	TOTAL	PER UNIT	TOTAL	COST	
Condensate Return								
Piping								
Plant								
2" Steel Pipe	400	LF	16.81		CSI	15060	$6,724.00	
Hangers & Hardware	10%	of	Pipe	Cost			672.00	
2" Gate Valves	4	EA	1,550		CSI	15100	6,200.00	
Warehouse								
2" Steel Pipe	400	LF	16.81		CSI	15060	6,724.00	
Hangers & Hardware	10%	of	Pipe	Cost			672.00	
2" Gate Valves	4	EA	1,550		CSI	15100	6,200.00	
							$27,192.00	
			20% Overhead		and	Profit	5,438.00	
						TOTAL	$32,630.00	

Figure 5-19 Estimate PCO #2.

Research shows that most (if not all) of this equipment is on-site for base contract work on a weekly or monthly basis.

Split the difference.   This approach is obvious: when final positions are reached, one party (either contractor or PM) suggests, "Why don't we split the difference?" The approach usually works when the spread between positions is not too great.

Win–win.   This approach is clearly the best. In the partnering process, each party tries to establish its cost position in a logical, rational approach. Each, then, looks for a way to help the other. For instance, the PM/CM will be in favor of a time extension when appropriate. The contractor, in turn, will agree (ahead of time) to forgo any claim for acceleration or disruption due to the change.

Contract Structure to Create Level Playing Field

The contract should include usual, tested provisions that set guidelines to keep the contractor within usual bounds:

Overhead and profit.   It is usual for the contract to limit contractor's overhead and profit to 20 percent of the labor (including fringes) and materials in the change. Conversely, if the contractor can identify specific charges to the change that are normally overhead, such as surveying, engineering, or dedicated supervision, these changes should be allowed (if supported).

Home office overhead.   This is to be included in the 20 percent overhead and profit limit.

Time and material.   In appropriate situations, such as when negotiations are bogged down or contractor proposals are unreasonable, the PM should have the prerogative to direct the change work to be done on a time and material basis.

Unilateral changes.   When the change work should proceed and time and material is not appropriate, the PM should be able to direct that the change proceed. In this mode, the PM unilaterally sets the change order amount so that progress payments can be made on the change work. (Even after issuance of a unilateral change order price, the final price is open to negotiation and/or claim.)

Right to audit.   If a PM is suspicious of cost claims made in a change order, the PM can include the intent to audit as part of the change order. If an audit shows lower costs, the change order can be modified accordingly.

Credit versus debit change orders.   There are two schools of thought in regard to credit change orders:

1. The contractor "owns" the overhead and profit in the base bid. Therefore, credit changes, including VECPs, do not include overhead and profit.

2. Overhead is a real cost and should therefore apply to a credit change, including VECPs. Profit as bid is not included.

Both approaches have merit. The contract should state which is to be used.

PCO #1—Sinkhole

Figure 5-20 presents the XYZ company proposal to correct the sinkhole. The proposed cost is $14,388 versus the estimated cost of $4590. In negotiating sessions, the PM/CM tells XYZ that the proposal is "way high." In response, XYZ points out that assembling a spread of equipment on a relatively small assignment is inherently inefficient. The PM/CM says that the change has to be rethought. After time, a second proposal is submitted (Fig. 5-21). The time frame is reduced from four calendar days to two calendar days. Note also that the estimate dropped to $9300.

The PM/CM compliments the contractor for the cost reduction and agrees that the equipment spread was reasonable and well planned. The PM/CM notes that the overhead elements were covered in the equipment and superintendent hourly rates and suggests that the change order be at base cost plus profit:

Base cost (from Fig. 5-21)	$7750
Profit 10%	+ 775
Total	$8525

After some grumbling, the contractor agrees to the $8525 figure. The PM is paying more than expected, but the negotiations have demonstrated two things: (1) with careful planning, the contractor reduced the cost substantially; and (2) a relatively small operation is very inefficient when multiple stages and multiple equipment are required.

PCO #2—Condensate Return

Figure 5-22 shows the XYZ proposal for the condensate return. Most of the work is by a subcontractor. The general contractor marks up the subcontract price by the 10 percent profit level. The proposal is $21,569, which compares favorably with the PM estimate of $32,630. The PM, wisely, does not quibble with a favorable proposal, and the deal is done.

Setting the Scene

The early negotiations are very important. The various players are "sizing up" each other. A few careless negotiating sessions for any party can set a disastrous pattern. Everyone should perform at the top of his or her potential throughout,

Figure 5-20   Initial XYZ Company cost proposal for PCO #1—sinkhole.

Plan: Sinkhole

Backhoe excavates for two days. On Day 1, excavated material is stockpiled. On Day 2, front-end loader loads 10 c.y. dump truck. Dump truck dumps on site at spoil location. Bulldozer spreads spoil, and "walks it down." On Day 2, backhoe completes excavation.

On Day 3, three 10 c.y. dump trucks begin importing fill from borrow pit 20 miles away (40-mile round-trip). Bulldozer pushes fill into place in 1-foot layers; compactor follows up. The fill operation, driven by the truck capacity, takes two days.

Equipment Rental Schedule

(Hourly cost includes equipment and operator or drive)

Equipment basis:	Daily	Weekly	Monthly
Cost per hour:			
Backhoe (2 c.y.)	$190	$130	$110
(1 c.y.)*	140	110	90
Front-end loader, 100 HP	110	80	70
10 c.y. dump truck	55	50	45
Bulldozer, 300 HP*	200	160	130
Compactor	50	45	42

*Bulldozer and 1 c.y. backhoe are on a monthly basis.

Borrow cost (loaded on truck) $5/c.y.

(Continued)

but especially in the early sessions. If unfavorable precedents are set, they are difficult to change.

VECP Negotiating

In almost all change negotiating, the owner is trying to control the cost of a change. The contractor, on the other hand, is trying to do two things: (1) maximize the cost; and (2) get the owner to carry part of the contractor's overhead cost. Every dollar of overhead (above that deserved) carried by a change is as good as a dollar of profit.

In the VECP negotiating world, the roles seem to be reversed. The change is a savings, part of which will become a credit. The contractor is trying to increase the size of the savings beyond its true worth. For every dollar that the contractor can inflate the savings, $0.50 comes back to the contractor; the other $0.50 doesn't cost the contractor anything since it relates to contract work not performed.

There are two elements to the VE savings:

$$\text{VE savings} = \text{base contract cost} - \text{cost of VECP}$$

$$\text{Cost of VECP} = \text{VE replacement work} + \text{costs related*}$$

---

*Costs related to the VE equals redesign, shipping, etc.

**O'BRIEN-KREITZBERG
& ASSOCIATES, INC.**

Project

John Doe

Location

WHATCO

Architect Engineer

CONSTRUCTION COST ESTIMATE				DATE PREPARED				SHEET 1 OF 1	
DRAWING NO.				ESTIMATOR			CHECKED BY:		
SUMMARY	QUANTITY		LABOR		MATERIAL		TOTAL COST		
	NO. UNITS	UNIT MEAS.	PER UNIT	TOTAL	PER UNIT	TOTAL			
PCO #1 Proposal #1									
Foreman	4	Days	$400				$1,600.00		
Backhoe (1 c.y.)	16	Hr	110				1,760.00		
Frontend Loader	8	Hr	110				880.00		
10 c.y. Dump Truck	56	Hr	55				3,080.00		
Bulldozer	24	Hr	130				3,120.00		
Compactor	16	Hr	50				800.00		
Backfill	150	c.y.	5.00				750.00		
							$11,990.00		
				20% Overhead	and	Profit	2,398.00		
						TOTAL	$14,388.00		

Figure 5-20 _(Continued)_

VECP: PCO #3—Long Span Joists for Roof Framing

This PCO was suggested by a letter proposal (Fig. 5-12). The actual proposal (VECP) is presented in Fig. 5-23. The thrust of this VECP is the use of long span joists (34) to replace heavier girders in the plant. The joists (56 inches) are approximately the same height as the 57-inch-high girders. Spaced at 10 feet on center, the roof deck panels ($1\frac{1}{2}$) can be supported directly on the long span joists. This results in a substantial savings in the weight of the secondary joists (i.e., they are deleted).

Plan 2: Sinkhole

Use larger backhoe (2 c.y.) and excavate on Day 1. Equipment on Day 1:
backhoe, bulldozer, and one 10 c.y. dump truck.
    Day 2: Use closer borrow pit and import 150 c.y. with two 10 c.y. dump
trucks. Equipment on Day 2: bulldozer, compactor, and two 10 c.y. dump trucks.

Project

     John Doe

O'BRIEN-KREITZBERG
& ASSOCIATES, INC.

Location

     WHATCO

Architect Engineer

CONSTRUCTION COST ESTIMATE			DATE PREPARED				SHEET 1 OF 1
DRAWING NO.			ESTIMATOR			CHECKED BY:	
	QUANTITY		LABOR		MATERIAL		TOTAL
SUMMARY	NO. UNITS	UNIT MEAS.	PER UNIT	TOTAL	PER UNIT	TOTAL	COST
PCO #1 Proposal #2							
Foreman	2	Days	$400				$800.00
Backhoe (2 c.y.)	8	Hr	190				1,520.00
Frontend Loader	8	Hr	110				880.00
10 c.y. Dump Truck	24	Hr	55				1,320.00
Bulldozer	16	Hr	130				2,080.00
Compactor	8	Hr	50				400.00
Backfill	150	c.y.	5.00				750.00
							$ 7,750.00
			20% Overhead	and	Profit		1,550.00
						TOTAL	$ 9,300.00

Figure 5-21 Second XYZ Company cost proposal for PCO #1—sinkhole.

Project
John Doe
Location
WHATCO
Architect Engineer

O'BRIEN-KREITZBERG & ASSOCIATES, INC.

CONSTRUCTION COST ESTIMATE				DATE PREPARED				SHEET 1 OF 1	
DRAWING NO.				ESTIMATOR				CHECKED BY:	
SUMMARY	QUANTITY		LABOR		MATERIAL			TOTAL	
	NO. UNITS	UNIT MEAS.	PER UNIT	TOTAL	PER UNIT	TOTAL		COST	
PCO #2 Proposal #1					&	Equipment			
Condensate Return									
Plant									
2" Steel Pipe	400	LF	7.10		5.16				
Sch. 40 A-53			x 1.35		x 1.35				
Welded W. Hangers			9.59	$3,836.00	6.97	2,788.00		$6,624.00	
2" Gate Valves	4	EA	600		100	15100		6,200.00	
			x 1.10		x 1.35				
			660	2,640.00	135	540.00		3,180.00	
Warehouse									
2" Steel Pipe	400	LF	9.59	3,836.00	6.97	2,788.00		6,624.00	
2" Gate Valves	4	EA	660	2,640.00	135	540.00		3,180.00	
				Subcontract	Total			$19,608.00	
						Profit 10%		1,961.00	
						TOTAL		$21,569.00	

Figure 5-22   Initial XYZ Company cost proposal for PCO #2—condensate return.

Gross savings = $317,184

Cost (redesign) =   30,000

Net savings = $287,184

On this basis the savings to the owner and the contractor is $143,592 each.

Figure 5-23   VECP for roof framing.

VE Recommendation

Plant

Replace 11 G57-173# girders with 34 long-span 56" joists DLH 13 (33#/lf).

Girders:	$11 \times 173\# \times 100' = 190{,}300\#$	
Joists:	$34 \times 33\# \times 100' = \underline{102{,}300}$	
	Savings 88,000#	

Delete secondary joists = 48 tons

Warehouse

Replace 1 G45-115# girder with 3 long-span joists 60' × 25#

2 girders:	$2 \times 115\# \times 60 = 13{,}800\#$	
Joists:	$6 \times 25\# \times 60 = \underline{9{,}000}$	
	Savings 4,800#	

Delete secondary joists = 22 tons

Office

Replace 4 WF 18 × 58# with 9 long-span joists 70' × 30#

WF:	$4 \times 58\# \times 70' = 16{,}240\#$	
Joists:	$9 \times 30\# \times 70' = \underline{18{,}900}$	
	Increase 2,660#	

Delete secondary joists = 21 tons

(Continued)

In negotiating, the contractor might well argue that the overhead profit was bid competitively. On that basis, the contractor claims that the overhead and profit as bid belong to the contractor. (Contractors, often successfully, make that same argument for credit change orders.) If the overhead and profit was not included in the savings, the VE savings would be as follows:

$$\text{Gross savings} = \$264{,}320$$
$$\text{Cost (redesign)} = \underline{30{,}000}$$
$$\text{Net savings} = \$234{,}320$$

In this case, the savings to the owner and the contractor is $117,160 each. However, the contractor's real gain (not increase) would be as follows:

$$\text{VE share} = \$117{,}160$$
$$\text{Overhead and profit} = \underline{52{,}864}$$
$$\text{Real gain} = \$170{,}024$$

There are strong arguments on either side as to whether the bid overhead and profit "belongs" to the contractor or whether a proportional part of the

COST WORKSHEET

Item		Units	No. Units	ORIGINAL ESTIMATE		NEW ESTIMATE	
				Cost/ Unit	Total	Cost/ Unit	Total
Structural Steel							
Plant	88,000#						
Warehouse	4,800						
Office	(2,660)						
	90,140#						
Say	45 Tons			$2240	$100,800		0
Short Span Joists							
Plant	48 Tons						
Warehouse	22 Tons						
Office	21 Tons						
Say	91 Tons			$1504	$136,864		0
Paint	136 Tons			$196	$22,656		0
					$264,320		
OH/Profit @ 20%					52,864		
Gross Savings					$317,184		
Cost (Redesign)					30,000		
Net Savings					$287,184		

Figure 5-23 *(Continued)*

overhead and profit goes with a credit or a VECP savings. The best place to resolve the questions is in the contract specifications.

Another contractor negotiating approach is to maximize the cost related to the VE. For every dollar this cost can be maximized, that dollar goes to the contractor (less the $0.50 it takes off the VE share).

Change Order Documentation

Documents identified thus far in the change order process are RFIs, RFI responses, RFI log, proposal to submit VECP, VECP, PCO evaluation, PCO log, PCO proposal, and PM/CM estimate. All of these documents have been used to identify a PCO, estimate its cost, and negotiate an agreed-upon price. After the negotiations, there should be a record of their content. Figure 5-24 shows a USPS guideline for preparing a record of the negotiation. Figure 5-25 shows a sample form to be signed by the parties for PCO #1.

Date_____

11.9 NEGOTIATION PROCEEDINGS
MODIFICATION TO CONSTRUCTION/A-E CONTRACT,
WORK REQUEST TO CONSTRUCTION/TERM A-E CONTRACT

1. A telephone/field conference was held on the above date with the firm of (_____), for the purpose of establishing the terms of a contract modification, subject to final approval of the Contracting Officer, for services in connection with (_____).

2. Participating in the negotiations representing the A-E Firm/Contractor was (_____), and representing the U.S. Postal Service was (_____).

3. This contract modification was initiated by (_____) for the following reasons: (_____ _____ _____).

4. Following a discussion of the scope of work, the contract modification and related material , the proposal submitted by the A-E/Contractor was considered. The proposal was (accepted as being fair and reasonable) (negotiated downward/upward) as indicated in the summary and comments below. The proposal was not accepted at this time; further negotiation necessary. Finding of Facts, dated _____ are (are not) attached to support recommendations.

5. Period of Service: (_____).

6. Time Extension Requested: ( ) yes   ( ) no   _____calendar days

7. Justification for extending or not extending Contractors time: (_____)

8. Cost Summary:

    <u>Government Estimate</u>   <u>Proposal</u>   <u>Negotiated Proposal</u>

9. Comments: (_____ _____ _____ _____)

Location: _____   Project: _____

MOD./WORK REQ. NO.: _____

Approval Recommended: Concurrence:

_____   _____
Project Manager   Date   Manager, D&C   Date

Figure 5-24 Record of negotiation proceedings. (*USPS, used with permission.*)

September 25, 1998

XYZ Company
Field Office

Project: John Doe
        WHATCO

Gentlemen:

This letter is to confirm the negotiation meeting that took place on the 25th day
of September 1998 with the following parties:

Name   *Firm*
John Smith   XYZ
William A. Jones   OK

   All parties agreed to the final settlement amount of $8,525 and any additional
time (TBD later) in calendar days.
      This document will serve as a binding agreement for all parties. Please return
signed copy.

Sincerely,

William A. Jones
Project Manager (OK)

Agreed: _____   Date: _____
                     John Smith, XYZ

Figure 5-25   Confirmation of negotiation.

   Note in Fig. 5-25 that the figure for days of delay is filled in as "TBD later"
(to be determined later). It is more usual than not that the time impact evaluation will follow determination of the scope and the cost of the change.

Paper Trail of a Change Order

   Figures 5-26 to 5-34 show a nine-step paper trail of a change order in a con-
tract involving a force main for the Camden County Municipal Utilities
Authority (CCMUA) in New Jersey. This trail is typical for a public authority.
The steps are as follows:

1. Daily report identifies PCO (requiring force account work); Fig. 5-26; dated
   September 21, 1990

2. Repair bill from subcontractor for pipe repair; Fig. 5-27; dated September
   25, 1990

3. Repair bill from paving subcontractor; Fig. 5-28; dated November 8, 1990

4. Cost estimate by CM; Fig. 5-29

5. CM requests authorization from CCMUA to issue change order; Fig. 5-30

6. CCMUA executes change proposal; Fig. 5-31; dated June 4, 1991

7. Contract modification and acceptance; Fig. 5-32; dated October 3, 1991

8. Modification to contract for change order; Fig. 5-33

9. Formal ratification of the modification and authorization to execute the modification; Fig. 5-34; dated December 23, 1991

Change Order Log

The potential change order log follows the change order process from request for a price quote from the contractor, through negotiations, and finally to the approved change order. It is cross-referenced to the RFI, C/E, and the change order logs. Change orders are numbered consecutively. The following information is logged:

Field	Description
CO #	The change order number assigned to this change order
CO Date	The date the change order is generated
PCO #	The potential change order number that initiated this change order
Amount	The total amount of this change order
Days	The number of days allotted for this change, if applicable
To Contractor	The date the change order is sent to the contractor for signature
Signed	The date the contractor signed the change order
To Owner	The date the change order is sent to the owner
Executed	The date the owner signs the change order
Work Started	If tracking the number of days on this change, the date the work was started
Completed	The date the work is finished
Status	The choices are:
	■ Pending—a CO that is pending will be added to the forecasted amount
	■ Approved—a CO that is approved will be added to the contractual amount
	■ Canceled—a CO that is canceled is not added to anything

PROJECT DAILY REPORT

O'BRIEN-KREITZBERG & ASSOCIATES, INC.
CAMDEN COUNTY MUNICIPAL UTILITIES AUTHORITY
DISTRICT II INTERCEPTOR PROGRAM

DAY: S M T W T F S
DATE: ___Sept. 21, 1990___
DAY NUMBER: _____717_____
DAYS REMAINING:_____

CHERRY HILL FORCE MAINS
AND INTERCEPTORS
CONTRACTOR'S HOURS: FROM __7__ TO __1 pm__

Weather _____
Temperature _____

LABOR & EQUIPMENT	CONTRACTORS: ☐ George Tripp, Inc. SUBCONTRACTOR: ☐ Crew: 1-FM 1-Open., 1-Flagman. and 2-Laborers, 1-Teamster 1-Two Ton Truck, 2-P/U Trucks, 1-Compressor, 1-Jack Hammer, 1-Power Saw, 1-Tamper, 1-B. Hoe, 1-Flat Bed Trailer, 1-Dump Truck. 13" Pump
	PLAN SHEET: EXISTING SURFACE CONDITION Location junction of Church Rd.. and E. Lake Dr.
COMMENTS	Work force made repair to leak in 8" transit sewerage area with 27,220 lbs of bituminous material. NOTE: Break in pipe was 2'0" beyond selected fill area.
	SAFETY:

	ITEM	DESCRIPTION	QUANTITY	LOCATION
PAY ITEMS		8" Dresser		

OKA FORM 5A-2
PAGE 1 OF 3

SIGNATURE _____

INSPECTOR
(PRINT NAME) _____Fred Brown_____

Figure 5-26 Daily report that identifies PCO.

PIONEER PIPE CONTRACTORS, INC.
Underground Utilities

September 25, 1990

GEORGE TRIPP, INC.
P.O. Box 283
Colmar, Pennsylvania 18915

ATTENTION: Mr. Paul Leonard

Reference: Force Main Repair
              Job #90-099

INVOICE #90-143

BILLING FOR THE REPAIRING OF THE EXISTING FORCE MAIN ADJACENT TO CHURCH ROAD AND COLUMBIA AVENUE IN CHERRY HILL, ON SEPTEMBER 21, 1990.

CHERRY HILL (SEE ATTACHED)	$ 745.74
7% SALES TAX	52.20
TOTAL AMOUNT DUE THIS INVOICE	$ 797.94

THANK YOU,

PIONEER PIPE CONTRACTORS, INC.

Figure 5-27 Pipe repair bill.

PIONEER PIPE CONTRACTORS, INC.
Underground Utilities

November 8, 1990

GEORGE TRIPP, INC.
P.O. Box 283
Colmar, Pennsylvania 18915

Attention: Mr. Paul Leonard

Reference: Top Paving
              Job #90-099

INVOICE #90-169

BILLING FOR THE TOP PAVING OF THE FORCE MAIN PATCH ADJACENT TO CHURCH ROAD AND COLUMBIA AVENUE IN CHERRY HILL, ON NOVEMBER 7, 1990:

CHERRY HILL (SEE ATTACHED)	$ 586.00
7% SALES TAX	41.02
TOTAL AMOUNT DUE THIS INVOICE	$ 627.02

THANK YOU,

PIONEER PIPE CONTRACTORS, INC.

Figure 5-28   Paving repair bill.

Project _____

O'BRIEN-KREITZBERG
& ASSOCIATES, INC.

Location _____

Architect Engineer _____

CONSTRUCTION COST ESTIMATE		DATE PREPARED		SHEET 1 OF 1			
DRAWING NO.		ESTIMATOR			CHECKED BY:		
SUMMARY	QUANTITY		LABOR		MATERIAL		TOTAL COST
	NO. UNITS	UNIT MEAS.	PER UNIT	TOTAL	PER UNIT	TOTAL	
Extra Work							
Sept. 21, 1998							
Labor:							
Foreman 1 ea.	7.5	MH	24.25	$181.88			
Laborers 3 ea.	22.5	MH	19.50	438.75			
			S/T	$620.63			
42% Tax & Ins.				260.65			
			S/T	$881.28			$881.28
Materials:							
8" AC Repair-	Band				EA	98.00	98.00
						S/T	$979.28
						10% OH	97.93
						S/T	$1,077.21
						10% Profit	107.72
						S/T	$1,184.93
Subs: Invoice #90-143					EA	745.75	
#90-169					EA	586.00	
						S/T	$1,331.75
				GC Markup		133.18	
						S/T $1,464.93	$1,464.93
Note: Break was 2' out-						TOTAL	$2,649.86
side of original trench.							

Figure 5-29 Cost estimate by CM.

O'BRIEN-KREITZBERG & ASSOCIATES, INC.

May 24, 1991
221-2 OK-UA-261

Paul Jones, P.E.
Chief Engineer
Camden County Municipal Utilities Authority
1645 Ferry Avenue
P.O. Box 1432
Camden, NJ 08101

REF: CAMDEN COUNTY MUNICIPAL UTILITIES AUTHORITIES
       District II Interceptor Program

Dear Mr. Jones:

Enclosed for your review and approval is Form 30A-1, "Request for Authorization to Issue Change, for Contract 221/2, Change Proposal No. 20 for repair of 8" force main - Church and E. Lake Drive.

This change is estimated to cost approximately $2,649.86, and is intended to cover costs associated with the extra work to repair an 8" force main broken outside the trench area.

The Construction Manager has reviewed the information relating to this Change Proposal, and recommends to you the approval of this Request for Authorization to Issue Change.

Sincerely yours,

Lawrence R. Veit, P.E.
Senior Project Manager

LRV/RSY/cs

Figure 5-30   Letter from CM requesting authorization to issue change.

THE CAMDEN COUNTY MUNICIPAL UTILITIES AUTHORITY

June 4, 1991

Mr. Lawrence Veit, P.E.
O'Brien-Kreitzberg & Associates, Inc.
Cuthbert Boulevard and Route 70
Suite 108
Cherry Hill, NJ 08002

RE: DISTRICT II INTERCEPTOR PROGRAM

Dear Mr. Veit:

Attached please find executed Change Proposal No. 221/2-20 for Contract 221/2. The estimated cost of the Change Proposal is $2,649.86.

Based on the recommendation of your letter dated May 24, 1991, you are hereby authorized to negotiate and issue these change to George Tripp, Inc.

If you should have any questions, please do not hesitate to contact this office.

Very truly yours,

Paul Jones, P.E.
Chief Engineer

PJ/ls

Figure 5-31 Letter from CCMUA executing change proposal.

CONTRACT MODIFICATION PROPOSAL AND ACCEPTANCE			(PAGE 1) (Ref: CP 221/2-20)
1. ISSUING OFFICE: Camden County Municipal Utilities Authority	2. GRANT NO. C-34-524-03	3. CONTRACT NO. 221/222	4. MODIFICATION NO. SIXTEEN
5. TO (CONTRACTOR) George Tripp, Inc.		6. PROJECT AND LOCATION DESCRIPTION: CAMDEN COUNTY MUNICIPAL UTILS. AUTH. DISTRICT II INTERCEPTOR PROGRAM, CAMDEN COUNTY, NJ	

7. A Proposal is required for making the hereinafter described Change in accordance with Specification and drawing revisions cited herein or listed in attachment hereto. Submit your Proposal in space indicated on Page 2, attach detailed breakdown of Prime and Sub-contract costs. (See the clause of this Contract entitled, "Changes". DO NOT start work under this Proposed Change until you receive a copy signed by the Contracting Officer or a Directive to proceed.)

	Camden County	
__10/3/91__	__Municipal Utilities Authority__	_____
DATE	TYPE NAME AND TITLE	SIGNATURE

8. DESCRIPTION OF CHANGE: Pursuant to the clause of this Contract entitled "Change", the Contractor shall furnish all plant labor and material, and perform all work necessary to accomplish the following described work:

Repair existing 8" force main outside trench excavation - E. Church and E. Lake Drive.

As a result of the above, the Contract price is revised as follows:

ITEM NO.	ITEM DESCRIPTION (See Page 2)	UNIT PRICE	ESTIMATED QUANTITY	TOTAL COST

TOTAL COST OF THIS MODIFICATION $2,649.86

The Contract Time is hereby (increased) (decreased) (remains the same) by ___0___ calendar days as a result of this Modification.

The foregoing Modification is hereby accepted:

George Tripp, Inc.	CAMDEN COUNTY MUNICIPAL UTILITIES AUTHORITY	(SEAL) Malcolm Pirnie
CONTRACTOR	OWNER	ENGINEER
BY: _____	BY: _____	BY: _____
DATE: _____	DATE: _____	DATE: _____
APPROVAL:		O'Brien-Kreitzberg & Associates, Inc. CONSTRUCTION MANAGER
		BY: _____
_____		
STATE OF NEW JERSEY REPRESENTATIVE		DATE: _____
_____		
USEPA - REPRESENTATIVE		

21 SEP 84   CHANGE ORDER PROCEDURES 01153-2

Figure 5-32 Contract modification proposal and acceptance.

CONTRACT MODIFICATION PROPOSAL AND ACCEPTANCE	(PAGE 2) (Ref: CP 221/2-20)		
9. ISSUING OFFICE: Camden County Municipal Utilities Authority	10. GRANT NO. C-34-524-03	11. CONTRACT NO. 221/222	12. MODIFICATION NO. SIXTEEN

13.   CONTRACTOR'S PROPOSAL-CHANGE IN CONTRACT PRICE
      (Detailed breakdown, attach additional sheets as necessary)

NOTE: SIGN AND RETURN ORIGINAL AND COPIES; RETAIN ONE COPY FOR YOUR FILE.

NET INCREASE: $_____	NET DECREASE: $_____	CALENDAR DAYS INCREASE: _____-0-_____ DAYS

(PROPOSED)
CP 221/2-20
The following is a summary of George Tripp cost breakdown.

Prime's Labor	$ 881.28
Prime's Material/Equip.	98.00
	$ 979.28
10% Overhead	97.93
	$1,077.21
10% Profit	107.72
	$1,184.93
Subcontractor	$1,464.93
	$2,649.86

In consideration for the additional work associated with this Change Order, the Contract amount is (increased) (decreased) (remains the same) by the total lump sum $2,649.86 with 0 days extension in Contract duration. It is stipulated that this agreement represents complete and total settlement for all costs associated with this additional work.

SIGNATURE:	DATE:	TYPE NAME AND TITLE OF CONTRACTOR'S REPRESENTATIVE George Trip, President

21 SEP 84   CHANGE ORDER PROCEDURES 01153-3

Figure 5-32   *(Continued)*

CONTRACT MODIFICATION PROPOSAL AND ACCEPTANCE	(PAGE 3) (Ref: CP 221/2-20)	
14. GRANTEE & GRANT NO. Camden County Municipal Utilities Authority C-34-524-03	15. CONTRACT NO. 221/222	16. MODIFICATION NO. SIXTEEN

17.   ORIGINAL CONTRACT BID PRICE - $ 4,911,489.00

TOTAL OF PREVIOUS CHANGE ORDER - $ 192,886.18

TOTAL CHANGE ORDERS TO DATE - $ 195,536.04

18.   NECESSITY FOR CHANGE AND REASON FOR OMISSION FROM PLANS AND SPECS.
During excavation of trench, an existing 8" force main broke outside of trench excavation, requiring repair.

19.   OTHER IMPACTS RESULTANT OF THIS CHANGE
None.

20.   RESUME OF NEGOTIATIONS OR RECOMMENDATIONS (Grantee's representative)

CP 221/222-20

09-21-90   Project Daily Report - 8" Force Main Break & Repair - (Page 1 of 3).

09-21-90   George Tripp Cost Summary to O'Brien-Kreitzberg outlining additional costs. (1 page w/2 attach.).

03-19-91   George Tripp letter to O'Brien-Kreitzberg request for reimbursement. (Page 2 only)

05-24-91   O'Brien-Kreitzberg letter to Authority requesting authorization to issue change.

06-04-91   Authority letter to O'Brien-Kreitzberg authorizing negotiations to issue change.

SIGNATURE:	DATE:	TYPE NAME AND TITLE OF GRANTEE'S REPRESENTATIVE Lawrence R. Veit, P.E., Senior Project Manager CONSTRUCTION MANAGER

21 JUL 1987   CHANGE ORDER PROCEDURES (3) 01153-4

Figure 5-32 (Continued)

THE CAMDEN COUNTY MUNICIPAL UTILITIES AUTHORITY

MODIFICATION TO CONTRACT   (REF: CP 221/2-20)
FOR
CHANGE ORDER

PROJECT:	Camden County MUA District II Interceptor Program	MODIFICATION NUMBER: Sixteen

TO (CONTRACTOR/CONSULTANT):
George Tripp, Inc.
P.O. Box 283
100 Trewigtown Road
Colmar, PA 18915

AUTHORITY'S PROJECT NO: DIII-2221/2222

CONTRACT FOR: Contract 221/222
Cherry Hill Force Main and Interceptor

CONTRACT DATE: October 5, 1989

You are directed to make the following changes in this Contract:

CP 221/2-20
Repair existing 8" force main outside trench excavation - E. Church and E. Lake Drive.

This change will increase the contract sum by $2,649.86 with no change in contract time.

The original Contract Sum was	$4,911,489.00
Net change by previous Change Order	$ 192,886.18
The Contract Sum prior to this Change Order was	$5,104,375.18
The Contract Sum will be (increased) (decreased)(unchanged) by this Change Order	$ 2,649.86
The new Contract Sum including this Change Order will be	$5,107,025.04
The Contract Time will be (increased)(decreased)(unchanged) by	(0) Days
The Date of Completion as of the date of this Change Order therefore is	January 3, 1990

Malcolm Pirnie, Inc.
ENGINEER
100 Eisenhower Drive
Address
Paramus, NJ 07683

BY Dennis J. Hayes, P.E., Vice President
   (Officer's Name)

(Signature)   (Date)

George Tripp, Inc.
CONTRACTOR
100 Trewigtown Road, P.O. Box 283
Address
Colmar, PA 18915

BY Ross Ludwick
   (Officer's Name)

(Signature)   (Date)

O'Brien-Kreitzberg & Associates, Inc.
CONSTRUCTION MANAGER
4350 Haddonfield Road, Suite 300
Address
Pennsauken, NJ 08109

BY Lawrence Veit, P.E., Senior Project Manager
   (Officer's Name)

(Signature)   (Date)

Camden County Municipal Utilities Authority
OWNER
P.O. Box 1432
Address
Camden, NJ 08101-1432

BY Stephen J. Kessler
   (Officer's Name)

(Signature)   (Date)

                    (RESOLUTION)   DATE

This Document must be fully executed to constitute a Modification to the original Contract.

Figure 5-33   Modification to contract for change order.

ℜesolution
of

THE CAMDEN COUNTY MUNICIPAL UTILITIES AUTHORITY
APPROVING MODIFICATION NO. SIXTEEN (CP 221/222-20) AND AMENDING
CONTRACT 221/222 WITH GEORGE TRIPP, INC.

#R-91:12-613

WHEREAS, the Authority has previously entered into a Contract with the Contractor for Construction Services for the District II Wastewater Facilities, which is a portion of the regional system; and

WHEREAS, the Change Order listed below has been submitted by the Contractor and is recommended for approval by the Construction Manager and the Camden County Municipal Utilities Authority staff; and

WHEREAS, a Change Order to the Contract is required as follows: CP 221/2-20 - Repair existing 8" force main outside trench excavation - E. Church and E. Lake Drive. This change increased the contract by $2,649.86, with no change in contract time.

WHEREAS, the above mentioned Change Order is attached to this resolution, which is made a part hereof, and fully describes the change of Contract being approved by this Resolution.

WHEREAS, Change Proposal No. 221/2-20 which was authorized by Project Committee on June 10, 1991, was found to contain erroneous information; and

WHEREAS, revised Change Proposal No. 221/2-20 was authorized by Project Committee on December 9, 1991; and

WHEREAS, there is attached to this resolution a certificate showing there are funds available to pay for the expenditures authorized herein.

NOW, THEREFORE, BE IT RESOLVED by the Camden County Municipal Utilities Authority and the members thereof that the Modification as described above is hereby ratified by the Authority.

BE IT FURTHER RESOLVED, that the Chairman of the Authority is hereby authorized to execute said amended contract incorporating this Modification.
ADOPTED: DECEMBER 23, 1991

_____
JESSICA S. SNYDER, SECRETARY

I hereby certify that the foregoing is a true copy of the Resolution adopted by the members of the Camden County Municipal Utilities Authority at a meeting held on December 23, 1991.

_____

Figure 5-34 Formal ratification by CCMUA of the modification and authorization to proceed.

6

Change Order Impact on Project Schedule

To this point, the impact of change orders on the project schedule has not been considered. This is analogous to what often happens. It takes time to identify the scope, prepare the estimate, and negotiate the effort required. Owners (and therefore their agents) are reluctant to grant time extensions. Contractors usually prefer to see how the events unfold to be sure that they request enough time.

The change order's "boilerplate" material usually has words to the effect that this change order covers all impacts related to this change. If cost has been agreed on, but time has not, the contractor should require that TBD (to be determined) be placed in the "___ days" space. This reserves the contractor's right to amend the change order for time impact.

Some owners refuse to do this and, unilaterally, place a zero in the time space. The contractor may waive the right to later claim time if it fails to disclaim the unilateral entry by the owner. The legal argument, if the contractor signs the change order with 0 days, is that it agreed to no time extension. In signing a change order with 0 days, the contractor's position is that the change order directed that the work be done, and the contractor can get paid only if the change order is executed.

At the least, the contractor must put the owner on notice that time will be claimed by a contemporaneous letter. Failure to provide such notice precluded any claim for additional time due to weather and changes in a 1996 state court case in New Jersey (Straus Group of Old Bridge v. Quaker Construction Management, Inc., Docket L-2741-94). In fact, the judge directed that no such testimony be placed before the jury.

Schedule Requirements

The owner uses advice from the designer and other confidants to set the schedule but generally undertakes establishment of the schedule personally. The typical schedule is a tight one, either intentionally or accidentally. The intentionally tight schedule (aggressive) is a reflection of the requirements that the project will fulfill for the owner. (Often by the time the design is completed, much of the time originally available has been used in preconstruction stages.) The accidentally short construction schedule occurs when the owner is not knowledgeable of or realistic about the time necessary to construct the project at hand.

Time of delivery of a construction project is a key factor to the owner. In terms of the cost of the project, it is important to the contractor also. Therefore, the contractor must have some definite opinion about the overall length of the project in order to make a meaningful bid on it. Some of the contractor's overhead can be spread throughout the job, but there will inevitably be additional costs to the contractor. Examples include price escalation that is due to wages, and a basic overhead cost that is tied to the length of the job rather than to the specific level of field activity the job will entail.

Preconstruction analysis

The owner who includes only the completion date in the contract has very little control during the progress of the job. To establish feasible schedules, many owners are turning to a preconstruction evaluation by their staff, consultants, or the construction manager (if one has already been assigned). The purpose of this prebid analysis is the development of a construction plan by knowledgeable people who can be used as the basis of the owner's schedule. The preconstruction study may well inform the owner that a reasonable contractor under normal circumstances cannot meet the owner's dates.

The owner will then have a number of alternatives. One of them is to describe the contract time as a tight one and insist that overtime be programmed into the project on a preset basis, such as six or seven days a week. Another is to require that the contractor work double shifts, although there are, of course, severe budget impositions as a result. Also, such an approach must be evaluated in terms of area work practice. Some labor unions require a full premium for double shifts, whereas others impose only a nominal increase.

Contractor preconstruction analysis

In most cases, bidding contractors do not make a serious evaluation of the contractual time requirements unless the requirements are unusually and obviously stringent. Twenty years ago, liquidated damages assigned by engineers were usually a wrist slap of $100 per day. [Compare this with the hospital that had a $200,000 per month (or $6700 per calendar day) time damage.]

Even today, liquidated damages are generally set fairly low. The contractor who includes a condition in the bid response will definitely be found nonresponsive by public agencies and may be found nonresponsive by private organizations. The bid of the contractor who has questioned or conditioned the time frame of a contract usually must be rejected. Therefore, most contractors will not do so, but they may state their reservations about the projected dates after the award of the contract.

Experienced contractors know that there will be unforeseen conditions and unexpected situations for which time extensions will be allowed. The contractors also expect changes on the part of owners and anticipate that either the owners will relax end dates, or, if need be, they will successfully handle any delay claims by the owners. Further, liquidated damages have traditionally been set too low by owners who are unaware that their claims for damages are usually limited to the liquidated damages specified.

Milestones

The preconstruction schedule can be used to develop something more than an end date. By means of the network evaluation, key milestone points can be identified. The analysis tells the owner that if certain things do not occur by certain stages of the project, there is no way in which the end date can be met. Therefore, the section in the contract on scheduling can establish the milestones as specific days following the notice to proceed.

Normally, the only scheduling requirement included in a contract is the end date by which the contract agrees to complete the project. There usually is general language to the effect that the contractor shall keep on schedule. However, when running behind schedule, contractors can always allege that they are going to put on more workforce, will work overtime when required, or are bringing more subcontractors onto the project. There are usually no definite means of establishing that they have failed to meet their contractual obligations.

The establishment of the milestone as a contractual requirement helps ensure that the owner has a means of controlling the project's progress, and it provides a definite area in which to require timely performance by the contractor. The contract language should include some flexibility to permit the owner to adjust milestone dates if a contractor requests such a change and can demonstrate a realistic means of readjusting. Such a request should be in writing, and its approval should require the signed concurrence of the owner.

Typical milestone points include completion of foundations, completion of structure, close-in and watertightness of structure, start of temporary heat, completion of basic air-handling system, completion of permanent heat, and completion of lighting system. Milestone dates can also be established by area. Thus, in a hospital, certain areas may be designated for acceptance by the owner in stages. Typical initial areas are the ambulatory care and staff administrative spaces. If the owner intends to take phased occupancy, the decision should be made early in the design stage so the layout of the facility will reflect the incremental occupancy intended. Also, the mechanical and electrical systems may require controls by local area.

Delay

The principal dimension measured by schedules is delay. In years past, delay in the completion of construction used to be a mutually accepted condition; even the courts on occasion recognized that delay was a normal situation in the construction process. Today, however, with tight budgets on the part of the owners, who usually want to expend their funds right up to the limits of their budgets but no further, coupled with the real costs contractors encounter in staying on a job longer than planned, delay is a very problematic area. When delays occur during construction, the parties involved attempt to shift the costs that result onto each other. If litigation results after negotiations fail, the lawsuits are between two or more losers—all of whom are attempting to mitigate their losses. There are no winners in a delay situation.

To the private owner, delay can mean a loss of revenues through the resulting lack of production facilities and rentable space, as well as through a continuing dependence on present facilities. To the public owner, it can mean that a building or facility is not available for use at the planned time. The service revenues lost through delay can never be recovered. To the contractor, delay means higher overhead costs that result from the longer construction period, higher prices for materials resulting from inflation, and escalation costs that are due to labor cost increases. Further, working capital and bonding capacity are so tied up that other projects cannot be undertaken.

Responsibility for delay

The assignment of responsibility for delay after the fact is often difficult, and the courts have often remarked that delay should be anticipated in any construction project. Traditionally, the courts have protected owners more than contractors. In recent years, no-damages-for-delay clauses have often been enforced in many states, with contractors receiving only time extensions when delays occurred. However, the granting of time extensions evades another owner-oriented remedy for problems connected with delay: liquidated damages. Even when courts are inclined to consider recovery of damages for owner-caused delays, the burden is on the contractor to prove active interference on the part of the owner to receive a favorable decision.

There are four general categories of responsibility:

1. Owner (or owner's agents) is responsible.

2. Contractor or subcontractors are responsible.

3. Neither contractual party is responsible.

4. Both contractual parties are responsible.

When the owner or owner's agents have caused the delay, the courts may find that the language of the contracts, in the form of the typical no-damages-for-delay clause, protects the owner from having to pay damages but requires an offsetting time extension to protect the contractor from having to pay liquidated damages. If the owner can be proven guilty of interfering with the contractor's progress on the project or has committed a breach of contract, however, the contractor can probably recover damages from the owner.

If the contractor or subcontractors cause the delay, the contract language does not generally offer the protection against litigation on the part of the owner to recover damages. If the delay is caused by forces beyond the control of either party to the contract, the finding generally is that each party must bear the brunt of its own damages. If both parties to the contract contribute to the delay or cause concurrent delays, the usual finding is that the delays offset one another. An exception would be instances in which the damages can be clearly and distinctly separated.

Types of delay

There are three basic types of delay: classic, concurrent, and serial. *Classic delay* occurs when a period of idleness and/or uselessness is imposed on the contracted work. In Grand Investment Co. v. United States (102 Ct. Cl. U.S. 40; 1944), the government issued a stop order by telegraph to the contractor that resulted in a work stoppage of 109 days. The contractor sued for damages caused by the delay, basing the suit on a claim of breach of contract. The court allowed, among other things, a damage due to the loss of utilization of equipment on the job site, finding inability to use equipment on the job site and stating, "When the government in breach of its contract, in effect, condemned a contractor's valuable and useful machines for a period of idleness and uselessness. . .it should make compensation comparable to what would be required if it took the machines for use for a temporary period."

Johnson v. Fenetra (305 F. 2d 179, 181, 3d Cir.; 1962) also involved a classic delay: workers were idled by the failure of the general contractor to supply materials. That type of delay, to be legally recognized as such, must be substantial, involve an essential segment of the work to be done, and remain a problem for an unreasonable amount of time.

Concurrent delay occurs when work stoppages happen at the same time to separate, parallel activities. Generally, if two parties claim concurrent delays, the court will not try to unravel the factors involved and will disallow the claims by both parties. In United States v. Citizens and Southern National Bank (367 F. 2d 473; 1966), a subcontractor was able to show delay damages caused by the general contractor. However, the general contractor, in turn, was able to demonstrate that portions of the damages were caused by factors for which he was not responsible. In the absence of clear evidence separating the two claims, the court rejected both claims, stating: "As the evidence does not provide any reasonable basis for allocating the additional costs among those contributing factors, we conclude that the entire claim should have been rejected."

Similarly, in Lichter v. Mellon-Stuart (305 F. 216, 3d Cir.; 1962), the court found that the facts supported evidence of delay imposed on a subcontractor by a general contractor. It also found that the work had been delayed by a number of other factors, including change orders, delays caused by other trades, and strikes. The subcontractor had based its claim for damages solely on the delay imposed by the general contractor, and both the trial court and the appeals court rejected the claim on the following basis:

> Even if one could find from the evidence that one or more of the interfering contingencies was a wrongful act on the part of the defendant, no basis appears for even an educated guess as to the increased costs. . . due to that particular breach . . .as distinguished from those causes from which defendant is contractually exempt.

It should be noted, however, that in recent decisions, the courts increasingly have demonstrated a willingness to allocate responsibility for concurrent delays.

Serial delay is a linkage of delays (or sometimes of different causes of a delay). Thus the effects of one delay might be amplified by a later delay. For instance, if an owner's representative delays review of shop drawings and the delay causes the project to drift into a strike or a period of severe weather, resulting in further delays, a court might find the owner liable for the total serial delay.

Force majeure causes

Force majeure causes include what are known as "acts of God." The general contact usually provides a list of such events: fires, strikes, earthquakes, tornadoes, floods, and so on. Should such an event occur, the contract provides for mutual relief from demands for damages that are due to delay, and the owner is obligated to provide a reasonable (usually a day-for-day) time extension.

In the case of weather-related delays, usually only the occurrences shown to be beyond the average weather conditions expected for the area based on past records can be considered as reasons for time extensions. These can, however, vary with contract language. A number of states and cities allow a day-for-day time extension (noncompensable) for all bad weather.

Many contracts have clauses stating that the time extensions for delay caused by acts of God shall be granted only to the portions of the projects specifically affected by such events. Thus a severe downpour after a site has been graded and drained and the building closed in may cause no actual delay; therefore, claims for time extensions would not be accepted even though it would qualify under other methods of evaluation as a force majeure act.

As-Planned CPM

The *critical path method* (CPM) is the predominant way of developing construction schedules. CPM is a logic network that defines the planned sequence of activities. The activities are analyzed by computer to identify the controlling or critical ones.

CPM can be useful in establishing the facts and also the intentions of the parties to a contract. The most important part of the CPM work in this respect is the initially approved CPM network, because it describes the manner in which the contractor intended to meet the requirements of the contract at the start of the project. The network can be used by the owner to demonstrate areas of failure on the part of the contractor, and it can be used by the contractor to demonstrate points of interference on the part of the owner or owner's agents.

A project involving regular (usually monthly) reviews or updatings of the CPM plan should provide a good basis, through the CPM reports, for evaluating the progress of the work done on it. Unfortunately, many such projects have only a collection of CPM diagrams and computer runs to show for the reviews. The CPM reports are far more valuable if each update is accompanied by a comprehensive narrative. Narratives, which should be normal portions of the project documentation, are prepared in the usual order of business and therefore can be accepted later at face value, with due weight given to their origins.

It is not unusual for the CPM scheduling team to periodically readjust the schedule of a project to attempt to maintain the end date or to accommodate problems and unexpected situations. When looking at those periods of rescheduling, it can appear that the project is either on schedule or has not fallen further behind schedule, whereas in reality, the dates are being revised in terms of the overall plan but do not necessarily reflect the true progress on the project.

A first step in using CPM to analyze what has happened on a project is to set up the initially approved plan in network form. If the original network was small (1000 or fewer activities), it is merely recomputerized to confirm the initially scheduled dates. If the network was larger, particularly in the range of 5000 to 10,000 activities, milestone points should be identified and a summary of activities should be prepared. A summary CPM network of 1000 or fewer activities equivalent to the detailed major network should then be developed. Finally, this equivalent summary network should be computerized to confirm that it gives the correct initial dates and that it is, indeed, equivalent to the original, larger network.

In addition to the above steps, a summary network should be redrawn to a time grid. A typical scale would be 2 inches for each month, so that a three-year project would be represented by a 6-foot-long diagram. The vertical dimension is a function of the arrangement of the schedule and the number of activities. If a more convenient size is preferred, the larger network can be reduced by photocopy to half-size, or a scale of 1 inch for each month can be used. Keep in mind, however, that too small a scale precludes the opportunity to use the as-planned network for demonstrating the effects of schedule changes. An alternative is a time-scaled plot using the graphics packages available with almost all high-level PC scheduling systems, although experience suggests they are not as useful in a courtroom as the drafted version.

As-Built CPM

When the activities on the as-planned network have been identified, work can start on an as-built network. The second network should include the same activities as the first for comparison purposes; however, it should be based upon actual performance dates. Those dates are researched from the updatings of the original CPM plan, the progress reports, and any other documentation available. Sparse or faulty project documentation makes development of an accurate as-built network difficult. (For that reason, in CPM updatings actual dates should be plugged in for all activities as they start and are completed.) The as-built network is drawn to the same time scale and organized in the same arrangement as the as-planned network. The two can now be compared directly.

Causative Factors

With the completion of the as-planned and as-built schedules, a uniform format for the evaluation of causative factors in the delay is now available. (Even before the completion of the networks, a separate group under the direction of the scheduling professional can begin that evaluation.) The identity of most of the causative factors should be readily apparent, but the specific impact of different factors may not be obvious.

One of the first areas to be identified is force majeure. The most common reasons for this area in construction projects are strikes and bad weather. Strikes should be documented in terms of their lengths, the remobilization time it takes when they are over, and the trades and areas of work affected by them. Most contracts provide for time extensions because of strikes but not for compensation. In the case of a contractor making a claim, demonstrating that a strike had little or no impact on the critical path of a project would be imperative so that other compensable factors could be shown to be the cause of the damages being claimed. Conversely, an owner defending against claims would try to demonstrate that strikes did indeed cause the delays and other problems were, at worst, concurrent.

Change orders are evaluated in terms of the specific impacts they have on the progress of a project. This is done in two ways. First, a determination is made at what point in the network a particular change order impacted the fieldwork. In addition, activities that were preparatory for implementing the change order are identified. Examples are change order proposals, ordering material, mobilization, and any other preimplementation factors.

Next, the change order's impact is identified in terms of the amount of labor power required to accomplish it. The size of a typical work crew can be derived either from standard estimating sources or from the labor portion of the work activity being evaluated if it is identified in the bid estimate or the approved progress payment breakdown. The worker-hours involved in implementing the change are then determined by multiplying the typical crew size by the number of hours it took to complete the work item.

A separate evaluation is done for every change order in the project. In addition to identifying the basic impact each has had on the plan, the analysis should also identify the times of issue of the individual change orders' notices to proceed. In each case, if the issues times are later than the late start date of the affected activity, obviously the change order had the potential to delay the project and, in fact, probably did delay it. An exception would be if methods were used to work around the change. In this case, the methods themselves must be demonstrated to have been used.

Another area to be researched is stop orders or suspensions. They are applied to a network in the form of actual dates or as activities inserted in the stream of activities affected.

Time Impact Evaluations

When all the causative factors have been identified, a time impact evaluation (TIE) is prepared for each factor. The information is assembled as described previously, and it is prepared in a format such that the impact of each factor on the as-planned network can be determined and applied to it. When the impacts of all the causative factors have been correctly determined and applied, the result should be an approximation of the as-built network. Then the impacted as-planned network should be compared with the as-built one, and any major disparities between them should be examined to identify whether TIEs were incorrectly applied or there were additional causative factors not identified.

The theoretical effects of the impacting factors on the as-planned network must be explainable in terms of the as-built network, or the proposed analysis is probably incorrect. Some professionals take a different position, however. One well-known scheduling consultant expounds the theory of the 500 bolts: If an owner is to provide 500 bolts and has delivered only 499, in the consultant's opinion, the activity involved will be impacted until the 500th bolt has been delivered. However, it appears more logical to examine the function of the 500th bolt. For instance, if the bolt is a spare or there is a readily acceptable substitute that permits construction to proceed, then it is not, theoretically speaking, proper to claim that the as-planned network has been impacted by its absence.

Another position, often taken by schedulers who conduct impact analyses on as-planned networks for contractor evaluations, is that all float (scheduling flexibility) belongs to the contractor. This has been a continuing argument in the profession. In fact, some recent owner's specifications, in order to counteract such claims, state outright that "all float belongs to the owner." Neither position is tenable, however. Float is a shared commodity. Like a natural resource, it must be used with common sense. The owner should be permitted to use float for change orders, shop drawing reviews, and other owner-responsible areas.

On the other hand, owners should not, of course, use float to the point that the entire project becomes totally critical. This would be an overreach on the part of the owners. Conversely, contractors should be expected to use float only to balance their workforces and to work efficiently to complete projects on time and at optimum budgets.

When all of the TIE information has been imposed on the as-planned network, a standard CPM calculation is made. The calculation should correlate, as discussed previously, with the as-built network. When such a correlation is observed, the TIEs are selectively zeroed out by category. For instance, the force majeure changes are zeroed out and a run is made to determine the overall impact of their absence on the network. Similarly, contractor-related TIEs are zeroed out, and whatever further improvement their absence makes in the schedule is noted. Then the owner-related TIEs, involving change orders, hold orders, and so on, are zeroed out, and the final result should bring the network back to its as-planned status. Since each category of change is zeroed out step-by-step, the effects of concurrency can be observed from the results of the three separate runs. This can provide an arbitrator or a court with the means to allocate delay damages and impacts caused by the various parties.

One of the first applications of this approach was to a major airport project. The airport authority had contracted for the installation of a $15 million underground fueling system. The contractor for the work, who was the low bidder by several million dollars, prepared a construction CPM plan that was never accepted by the owner, and all the milestone dates were completely missed. The airport authority took under advisement whether to enter suit for delay damages that were due to losses in interest on money and in airport operating efficiency, as well as for other direct delay damages. When the contractor filed a $6 million delay suit against the authority, the authority promptly filed a counterclaim and litigation ensued.

In the absence of a mutually acceptable as-planned CPM, the owner directed that an as-built CPM be prepared to evaluate the real causes of the delays. The daily, weekly, and monthly reports, as well as personal observations by the owner's field team and the CPM consultant, were used in developing the comprehensive plan. It contained milestone points reflecting actual dates of accomplishment for various activities. Between the milestone points, the estimates for the time that the work should have taken were inserted, and the CPM team then divided the delay proportionally by its causes. The causes were either by contractor, by owner, or combined, or none of these.

The first computer run of the network showed the actual dates for all the events. The next computation established the amount of delay due to the contractor alone. The third established the amount of delay caused by the owner alone. The fourth identified the amount of delay due to both. But the total actual delay was less than the combined total when the amounts caused by the owner alone and the contractor alone were added together.

Using this very specific information, the managing engineer for the owner was able to direct efforts toward an out-of-court settlement that took more than a year to negotiate. (Part of the willingness to negotiate on the part of the owner's management personnel arose out of a recognition of the very real delays caused by slow shop drawing review. Many of these delays were due to the high workload that the owner's engineering department was carrying at the time, but many were also identified as coming from attempts by the owner's engineers to redesign the shop drawing submissions, a common mistake made in the course of many shop drawing reviews.)

CPM Plan and TIE Examples—John Doe Project

Figure 6-1 is the i-j (dictionary) CPM plan for the John Doe Project for WHATCO. The start of construction date is July 1, 1998 and the end date is April 22, 1999. Figure 6-1 gives the schedule data for the project in i-j order. The network underlying this output is found in App. C.

Figure 6-2 (pp. 124 to 130) is the total float–early start CPM output for the project. In this sort (or edit), the output is first sorted by total float (TF, low to high) and then sorted by early start (ES). Thus, on page 1, the critical path (Total Float = 0) can be read chronologically from Activity 0-1 Clear Site to 94-80 Seed and Plant.

Electric duct bank

Figure 6-3 (p. 131) shows the TIE for an electric duct bank in the way of plant excavation (discovered July 31, 1998). See Activity 15-16 in Fig. 6-4 (pp. 132 to 133). The impact is Event 15, the start of excavation.

PCO #1—sinkhole

Figure 6-5 (p. 134) is the TIE for a sinkhole in the footprint of the John Doe plant (discovered August 21, 1998 at column lines A and 11). The impact is on the same activity (15-16), but this can be done in parallel with excavation of the plant. Accordingly, the impact is on Event 16.

Figure 6-6 (p. 135) combines the "fragnets" (fragments of a network) from TIEs for RFI 001 and RFI 002/PCO #1. This figure is drawn to a time scale. The delay due to the duct bank is a delay in starting Activity 15-16 of 13 calendar days. The delay caused by the sinkhole is a delay in finishing the excavation of 17 calendar days. The start and finish delays are in parallel. Therefore, the combined delays are *not* the total, or 30 calendar days. The "combination" is 17 calendar days, the "greater" of the two.

As a check:

> *Delayed finish*
>
> 9-25 complete
>
> <u>9-15</u> delayed start
>
>   10 calendar days (c.d.)
>
>   6 c.d. work (8 Sep. planned finish)
>
>   (2 Sep. planned start)
>
>   <u> 4</u> c.d. delayed finish
>
>   10 c.d. *(Text continues on p. 123)*

(Text continues on p. 123)

O'BRIEN KREITZBERG INC.

FINEST HOUR

JOHN DOE 1

REPORT DATE 29JAN97   RUN NO. 29
12:07

SCHEDULE REPORT BY I J

JOHN DOE

START DATE 1JUL98   FIN DATE 22APR99

DATA DATE 1JUL98   PAGE NO. 1

PRED	SUCC	ORIG DUR	REM DUR	%	CODE	ACTIVITY DESCRIPTION	EARLY START	EARLY FINISH	LATE START	LATE FINISH	TOTAL FLOAT
0	1	3	3	0	1	CLEAR SITE	1JUL98	3JUL98	1JUL98	3JUL98	0
0	210	10	10	0	31	SUBMIT FOUNDATION REBAR	1JUL98	14JUL98	29JUL98	11AUG98	20
0	212	20	20	0	31	SUBMIT STRUCTURAL STEEL	1JUL98	28JUL98	27JUL98	21AUG98	18
0	214	20	20	0	31	SUBMIT CRANE	1JUL98	28JUL98	3AUG98	28AUG98	23
0	216	20	20	0	31	SUBMIT BAR JOISTS	1JUL98	28JUL98	7SEP98	2OCT98	48
0	218	20	20	0	31	SUBMIT SIDING	1JUL98	28JUL98	1SEP98	28SEP98	44
0	220	20	20	0	31	SUBMIT PLANT ELECTRICAL LOAD CENTER	1JUL98	28JUL98	7JUL98	3AUG98	4
0	222	20	20	0	31	SUBMIT POWER PANELS-PLANT	1JUL98	28JUL98	13AUG98	9SEP98	31
0	224	20	20	0	31	SUBMIT EXTERIOR DOORS	1JUL98	28JUL98	21AUG98	17SEP98	37
0	225	30	30	0	31	SUBMIT PLANT EXECTRICAL FIXTURES	1JUL98	11AUG98	21AUG98	1OCT98	37
0	227	20	20	0	31	SUBMIT PLANT HEATING AND VENTILATING FANS	1JUL98	28JUL98	28AUG98	24SEP98	42
0	229	20	20	0	31	SUBMIT BOILER	1JUL98	28JUL98	16OCT98	12NOV98	77
0	231	20	20	0	31	SUBMIT OIL TANK	1JUL98	28JUL98	8DEC98	4JAN99	114
0	235	30	30	0	31	SUBMIT PACKAGED A/C	1JUL98	11AUG98	5OCT98	13NOV98	68
0	237	40	40	0	31	SUBMIT PRECAST	1JUL98	25AUG98	7SEP98	30OCT98	48
1	2	2	2	0	2	SURVEY AND LAYOUT	6JUL98	7JUL98	6JUL98	7JUL98	0
2	3	2	2	0	1	ROUGH GRADE	8JUL98	9JUL98	8JUL98	9JUL98	0
3	4	15	15	0	9	DRILL WELL	10JUL98	30JUL98	10JUL98	30JUL98	0
3	6	4	4	0	3	WATER TANK FOUNDATIONS	10JUL98	15JUL98	13JUL98	16JUL98	1
3	9	10	10	0	1	EXCAVATE FOR SEWER	10JUL98	23JUL98	16JUL98	29JUL98	4
3	10	1	1	0	1	EXCAVATE ELECTRICAL MANHOLES	10JUL98	10JUL98	29JUL98	29JUL98	13
3	12	6	6	0	4	OVERHEAD POLE LINE	10JUL98	17JUL98	3AUG98	10AUG98	16

116

Figure 6-1 is a tabular CPM plan (i-j sort). The columns, left to right, are: i-node, j-node, duration (shown in two columns), and additional planning columns, followed by the activity description, the early start, early finish, late start, late finish dates, and the total float.

I	J	DUR	DUR					DESCRIPTION	EARLY START	EARLY FINISH	LATE START	LATE FINISH	TOTAL FLOAT
4	5	2	2	0	1	1	9	INSTALL WELL PUMP	31JUL98	3AUG98	31JUL98	3AUG98	0
5	8	8	8	0	1	1	5	UNDERGROUND WATER PIPING	4AUG98	13AUG98	4AUG98	13AUG98	0
6	7	10	10	0	1	1	10	ERECT WATER TOWER	16JUL98	29JUL98	17JUL98	30JUL98	1
7	8	10	10	0	1	1	10	TANK PIPING AND VALVES	30JUL98	12AUG98	31JUL98	13AUG98	1
8	13	2	2	0	1	1	10	CONNECT WATER PIPING	14AUG98	17AUG98	14AUG98	17AUG98	0
9	11	5	5	0	1	1	5	INSTALL SEWER AND BACKFILL	24JUL98	30JUL98	30JUL98	5AUG98	4
10	11	5	5	0	1	1	4	INSTALL ELECTRICAL MANHOLES	13JUL98	17JUL98	30JUL98	5AUG98	13
11	12	3	3	0	1	1	4	ELECTRICAL DUCT BANK	31JUL98	6AUG98	4AUG98	10AUG98	4
12	13	5	5	0	1	1	4	PULL IN POWER FEEDER	5AUG98	11AUG98	11AUG98	17AUG98	4
13	14	1	1	0	1	1	2	BUILDING LAYOUT	18AUG98	18AUG98	18AUG98	18AUG98	0
14	15	10	10	0	2	1	11	DRIVE AND POUR PILES	19AUG98	1SEP98	19AUG98	1SEP98	0
14	23	3	3	0	2	1	1	EXCAVATE FOR OFFICE BUILDING	19AUG98	21AUG98	25NOV98	27NOV98	70
15	16	5	5	0	2	1	1	EXCAVATE FOR PLANT WAREHOUSE	2SEP98	8SEP98	2SEP98	8SEP98	0
16	17	5	5	0	2	1	3	POUR PILE CAPS PLANT-WAREHSE	9SEP98	15SEP98	9SEP98	15SEP98	0
17	18	10	10	0	2	1	3	FORM AND POUR GRADE BEAMS P-W	16SEP98	29SEP98	16SEP98	29SEP98	0
18	19	3	3	0	2	1	1	BACKFILL AND COMPACT P-W	30SEP98	2OCT98	30SEP98	2OCT98	0
18	21	5	5	0	2	1	3	FORM AND POUR RR LOAD DOCK P-W	30SEP98	6OCT98	12OCT98	16OCT98	8
18	22	5	5	0	2	1	3	FORM AND POUR TK LOAD DOCK P-W	30SEP98	6OCT98	12OCT98	16OCT98	8
19	20	5	5	0	2	1	5	UNDERSLAB PLUMBING P-W	5OCT98	9OCT98	5OCT98	9OCT98	0
20	22	5	5	0	2	1	4	UNDERSLAB CONDUIT P-W	12OCT98	16OCT98	12OCT98	16OCT98	0
21	22	0	0	0	0	0		RESTRAINT	7OCT98	6OCT98	19OCT98	16OCT98	8
22	29	10	10	0	2	1	3	FORM AND POUR SLABS P-W	19OCT98	30OCT98	19OCT98	30OCT98	0
23	24	4	4	0	2	1	3	SPREAD FOOTINGS OFFICE	24AUG98	27AUG98	30NOV98	3DEC98	70
24	25	6	6	0	2	1	3	FORM AND POUR GRADE BEAMS OFF	28AUG98	4SEP98	4DEC98	11DEC98	70
25	26	1	1	0	2	1	1	BACKFILL AND COMPACT OFFICE	7SEP98	7SEP98	14DEC98	14DEC98	70
26	27	3	3	0	2	1	5	UNDERSLAB PLUMBING OFFICE	8SEP98	10SEP98	15DEC98	17DEC98	70
27	28	3	3	0	2	1	4	UNDERSLAB CONDUIT OFFICE	11SEP98	15SEP98	18DEC98	22DEC98	70
28	99	3	3	0	2	1	3	FORM AND POUR OFFICE SLAB	16SEP98	18SEP98	23DEC98	25DEC98	70
29	30	10	10	0	3	1	6	ERECT STRUCT STEEL P-W	2NOV98	13NOV98	2NOV98	13NOV98	0
30	31	5	5	0	3	1	6	PLUMB STEEL AND BOLT P-W	16NOV98	20NOV98	16NOV98	20NOV98	0
31	32	5	5	0	3	1	6	ERECT CRANE WAY AND CRANE P-W	23NOV98	27NOV98	23NOV98	27NOV98	0
31	33	3	3	0	3	1	6	ERECT MONORAIL TRACK P-W	23NOV98	25NOV98	25NOV98	27NOV98	2
32	33	0	0	0	0	0		RESTRAINT	30NOV98	27NOV98	30NOV98	27NOV98	0
33	34	3	3	0	3	1	6	ERECT BAR JOISTS P-W	30NOV98	2DEC98	30NOV98	2DEC98	0

Figure 6-1 CPM plan (i-j sort).

PRED	SUCC	ORIG DUR	REM DUR	%	CODE		ACTIVITY DESCRIPTION	EARLY START	EARLY FINISH	LATE START	LATE FINISH	TOTAL FLOAT
34	35	3	3	0	3	7	ERECT ROOF PLANKS P-W	3DEC98	7DEC98	3DEC98	7DEC98	0
35	36	10	10	0	3	12	ERECT SIDING P-W	8DEC98	21DEC98	8DEC98	21DEC98	0
35	37	5	5	0	3	13	ROOFING P-W	8DEC98	14DEC98	15DEC98	21DEC98	5
36	37	0	0	0			RESTRAINT	22DEC98	21DEC98	22DEC98	21DEC98	0
37	80	10	10	0	5	15	PERIMETER FENCE	22DEC98	4JAN99	9APR99	22APR99	78
37	90	5	5	0	5	16	PAVE PARKING AREA	22DEC98	28DEC98	2APR99	8APR99	73
37	91	5	5	0	5	17	GRADE AND BALLAST RR SIDING	22DEC98	28DEC98	19MAR99	25MAR99	63
37	92	10	10	0	5	16	ACCESS ROAD	22DEC98	4JAN99	26MAR99	8APR99	68
37	93	20	20	0	5	4	AREA LIGHTING	22DEC98	18JAN99	26MAR99	22APR99	68
37	300	0	0	0			RESTRAINT	22DEC98	21DEC98	22DEC98	21DEC98	0
37	301	0	0	0			RESTRAINT	22DEC98	21DEC98	7JAN99	6JAN99	12
37	302	0	0	0			RESTRAINT	22DEC98	21DEC98	22JAN99	21JAN99	23
37	303	0	0	0			RESTRAINT	22DEC98	21DEC98	8JAN99	7JAN99	13
37	304	0	0	0			RESTRAINT	22DEC98	21DEC98	22JAN99	21JAN99	23
37	305	0	0	0			RESTRAINT	22DEC98	21DEC98	30MAR99	29MAR99	70
37	306	0	0	0			RESTRAINT	22DEC98	21DEC98	19FEB99	18FEB99	43
37	307	0	0	0			RESTRAINT	22DEC98	21DEC98	5FEB99	4FEB99	33
37	308	0	0	0			RESTRAINT	22DEC98	21DEC98	2APR99	1APR99	73
38	43	20	20	0	3	4	INSTALL POWER CONDUIT P-W	24DEC98	20JAN99	24DEC98	20JAN99	0
39	42	5	5	0	3	18	FRAME CEILINGS P-W	5JAN99	11JAN99	22JAN99	28JAN99	13
40	47	10	10	0	3	8	TEST PIPING SYSTEMS P-W	2FEB99	15FEB99	19MAR99	1APR99	33
41	47	5	5	0	3	8	PREOPERATIONAL BOILER CHECK	26JAN99	1FEB99	26MAR99	1APR99	43

							Activity					
42	44	10	10	0	3	19	DRYWELLL PARTITIONS P-W	12JAN99	25JAN99	29JAN99	11FEB99	13
43	49	15	15	0	3	4	INSTALL BRANCH CONDUIT P-W	21JAN99	10FEB99	21JAN99	10FEB99	0
44	45	0	0	0			RESTRAINT	26JAN99	25JAN99	18MAR99	17MAR99	37
44	46	0	0	0			RESTRAINT	26JAN99	25JAN99	12FEB99	11FEB99	13
44	48	10	10	0	3	20	CERAMIC TILE	26JAN99	8FEB99	19FEB99	4MAR99	18
44	58	10	10	0	3	21	HANG INTERIOR DOORS P-W	26JAN99	8FEB99	26MAR99	8APR99	43
45	51	5	5	0	3	4	ROOM OUTLETS P-W	11FEB99	17FEB99	18MAR99	24MAR99	25
46	52	25	25	0	3	8	INSTALL DUCTWORK P-W	26JAN99	1MAR99	12FEB99	18MAR99	13
47	58	5	5	0	3	8	LIGHTOFF BOILER AND TEST	16FEB99	22FEB99	2APR99	8APR99	33
48	53	5	5	0	3	22	PAINT ROOMS P-W	9FEB99	15FEB99	5MAR99	11MAR99	18
49	45	0	0	0			RESTRAINT	11FEB99	10FEB99	18MAR99	17MAR99	25
49	50	15	15	0	3	4	PULL WIRE P-W	11FEB99	3MAR99	11FEB99	3MAR99	0
50	54	5	5	0	3	4	INSTALL PANEL INTERNALS P-W	4MAR99	10MAR99	4MAR99	10MAR99	0
51	56	10	10	0	3	8	INSTALL ELECTRICAL FIXTURES	18FEB99	3MAR99	25MAR99	7APR99	25
52	58	15	15	0	3	22	INSULATE H&V SYSTEM P-W	2MAR99	22MAR99	19MAR99	8APR99	13
53	57	10	10	0	3	5	FLOOR TILE P-W	16FEB99	1MAR99	12MAR99	25MAR99	18
53	58	10	10	0	3	4	INSTALL PLUMBING FIXTURES P-W	16FEB99	1MAR99	26MAR99	8APR99	28
54	55	10	10	0	3	4	TERMINATE WIRES P-W	11MAR99	24MAR99	11MAR99	24MAR99	0
55	56	10	10	0	3	4	RINGOUT P-W	25MAR99	7APR99	25MAR99	7APR99	0
56	58	1	1	0	3	24	ENERGIZE POWER	8APR99	8APR99	8APR99	8APR99	0
57	58	10	10	0	3	27	INSTALL FURNISHING P-W	2MAR99	15MAR99	26MAR99	8APR99	18
58	80	5	5	0	5	27	ERECT FLAGPOLE	9APR99	15APR99	16APR99	22APR99	5
58	94	5	5	0	5	7	FINE GRADE	9APR99	15APR99	9APR99	15APR99	0
59	60	5	5	0	4	14	ERECT PRECAST ROOF OFFICE	28OCT98	3NOV98	4JAN99	8JAN99	48
60	61	10	10	0	4	21	EXTERIOR MASONRY OFFICE	4NOV98	17NOV98	11JAN99	22JAN99	48
60	98	0	0	0			RESTRAINT	4NOV98	3NOV98	5APR99	2APR99	108
61	62	5	5	0	4	13	EXTERIOR DOORS OFFICE	18NOV98	24NOV98	12FEB99	18FEB99	62
61	63	5	5	0	4	8	ROOFING OFFICE	18NOV98	24NOV98	12FEB99	18FEB99	62
61	64	10	10	0	4	4	INSTALL PIPING OFFICE	18NOV98	1DEC98	25JAN99	5FEB99	48
61	65	4	4	0	4	25	INSTALL ELEC BACKING BOXES	18NOV98	23NOV98	25JAN99	28JAN99	48
61	68	5	5	0	4	8	GLAZE OFFICE	18NOV98	24NOV98	12FEB99	18FEB99	62
61	77	15	15	0	4		DUCTWORK OFFICE	18NOV98	8DEC98	12MAR99	1APR99	82
62	63	0	0	0			RESTRAINT	25NOV98	24NOV98	19FEB99	18FEB99	62
63	68	0	0	0			RESTRAINT	25NOV98	24NOV98	19FEB99	18FEB99	62

Figure 6-1 (*Continued*)

119

O'BRIEN KREITZBERG INC.   FINEST HOUR   JOHN DOE 1

REPORT DATE 29JAN97 RUN NO. 29   JOHN DOE   START DATE 1JUL98   FIN DATE 22APR99
          12:07
SCHEDULE REPORT BY I J   DATA DATE 1JUL98   PAGE NO. 3

PRED	SUCC	ORIG DUR	REM DUR	%	CODE	ACTIVITY DESCRIPTION	EARLY START	EARLY FINISH	LATE START	LATE FINISH	TOTAL FLOAT
63	80	5	5	0	22	PAINT OFFICE EXTERIOR	25NOV98	1DEC98	16APR99	22APR99	102
64	67	4	4	0	8	TEST PIPING OFFICE	2DEC98	7DEC98	8FEB99	11FEB99	48
65	66	10	10	0	4	INSTALL CONDUIT OFFICE	24NOV98	7DEC98	29JAN99	11FEB99	48
66	67	0	0	0		RESTRAINT	8DEC98	7DEC98	12FEB99	11FEB99	48
66	74	10	10	0	4	PULL WIRE OFFICE	8DEC98	21DEC98	12FEB99	25MAR99	68
67	68	5	5	0	19	PARTITIONS OFFICE	8DEC98	14DEC98	12FEB99	18FEB99	48
68	69	10	10	0	19	DRYWALL	15DEC98	21DEC98	19FEB99	25FEB99	48
69	70	10	10	0	19	DRYWALL	22DEC98	4JAN99	26FEB99	11MAR99	48
69	73	10	10	0	20	CERAMIC TILE OFFICE	22DEC98	4JAN99	2APR99	15APR99	73
70	71	10	10	0	26	WOOD TRIM OFFICE	5JAN99	18JAN99	12MAR99	25MAR99	48
70	77	0	0	0		RESTRAINT	5JAN99	4JAN99	2APR99	1APR99	63
71	72	10	10	0	22	PAINT INTERIOR OFFICE	19JAN99	1FEB99	26MAR99	8APR99	48
71	80	5	5	0	21	HANG DOORS OFFICE	19JAN99	25JAN99	16APR99	22APR99	63
72	73	0	0	0		RESTRAINT	2FEB99	1FEB99	16APR99	15APR99	53
72	78	0	0	0		RESTRAINT	2FEB99	1FEB99	9APR99	8APR99	48
72	80	10	10	0	20	FLOOR TILE OFFICE	2FEB99	15FEB99	9APR99	22APR99	48
73	80	5	5	0	5	TOILET FIXTURES OFFICE	2FEB99	8FEB99	16APR99	22APR99	53
74	75	5	5	0	4	INSTALL PANEL INTERNALS OFFICE	22DEC98	28DEC98	26MAR99	1APR99	68
74	76	0	0	0		RESTRAINT	22DEC98	21DEC98	12APR99	9APR99	79
75	79	10	10	0	4	TERMINATE WIRES OFFICE	29DEC98	11JAN99	2APR99	15APR99	68
76	79	4	4	0	4	AIR CONDITIONING ELEC CONNECT	6JAN99	11JAN99	12APR99	15APR99	68
77	78	5	5	0	18	INSTALL CEILING GRID OFFICE	5JAN99	11JAN99	2APR99	8APR99	63

78	80	10	10	0	4	18	ACOUSTIC TILE OFFICE	2FEB99	15FEB99	9APR99	22APR99	48
79	80	5	5	0	4	4	RINGOUT ELECT	12JAN99	18JAN99	16APR99	22APR99	68
90	58	0	0	0			RESTRAINT	29DEC98	28DEC98	9APR99	8APR99	73
91	58	10	10	0	5	17	INSTALL RR SIDING	29DEC98	11JAN99	26MAR99	8APR99	63
92	58	0	0	0			RESTRAINT	5JAN99	4JAN99	9APR99	8APR99	68
93	80	0	0	0			RESTRAINT	19JAN99	18JAN99	23APR99	22APR99	68
94	80	5	5	0	5	27	SEED AND PLANT	16APR99	22APR99	16APR99	22APR99	0
98	76	5	5	0	4	8	INSTALL PACKAGE AIR CONDITONER	30DEC98	5JAN99	5APR99	9APR99	68
99	59	5	5	0	4	7	ERECT PRECAST STRUCT. OFFICE	21OCT98	27OCT98	28DEC98	1JAN99	48
210	211	10	10	0	6	32	APPROVE FONDATION REBAR	15JUL98	28JUL98	12AUG98	25AUG98	20
211	16	10	10	0	6	33	FAB&DEL FOUNDATION REBAR	29JUL98	11AUG98	26AUG98	8SEP98	20
212	213	10	10	0	6	32	APPROVE STRUCTURAL STEEL	29JUL98	11AUG98	24AUG98	4SEP98	18
213	29	40	40	0	6	33	FAB&DEL STRUCTURAL STEEL	12AUG98	7SEP98	7SEP98	30OCT98	18
214	215	10	10	0	6	32	APPROVE CRANE	29JUL98	11AUG98	31AUG98	11SEP98	23
215	31	50	50	0	6	33	FAB&DEL CRANE	12AUG98	20OCT98	14SEP98	20NOV98	23
216	217	10	10	0	6	32	APPROVE BAR JOISTS	29JUL98	11AUG98	5OCT98	16OCT98	48
217	33	30	30	0	6	33	FAB&DEL BAR JOISTS	12AUG98	22SEP98	19OCT98	27NOV98	48
218	219	10	10	0	6	32	APPROVE SIDING	22AUG98	11AUG98	29SEP98	12OCT98	44
219	35	40	40	0	6	33	FAB&DEL SIDING	29JUL98	6OCT98	13OCT98	7DEC98	44
220	221	10	10	0	6	32	APPROVE PLANT ELECTRICAL LOAD CENTER	12AUG98	11AUG98	4AUG98	17AUG98	4
221	300	90	90	0	6	33	FAB&DEL PLANT ELECTRICAL LOAD CENTER	29JUL98	15DEC98	18AUG98	21DEC98	4
222	223	10	10	0	6	32	APPROVE POWER PANELS-PLANT	12AUG98	11AUG98	10SEP98	23SEP98	31
223	301	75	75	0	6	33	FAB&DEL POWER PANELS-PLANT	29JUL98	24NOV98	24SEP98	6JAN99	31
224	225	10	10	0	6	32	APPROVE EXTERIOR DOORS	12AUG98	11AUG98	18SEP98	1OCT98	37
225	226	15	15	0	6	33	APPROVE PLANT ELECTRICAL FIXTURES	29JUL98	1SEP98	19NOV98	9DEC98	71
225	302	80	80	0	6	32	FAB&DEL EXTERIOR DOORS	12AUG98	1DEC98	20OCT98	21JAN99	37
226	51	75	75	0	6	33	FAB&DEL PLANT ELECTRICAL FIXTURES	2SEP98	15DEC98	10DEC98	24MAR99	71
227	228	10	10	0	6	32	APPROVE PLANT HEATING AND VENTILATING FANS	29JUL98	11AUG98	25SEP98	8OCT98	42
228	304	75	75	0	6	33	FAB&DEL PLANT HEATING AND VENTILATING FANS	12AUG98	24NOV98	9OCT98	21JAN99	42
229	230	10	10	0	6	32	APPROVE BOILER	29JUL98	11AUG98	13NOV98	26NOV98	77
230	306	60	60	0	6	33	FAB&DEL BOILER	12AUG98	3NOV98	27NOV98	18FEB99	77
231	232	10	10	0	6	32	APPROVE OIL TANK	29JUL98	11AUG98	5JAN99	18JAN99	114
232	305	50	50	0	6	33	FAB&DEL OIL TANK	12AUG98	20OCT98	19JAN99	29MAR99	114
235	236	10	10	0	6	32	APPROVE PACKAGED A/C	12AUG98	25AUG98	16NOV98	27NOV98	68

Figure 6-1 (*Continued*)

O'BRIEN KREITZBERG INC.

FINEST HOUR

JOHN DOE 1

REPORT DATE 29JAN97 RUN NO. 29
12:07

JOHN DOE

START DATE 1JUL98 FIN DATE 22APR99

SCHEDULE REPORT BY I J

DATA DATE 1JUL98 PAGE NO. 4

PRED	SUCC	ORIG DUR	REM DUR	%	CODE	ACTIVITY DESCRIPTION	EARLY START	EARLY FINISH	LATE START	LATE FINISH	TOTAL FLOAT
236	98	90	90	0	6	33	FAB&DEL PACKAGED A/C	26AUG98	29DEC98	30NOV98	2APR99	68
237	238	10	10	0	6	32	APPROVE PRECAST	26AUG98	8SEP98	2NOV98	13NOV98	48
238	99	30	30	0	6	33	FAB&DEL PRECAST	9SEP98	20OCT98	16NOV98	25DEC98	48
300	38	2	2	0	3	4	SET ELECTRICAL LOAD CENTER P-W	22DEC98	23DEC98	22DEC98	23DEC98	0
301	43	10	10	0	3	4	POWER PANEL BACKFILL BOXES P-W	22DEC98	4JAN99	7JAN99	20JAN99	12
302	42	5	5	0	3	6	ERECT EXTERIOR DOORS P-W	22DEC98	28DEC98	22JAN99	28JAN99	23
303	39	10	10	0	3	14	MASONRY PARTITIONS P-W	22DEC98	4JAN99	8JAN99	21JAN99	13
304	46	15	15	0	3	8	INSTALL H AND V UNITS P-W	22DEC98	11JAN99	22JAN99	11FEB99	23
305	47	3	3	0	3	8	INSTALL FUEL TANK P-W	22DEC98	24DEC98	30MAR99	1APR99	70
306	41	25	25	0	3	8	ERECT BOILER AND AUXILIARY P-W	22DEC98	25JAN99	19FEB99	25MAR99	43
307	40	30	30	0	3	8	FABRICATE PIPING P-W	22DEC98	1FEB99	5FEB99	18MAR99	33
308	58	5	5	0	3	6	INSTALL MONORAIL WAREHOUSE	22DEC98	28DEC98	2APR99	8APR99	73

Figure 6-1 (*Continued*)

Delayed start

9-15 delayed start

<u>9- 2</u> planned start

13 c.d.

Delayed start (13 c.d.) + delayed finish (4 c.d.) = 17 c.d. delay

PCO #2—hot-water return

Figure 6-7 (p. 136) is the TIE for the missing hot-water return from the plant and warehouse. The negotiations and piping material are projected to be available as early as November 13, 1998.

Under the baseline—early need 12/22/98

Under the revised baseline—early need 1/08/99

Further, the selection of Event 307 as the impact point is conservative. The work activity (307-40) is 30 working days (or 6 weeks).

PCO #3—VE joists versus girders

Contractor XYZ proposed a major cost savings through VECP #1 (PCO #3). This TIE is very important because a major time impact could cause the PM/CM for the owner to recommend against it. As shown in the TIE (Fig. 6-8, p. 137), the change to joists at this time has a −18 c.d. impact. The TIE recognizes two workarounds (two-week submittal versus four) and (one-week approval versus two weeks). These make sense since the joists are able to be manufactured faster, and this VECP deletes the secondary joists.

Accordingly, the PM/CM can readily agree to proceed with the VECP. (Also, the PM/CM is aware that the delays to excavating the plant are in the same order of magnitude.)

PCO #4—multiple chillers

This owner bulletin substitutes three chillers for one. Two can run at optimum (full) power with the third as a spare. The fragnet in Fig. 6-9 (p. 138) shows:

Bulletin	09/03/98		
Three-week negotiation	10/01/98	30 w.d. (42 c.d.) submittal	11/12/98

This is one day before the late finish; therefore, no impact.

(*Text continues on p. 135*)

O'BRIEN KREITZBERG INC.   FINEST HOUR   JOHN DOE 1

REPORT DATE 29JAN97   RUN NO. 30   JOHN DOE   START DATE 1JUL98   FIN DATE 22APR99
            12:08

SCHEDULE REPORT - SORT BY TF, ES   DATA DATE 1JUL98   PAGE NO. 1

PRED	SUCC	ORIG DUR	REM DUR	%	CODE	ACTIVITY DESCRIPTION	EARLY START	EARLY FINISH	LATE START	LATE FINISH	TOTAL FLOAT
0	1	3	3	0		CLEAR SITE	1JUL98	3JUL98	1JUL98	3JUL98	0
1	2	2	2	0		SURVEY AND LAYOUT	6JUL98	7JUL98	6JUL98	7JUL98	0
2	3	2	2	0		ROUGH GRADE	8JUL98	9JUL98	8JUL98	9JUL98	0
3	4	15	15	0		DRILL WELL	10JUL98	30JUL98	10JUL98	30JUL98	0
4	5	2	2	0		INSTALL WELL PUMP	31JUL98	3AUG98	31JUL98	3AUG98	0
5	8	8	8	0		UNDERGROUND WATER PIPING	4AUG98	13AUG98	4AUG98	13AUG98	0
8	13	2	2	0		CONNECT WATER PIPING	14AUG98	17AUG98	14AUG98	17AUG98	0
13	14	1	1	0		BUILDING LAYOUT	18AUG98	18AUG98	18AUG98	18AUG98	0
14	15	10	10	0		DRIVE AND POUR PILES	19AUG98	1SEP98	19AUG98	1SEP98	0
15	16	5	5	0		EXCAVATE FOR PLANT WAREHOUSE	2SEP98	8SEP98	2SEP98	8SEP98	0
16	17	5	5	0		POUR PILE CAPS PLANT-WAREHSE	9SEP98	15SEP98	9SEP98	15SEP98	0
17	18	10	10	0		FORM AND POUR GRADE BEAMS P-W	16SEP98	29SEP98	16SEP98	29SEP98	0
18	19	3	3	0		BACKFILL AND COMPACT P-W	30SEP98	2OCT98	30SEP98	2OCT98	0
19	20	5	5	0		UNDERSLAB PLUMBING P-W	5OCT98	9OCT98	5OCT98	9OCT98	0
20	22	5	5	0		UNDERSLAB CONDUIT P-W	12OCT98	16OCT98	12OCT98	16OCT98	0
22	29	10	10	0		FORM AND POUR SLABS P-W	19OCT98	30OCT98	19OCT98	30OCT98	0
29	30	10	10	0		ERECT STRUCT STEEL P-W	2NOV98	13NOV98	2NOV98	13NOV98	0
30	31	5	5	0		PLUMB STEEL AND BOLT P-W	16NOV98	20NOV98	16NOV98	20NOV98	0
31	32	5	5	0		ERECT CRANE WAY AND CRANE P-W	23NOV98	27NOV98	23NOV98	27NOV98	0
32	33	0	0	0		RESTRAINT	30NOV98	27NOV98	30NOV98	27NOV98	0
33	34	3	3	0		ERECT BAR JOISTS P-W	30NOV98	2DEC98	30NOV98	2DEC98	0
34	35	3	3	0		ERECT ROOF PLANKS P-W	3DEC98	7DEC98	3DEC98	7DEC98	0

124

I	J	OD	RD	%	Description	Early Start	Early Finish	Late Finish	TF
35	36	10	10	0	ERECT SIDING P-W	8DEC98	21DEC98	21DEC98	0
36	37	0	0	0	RESTRAINT	22DEC98	21DEC98	21DEC98	0
37	300	0	0	0	RESTRAINT	22DEC98	21DEC98	21DEC98	0
300	38	2	2	0	SET ELECTRICAL LOAD CENTER P-W	22DEC98	23DEC98	23DEC98	0
38	43	20	20	0	INSTALL POWER CONDUIT P-W	24DEC98	20JAN99	20JAN99	0
43	49	15	15	0	INSTALL BRANCH CONDUIT P-W	21JAN99	10FEB99	10FEB99	0
49	50	15	15	0	PULL WIRE P-W	11FEB99	3MAR99	3MAR99	0
50	54	5	5	0	INSTALL PANEL INTERNALS P-W	4MAR99	10MAR99	10MAR99	0
54	55	10	10	0	TERMINATE WIRES P-W	11MAR99	24MAR99	24MAR99	0
55	56	10	10	0	RINGOUT P-W	25MAR99	7APR99	7APR99	0
56	58	1	1	0	ENERGIZE POWER	8APR99	8APR99	8APR99	0
58	94	5	5	0	FINE GRADE	9APR99	15APR99	15APR99	0
94	80	5	4	0	SEED AND PLANT	16APR99	22APR99	22APR99	0
3	6	4	4	0	WATER TANK FOUNDATIONS	10JUL98	15JUL98	16JUL98	1
6	7	10	10	0	ERECT WATER TOWER	16JUL98	29JUL98	30JUL98	1
7	8	10	10	0	TANK PIPING AND VALVES	30JUL98	12AUG98	13AUG98	1
31	33	3	3	0	ERECT MONORAIL TRACK P-W	23NOV98	25NOV98	27NOV98	2
0	220	20	20	0	SUBMIT PLANT ELECTRICAL LOAD CENTER	1JUL98	28JUL98	3AUG98	4
3	9	10	10	0	EXCAVATE FOR SEWER	10JUL98	23JUL98	29JUL98	4
9	11	5	5	0	INSTALL SEWER AND BACKFILL	24JUL98	30JUL98	5AUG98	4
220	221	10	10	0	APPROVE PLANT ELECTRICAL LOAD CENTER	29JUL98	11AUG98	17AUG98	4
11	12	3	3	0	ELECTRICAL DUCT BANK	31JUL98	4AUG98	10AUG98	4
12	13	5	5	0	PULL IN POWER FEEDER	5AUG98	11AUG98	17AUG98	4
221	300	90	90	0	FAB&DEL PLANT ELECTRICAL LOAD CENTER	12AUG98	15DEC98	21DEC98	5
35	37	5	5	0	ROOFING P-W	8DEC98	14DEC98	21DEC98	5
58	80	5	5	0	ERECT FLAGPOLE	9APR99	15APR99	22APR99	5
18	21	5	5	0	FORM AND POUR RR LOAD DOCK P-W	30SEP98	6OCT98	16OCT98	8
18	22	5	5	0	FORM AND POUR TK LOAD DOCK P-W	30SEP98	6OCT98	16OCT98	8
21	22	0	0	0	RESTRAINT	7OCT98	6OCT98	16OCT98	8
37	301	0	0	0	RESTRAINT	22DEC98	21DEC98	6JAN99	12
301	43	10	10	0	POWER PANEL BACKFILL BOXES P-W	22DEC98	4JAN99	20JAN99	12
3	10	1	1	0	EXCAVATE ELECTRICAL MANHOLES	10JUL98	10JUL98	29JUL98	13
10	11	5	5	0	INSTALL ELECTRICAL MANHOLES	13JUL98	17JUL98	5AUG98	13
37	303	0	0	0	RESTRAINT	22DEC98	21DEC98	7JAN99	13

Figure 6-2 CPM plan (TF/ES sort).

O'BRIEN KREITZBERG INC.   FINEST HOUR   JOHN DOE 1

REPORT DATE 29JAN97 RUN NO. 30   JOHN DOE
            12:08   START DATE 1JUL98 FIN DATE 22APR99

SCHEDULE REPORT - SORT BY TF, ES   DATA DATE 1JUL98   PAGE NO.   2

PRED	SUCC	ORIG DUR	REM DUR	%	CODE	ACTIVITY DESCRIPTION	EARLY START	EARLY FINISH	LATE START	LATE FINISH	TOTAL FLOAT
303	39	10	10	0		MASONRY PARTITIONS P-W	22DEC98	4JAN99	8JAN99	21JAN99	13
39	42	5	5	0		FRAME CEILINGS P-W	5JAN99	11JAN99	22JAN99	28JAN99	13
42	44	10	10	0		DRYWELLL PARTITIONS P-W	12JAN99	25JAN99	29JAN99	11FEB99	13
44	46	0	0	0		RESTRAINT	26JAN99	25JAN99	12FEB99	11FEB99	13
46	52	25	25	0		INSTALL DUCTWORK P-W	26JAN99	1MAR99	12FEB99	18MAR99	13
52	58	15	15	0		INSULATE H&V SYSTEM P-W	2MAR99	22MAR99	19MAR99	8APR99	13
3	12	6	6	0		OVERHEAD POLE LINE	10JUL98	17JUL98	3AUG98	10AUG98	16
0	212	20	20	0		SUBMIT STRUCTURAL STEEL	1JUL98	28JUL98	27JUL98	21AUG98	18
212	213	10	10	0		APPROVE STRUCTURAL STEEL	29JUL98	11AUG98	24AUG98	4SEP98	18
213	29	40	40	0		FAB&DEL STRUCTURAL STEEL	12AUG98	6OCT98	7SEP98	30OCT98	18
44	48	10	10	0		CERAMIC TILE	26JAN99	8FEB99	19FEB99	4MAR99	18
48	53	5	5	0		PAINT ROOMS P-W	9FEB99	15FEB99	5MAR99	11MAR99	18
53	57	10	10	0		FLOOR TILE P-W	16FEB99	1MAR99	12MAR99	25MAR99	18
57	58	10	10	0		INSTALL FURNISHING P-W	2MAR99	15MAR99	26MAR99	8APR99	18
0	210	10	10	0		SUBMIT FOUNDATION REBAR	1JUL98	14JUL98	29JUL98	11AUG98	20
210	211	10	10	0		APPROVE FONDATION REBAR	15JUL98	28JUL98	12AUG98	25AUG98	20
211	16	10	10	0		FAB&DEL FOUNDATION REBAR	29JUL98	11AUG98	26AUG98	8SEP98	20
0	214	20	20	0		SUBMIT CRANE	1JUL98	28JUL98	3AUG98	28AUG98	23
214	215	10	10	0		APPROVE CRANE	29JUL98	11AUG98	31AUG98	11SEP98	23
215	31	50	50	0		FAB&DEL CRANE	12AUG98	20OCT98	14SEP98	20NOV98	23
37	302	0	0	0		RESTRAINT	22DEC98	21DEC98	22JAN99	21JAN99	23
37	304	0	0	0		RESTRAINT	22DEC98	21DEC98	22JAN99	21JAN99	23

I	J	Dur	Dur	Dur	Description					
302	42	5	5	0	ERECT EXTERIOR DOORS P-W	22DEC98	28DEC98	22JAN99	28JAN99	23
304	46	15	15	0	INSTALL H AND V UNITS P-W	22DEC98	11JAN99	22JAN99	11FEB99	23
45	51	5	5	0	ROOM OUTLETS P-W	11FEB99	17FEB99	18MAR99	24MAR99	25
49	45	0	0	0	RESTRAINT	11FEB99	10FEB99	18MAR99	17MAR99	25
51	56	10	10	0	INSTALL ELECTRICAL FIXTURES	18FEB99	3MAR99	25MAR99	7APR99	25
53	58	10	10	0	INSTALL PLUMBING FIXTURES P-W	16FEB99	1MAR99	26MAR99	8APR99	28
0	222	20	20	0	SUBMIT POWER PANELS-PLANT	1JUL98	28JUL98	13AUG98	9SEP98	31
222	223	10	10	0	APPROVE POWER PANELS-PLANT	29JUL98	11AUG98	10SEP98	23SEP98	31
223	301	75	75	0	FAB&DEL POWER PANELS-PLANT	12AUG98	24NOV98	24SEP98	6JAN99	31
37	307	0	0	0	RESTRAINT	22DEC98	21DEC98	5FEB99	4FEB99	33
307	40	30	30	0	FABRICATE PIPING P-W	22DEC98	1FEB99	5FEB99	18MAR99	33
40	47	10	10	0	TEST PIPING SYSTEMS P-W	2FEB99	15FEB99	19MAR99	1APR99	33
47	58	5	5	0	LIGHTOFF BOILER AND TEST	16FEB99	22FEB99	2APR99	8APR99	33
0	224	20	20	0	SUBMIT EXTERIOR DOORS	1JUL98	28JUL98	21AUG98	17SEP98	37
0	225	30	30	0	SUBMIT PLANT EXECTRICAL FIXTURES	1JUL98	11AUG98	21AUG98	1OCT98	37
224	225	10	10	0	APPROVE EXTERIOR DOORS	29JUL98	11AUG98	18SEP98	1OCT98	37
225	302	80	80	0	FAB&DEL EXTERIOR DOORS	12AUG98	1DEC98	2OCT98	21JAN99	37
44	45	0	0	0	RESTRAINT	26JAN99	25JAN99	18MAR99	17MAR99	37
0	227	20	20	0	SUBMIT PLANT HEATING AND VENTILATING FANS	1JUL98	28JUL98	28AUG98	24SEP98	42
227	228	10	10	0	APPROVE PLANT HEATING AND VENTILATING FANS	29JUL98	11AUG98	25SEP98	8OCT98	42
228	304	75	75	0	FAB&DEL PLANT HEATING AND VENTILATING FANS	12AUG98	24NOV98	9OCT98	21JAN99	42
37	306	0	0	0	RESTRAINT	22DEC98	21DEC98	19FEB99	18FEB99	43
306	41	25	25	0	ERECT BOILER AND AUXILIARY P-W	22DEC98	25JAN99	19FEB99	25MAR99	43
41	47	5	5	0	PREOPERATIONAL BOILER CHECK	26JAN99	1FEB99	26MAR99	1APR99	43
44	58	10	10	0	HANG INTERIOR DOORS P-W	26JAN99	8FEB99	26MAR99	8APR99	43
0	218	20	20	0	SUBMIT SIDING	1JUL98	28JUL98	1SEP98	28SEP98	44
218	219	10	10	0	APPROVE SIDING	29JUL98	11AUG98	29SEP98	12OCT98	44
219	35	40	40	0	FAB&DEL SIDING	12AUG98	6OCT98	13OCT98	7DEC98	44
0	216	20	20	0	SUBMIT BAR JOISTS	1JUL98	28JUL98	7SEP98	2OCT98	48
0	237	40	40	0	SUBMIT PRECAST	1JUL98	25AUG98	7SEP98	30OCT98	48
216	217	10	10	0	APPROVE BAR JOISTS	29JUL98	11AUG98	5OCT98	16OCT98	48
217	33	30	30	0	FAB&DEL BAR JOISTS	12AUG98	22SEP98	19OCT98	27NOV98	48
237	238	10	10	0	APPROVE PRECAST	26AUG98	8SEP98	2NOV98	13NOV98	48
238	99	30	30	0	FAB&DEL PRECAST	9SEP98	20OCT98	16NOV98	25DEC98	48

Figure 6-2 (*Continued*)

O'BRIEN KREITZBERG INC. FINEST HOUR JOHN DOE 1

REPORT DATE 29JAN97 RUN NO. 30 JOHN DOE START DATE 1JUL98 FIN DATE 22APR99

12:08 DATA DATE 1JUL98 PAGE NO. 3

SCHEDULE REPORT - SORT BY TF, ES

PRED	SUCC	ORIG DUR	REM DUR	%	CODE	ACTIVITY DESCRIPTION	EARLY START	EARLY FINISH	LATE START	LATE FINISH	TOTAL FLOAT
99	59	5	5	0		ERECT PRECAST STRUCT. OFFICE	21OCT98	27OCT98	28DEC98	1JAN99	48
59	60	5	5	0		ERECT PRECAST ROOF OFFICE	28OCT98	3NOV98	4JAN99	8JAN99	48
60	61	10	10	0		EXTERIOR MASONRY OFFICE	4NOV98	17NOV98	11JAN99	22JAN99	48
61	64	10	10	0		INSTALL PIPING OFFICE	18NOV98	1DEC98	25JAN99	5FEB99	48
61	65	4	4	0		INSTALL ELEC BACKING BOXES	18NOV98	23NOV98	25JAN99	28JAN99	48
65	66	10	10	0		INSTALL CONDUIT OFFICE	24NOV98	7DEC98	29JAN99	11FEB99	48
64	67	4	4	0		TEST PIPING OFFICE	2DEC98	7DEC98	8FEB99	11FEB99	48
66	67	0	0	0		RESTRAINT	8DEC98	7DEC98	12FEB99	11FEB99	48
67	68	5	5	0		PARTITIONS OFFICE	8DEC98	14DEC98	12FEB99	18FEB99	48
68	69	5	5	0		DRYWALL	15DEC98	21DEC98	19FEB99	25FEB99	48
69	70	10	10	0		DRYWALL	22DEC98	4JAN99	26FEB99	11MAR99	48
70	71	10	10	0		WOOD TRIM OFFICE	5JAN99	18JAN99	12MAR99	25MAR99	48
71	72	10	10	0		PAINT INTERIOR OFFICE	19JAN99	1FEB99	26MAR99	8APR99	48
72	78	0	0	0		RESTRAINT	2FEB99	1FEB99	9APR99	8APR99	48
72	80	10	10	0		FLOOR TILE OFFICE	2FEB99	15FEB99	9APR99	22APR99	48
78	80	10	10	0		ACOUSTIC TILE OFFICE	2FEB99	15FEB99	9APR99	22APR99	48
72	73	0	0	0		RESTRAINT	2FEB99	1FEB99	16APR99	15APR99	53
73	80	5	5	0		TOILET FIXTURES OFFICE	2FEB99	8FEB99	16APR99	22APR99	53
61	62	5	5	0		EXTERIOR DOORS OFFICE	18NOV98	24NOV98	12FEB99	18FEB99	62
61	63	5	5	0		ROOFING OFFICE	18NOV98	24NOV98	12FEB99	18FEB99	62
61	68	5	5	0		GLAZE OFFICE	18NOV98	24NOV98	12FEB99	18FEB99	62
62	63	0	0	0		RESTRAINT	25NOV98	24NOV98	19FEB99	18FEB99	62

Pred	Succ	Dur	FF	Description	Early Start	Early Finish	Late Start	Late Finish	TF
63	68	0	0	RESTRAINT	25NOV98	24NOV98	19FEB99	18FEB99	62
37	91	5	0	GRADE AND BALLAST RR SIDING	22DEC98	28DEC98	19MAR99	25MAR99	63
91	58	10	0	INSTALL RR SIDING	29DEC98	11JAN99	26MAR99	8APR99	63
70	77	0	0	RESTRAINT	5JAN99	4JAN99	2APR99	1APR99	63
77	78	5	0	INSTALL CEILING GRID OFFICE	5JAN99	11JAN99	2APR99	8APR99	63
71	80	5	0	HANG DOORS OFFICE	19JAN99	25JAN99	16APR99	22APR99	63
0	235	30	0	SUBMIT PACKAGED A/C	1JUL98	11AUG98	5OCT98	13NOV98	68
235	236	10	0	APPROVE PACKAGED A/C	12AUG98	25AUG98	16NOV98	27NOV98	68
236	98	90	0	FAB&DEL PACKAGED A/C	26AUG98	29DEC98	30NOV98	2APR99	68
66	74	10	0	PULL WIRE OFFICE	8DEC98	21DEC98	12MAR99	25MAR99	68
37	92	10	0	ACCESS ROAD	22DEC98	4JAN99	26MAR99	8APR99	68
37	93	20	0	AREA LIGHTING	22DEC98	18JAN99	26MAR99	22APR99	68
74	75	5	0	INSTALL PANEL INTERNALS OFFICE	29DEC98	28DEC98	2APR99	1APR99	68
75	79	10	0	TERMINATE WIRES OFFICE	30DEC98	5JAN99	5APR99	15APR99	68
98	76	5	0	INSTALL PACKAGE AIR CONDITONER	5JAN99	4JAN99	9APR99	9APR99	68
92	58	0	0	RESTRAINT	6JAN99	11JAN99	8APR99	8APR99	68
76	79	4	0	AIR CONDITIONING ELEC CONNECT	12JAN99	18JAN99	12APR99	15APR99	68
79	80	5	0	RINGOUT ELECT	19JAN99	18JAN99	16APR99	22APR99	68
93	80	0	0	RESTRAINT	19AUG98	21AUG98	23APR99	22APR99	70
14	23	3	0	EXCAVATE FOR OFFICE BUILDING	24AUG98	27AUG98	25NOV98	27NOV98	70
23	24	4	0	SPREAD FOOTINGS OFFICE	28AUG98	4SEP98	30NOV98	3DEC98	70
24	25	6	0	FORM AND POUR GRADE BEAMS OFF	7SEP98	7SEP98	4DEC98	11DEC98	70
25	26	1	0	BACKFILL AND COMPACT OFFICE	8SEP98	10SEP98	14DEC98	14DEC98	70
26	27	3	0	UNDERSLAB PLUMBING OFFICE	11SEP98	15SEP98	15DEC98	17DEC98	70
27	28	3	0	UNDERSLAB CONDUIT OFFICE	16SEP98	18SEP98	18DEC98	22DEC98	70
28	99	3	0	FORM AND POUR OFFICE SLAB	22DEC98	21DEC98	23DEC98	25DEC98	70
37	305	0	0	RESTRAINT	22DEC98	24DEC98	30MAR99	29MAR99	70
305	47	3	0	INSTALL FUEL TANK P-W	12AUG98	1SEP98	30MAR99	1APR99	71
225	226	15	0	APPROVE PLANT ELECTRICAL FIXTURES	2SEP98	15DEC98	19NOV98	9DEC98	71
226	51	75	0	FAB&DEL PLANT ELECTRICAL FIXTURES	22DEC98	28DEC98	10DEC98	24MAR99	73
37	90	5	0	PAVE PARKING AREA	22DEC98	21DEC98	2APR99	8APR99	73
37	308	0	0	RESTRAINT	22DEC98	4JAN99	2APR99	1APR99	73
69	73	10	0	CERAMIC TILE OFFICE	22DEC98	28DEC98	2APR99	15APR99	73
308	58	5	0	INSTALL MONORAIL WAREHOUSE	22DEC98	28DEC98	2APR99	8APR99	73

Figure 6-2 (Continued)

129

REPORT DATE 29JAN97   RUN NO. 30   JOHN DOE   START DATE 1JUL98   FIN DATE 22APR99
          12:08

SCHEDULE REPORT - SORT BY TF, ES   DATA DATE   1JUL98   PAGE NO.   4

                                                    JOHN DOE 1

PRED	SUCC	ORIG DUR	REM DUR	%	CODE	ACTIVITY DESCRIPTION	EARLY START	EARLY FINISH	LATE START	LATE FINISH	TOTAL FLOAT
90	58	0	0	0		RESTRAINT	29DEC98	28DEC98	9APR99	8APR99	73
0	229	20	20	0		SUBMIT BOILER	1JUL98	28JUL98	16OCT98	12NOV98	77
229	230	10	10	0		APPROVE BOILER	29JUL98	11AUG98	13NOV98	26NOV98	77
230	306	60	60	0		FAB&DEL BOILER	12AUG98	3NOV98	27NOV98	18FEB99	77
37	80	10	10	0		PERIMETER FENCE	22DEC98	4JAN99	9APR99	22APR99	78
74	76	0	0	0		RESTRAINT	22DEC98	21DEC98	12APR99	9APR99	79
61	77	15	15	0		DUCTWORK OFFICE	18NOV98	8DEC98	12MAR99	1APR99	82
63	80	5	5	0		PAINT OFFICE EXTERIOR	25NOV98	1DEC98	16APR99	22APR99	102
60	98	0	0	0		RESTRAINT	4NOV98	3NOV98	5APR99	2APR99	108
0	231	20	20	0		SUBMIT OIL TANK	1JUL98	28JUL98	8DEC98	4JAN99	114
231	232	10	10	0		APPROVE OIL TANK	29JUL98	11AUG98	5JAN99	18JAN99	114
232	305	50	50	0		FAB&DEL OIL TANK	12AUG98	20OCT98	19JAN99	29MAR99	114

Figure 6-2 *(Continued)*

130

TIME IMPACT EVALUATION
(T.I.E.)
JOHN DOE PROJECT / WHATCO

P.C.O. _____

R.F.I. _____ 001 (7/31/98) _____

Other _____

R.F.I. Response _____ 8/12/98 _____

Project Status: _____ On schedule _____

Date Issued _____ 7/31/98 _____

Description: Underground duct bank
(electrical) must be relocated.

Date Resolved _____ 9/15/98 _____

Action Required:
WHATCO will arrange for electric utility to relocate by 9/15/98.

Activity Affected: _____ 15-16 _____   **Event Affected:** _____ 15 _____

Description: Excavate for plant.

Duration	Early Start	Early Finish	Late Start	Late Finish	Finish
5	2 Sep 98	8 Sep 98	2 Sep 98	8 Sep 8	0

Impact: 13 calendar days

T.I.E. DATE: October 8, 1998   BY: J. O'Brien

Fragnet:

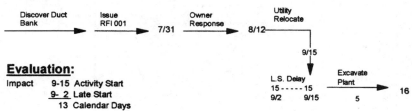

Evaluation:

Impact 9-15 Activity Start
        9- 2 Late Start
        13 Calendar Days

Responsibility:

TBD
Requires review of as-built information furnished to the architect-engineer.

Workaround:

• Owner response time (including negotiations with utility) 12 c.d. This is deemed reasonable.
• Utility response 34 c.d. (less Labor Day holiday) also reasonable.
• Work could not be delegated to contractor.

Figure 6-3 Time impact evaluation for electric duct bank.

O'BRIEN KREITZBERG INC.   FINEST HOUR   JOHN DOE 1

REPORT DATE 29JAN97   RUN NO. 30   JOHN DOE   START DATE 1JUL98   FIN DATE 22APR99
          12:08
SCHEDULE REPORT - SORT BY TF, ES   DATA DATE 1JUL98   PAGE NO.   1

PRED	SUCC	ORIG DUR	REM DUR	%	CODE	ACTIVITY DESCRIPTION	EARLY START	EARLY FINISH	LATE START	LATE FINISH	TOTAL FLOAT
0	1	3	3	0		CLEAR SITE	1JUL98	3JUL98	1JUL98	3JUL98	0
1	2	2	2	0		SURVEY AND LAYOUT	6JUL98	7JUL98	6JUL98	7JUL98	0
2	3	2	2	0		ROUGH GRADE	8JUL98	9JUL98	8JUL98	9JUL98	0
3	4	15	15	0		DRILL WELL	10JUL98	30JUL98	10JUL98	30JUL98	0
4	5	2	2	0		INSTALL WELL PUMP	31JUL98	3AUG98	31JUL98	3AUG98	0
5	8	8	8	0		UNDERGROUND WATER PIPING	4AUG98	13AUG98	4AUG98	13AUG98	0
8	13	2	2	0		CONNECT WATER PIPING	14AUG98	17AUG98	14AUG98	17AUG98	0
13	14	1	1	0		BUILDING LAYOUT	18AUG98	18AUG98	18AUG98	18AUG98	0
14	15	10	10	0		DRIVE AND POUR PILES	19AUG98	1SEP98	19AUG98	1SEP98	0
15	16	5	5	0		EXCAVATE FOR PLANT WAREHOUSE	2SEP98	8SEP98	2SEP98	8SEP98	0
16	17	5	5	0		POUR PILE CAPS PLANT-WAREHSE	9SEP98	15SEP98	9SEP98	15SEP98	0
17	18	10	10	0		FORM AND POUR GRADE BEAMS P-W	16SEP98	29SEP98	16SEP98	29SEP98	0
18	19	3	3	0		BACKFILL AND COMPACT P-W	30SEP98	2OCT98	30SEP98	2OCT98	0
19	20	5	5	0		UNDERSLAB PLUMBING P-W	5OCT98	9OCT98	5OCT98	9OCT98	0
20	22	5	5	0		UNDERSLAB CONDUIT P-W	12OCT98	16OCT98	12OCT98	16OCT98	0
22	29	10	10	0		FORM AND POUR SLABS P-W	19OCT98	30OCT98	19OCT98	30OCT98	0
29	30	10	10	0		ERECT STRUCT STEEL P-W	2NOV98	13NOV98	2NOV98	13NOV98	0
30	31	5	5	0		PLUMB STEEL AND BOLT P-W	16NOV98	20NOV98	16NOV98	20NOV98	0
31	32	5	5	0		ERECT CRANE WAY AND CRANE P-W	23NOV98	27NOV98	23NOV98	27NOV98	0
32	33	0	0	0		RESTRAINT	30NOV98	27NOV98	30NOV98	27NOV98	0
33	34	3	3	0		ERECT BAR JOISTS P-W	30NOV98	2DEC98	30NOV98	2DEC98	0
34	35	3	3	0		ERECT ROOF PLANKS P-W	3DEC98	7DEC98	3DEC98	7DEC98	0

I	J	Dur	Rem	%	Activity	Early Start	Early Finish	Late Start	Late Finish	TF
35	36	10	10	0	ERECT SIDING P-W	8DEC98	21DEC98	8DEC98	21DEC98	0
36	37	0	0	0	RESTRAINT	22DEC98	21DEC98	22DEC98	21DEC98	0
37	300	0	0	0	RESTRAINT	22DEC98	21DEC98	22DEC98	21DEC98	0
300	38	2	2	0	SET ELECTRICAL LOAD CENTER P-W	22DEC98	23DEC98	22DEC98	23DEC98	0
38	43	20	20	0	INSTALL POWER CONDUIT P-W	24DEC98	20JAN99	24DEC98	20JAN99	0
43	49	15	15	0	INSTALL BRANCH CONDUIT P-W	21JAN99	10FEB99	21JAN99	10FEB99	0
49	50	15	5	0	PULL WIRE P-W	11FEB99	3MAR99	11FEB99	3MAR99	0
50	54	5	10	0	INSTALL PANEL INTERNALS P-W	4MAR99	10MAR99	4MAR99	10MAR99	0
54	55	10	10	0	TERMINATE WIRES P-W	11MAR99	24MAR99	11MAR99	24MAR99	0
55	56	10	10	0	RINGOUT P-W	25MAR99	7APR99	25MAR99	7APR99	0
56	58	1	1	0	ENERGIZE POWER	8APR99	8APR99	8APR99	8APR99	0
58	94	5	5	0	FINE GRADE	9APR99	15APR99	9APR99	15APR99	0
94	80	5	4	0	SEED AND PLANT	16APR99	22APR99	16APR99	22APR99	0
3	6	10	10	0	WATER TANK FOUNDATIONS	10JUL98	15JUL98	13JUL98	16JUL98	1
6	7	10	10	0	ERECT WATER TOWER	16JUL98	29JUL98	17JUL98	30JUL98	1
7	8	10	3	0	TANK PIPING AND VALVES	30JUL98	12AUG98	31JUL98	13AUG98	1
31	33	3	3	0	ERECT MONORAIL TRACK P-W	23NOV98	25NOV98	25NOV98	27NOV98	2
0	220	20	20	0	SUBMIT PLANT ELECTRICAL LOAD CENTER	1JUL98	28JUL98	7JUL98	3AUG98	4
3	9	10	10	0	EXCAVATE FOR SEWER	10JUL98	23JUL98	16JUL98	29JUL98	4
9	11	5	5	0	INSTALL SEWER AND BACKFILL	24JUL98	30JUL98	30JUL98	5AUG98	4
220	221	10	10	0	APPROVE PLANT ELECTRICAL LOAD CENTER	29JUL98	11AUG98	4AUG98	17AUG98	4
11	12	3	3	0	ELECTRICAL DUCT BANK	31JUL98	4AUG98	6AUG98	10AUG98	4
12	13	5	5	0	PULL IN POWER FEEDER	5AUG98	11AUG98	11AUG98	17AUG98	4
221	300	90	90	0	FAB&DEL PLANT ELECTRICAL LOAD CENTER	12AUG98	15DEC98	18AUG98	21DEC98	4
35	37	5	5	0	ROOFING P-W	8DEC98	14DEC98	15DEC98	21DEC98	5
58	80	5	5	0	ERECT FLAGPOLE	9APR99	15APR99	16APR99	22APR99	5
18	21	5	5	0	FORM AND POUR RR LOAD DOCK P-W	30SEP98	6OCT98	12OCT98	16OCT98	8
18	22	5	5	0	FORM AND POUR TK LOAD DOCK P-W	30SEP98	6OCT98	12OCT98	16OCT98	8
21	22	0	0	0	RESTRAINT	7OCT98	6OCT98	19OCT98	16OCT98	8
37	301	0	0	0	RESTRAINT	22DEC98	21DEC98	7JAN99	6JAN99	12
301	43	10	10	0	POWER PANEL BACKFILL BOXES P-W	22DEC98	4JAN99	7JAN99	20JAN99	12
3	10	1	1	0	EXCAVATE ELECTRICAL MANHOLES	10JUL98	29JUL98	29JUL98	17AUG98	13
10	11	5	5	0	INSTALL ELECTRICAL MANHOLES	13JUL98	17JUL98	30JUL98	5AUG98	13
37	303	0	0	0	RESTRAINT	22DEC98	21DEC98	8JAN99	7JAN99	13

Figure 6-4 Activity 15-16 and progress to date at 2 Sep 98.

TIME IMPACT EVALUATION
(T.I.E.)
JOHN DOE PROJECT / WHATCO

P.C.O. _____#1_____   **Description:** Sinkhole at Cols. A/11

R.F.I. _____002_____

Other _____

R.F.I. Response___8/27/98____

Project Status:_____On schedule @ 8/21/98_____

Date Issued _____8/21/98_____   **Date Resolved** _____9/25/98_____

Action Required:
    Bulletin 9/11/98
    Negotiate 9/18/98
    Work complete 9/25/98
    Remove bad material, borrow, backfill, compact

Activity Affected: _____15-16_____   **Event Affected:** _____16_____

Description:

Duration	Early Start	Early Finish	Late Start	Late Finish	Finish
5	2 Sep 98	8 Sep 98	2 Sep 98	8 Sep 8	0

    Impact: 17 calendar days

T.I.E. DATE: October 8, 1998 **BY:** J. O'Brien

Fragnet:

Discover Sinkhole → Issue RFI 002 → 8/21 → Owner Response → 8/27 → A-E Bulletin → 9/11 → Negotiate → 9/18 → Approval & NTP → 9/23 → 9/25

Evaluation:

Impact 9-25 Actual Finish
       9- 8 Late Finish
       17 Calendar Days

L.F 16 -------------------- 16
9/8   Delay

Responsibility:

TBD
Requires review of geotechnical studies by the architect-engineer.

Workaround:

• Owner-A-E-contractor response times - all reasonable.
• Time could have been saved by going Time & Material, ASAP.
• Time & Material would have been more expensive.

Figure 6-5 TIE for sinkhole.

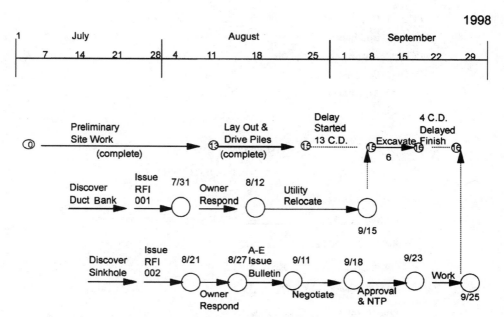

Figure 6-6   Combined fragnets from TIEs for RFI 001 and RFI 002—PCO #1.

Lessons learned

- Identify problems as early as possible.

- For problems on the critical path, decisive action (including time/material) is necessary to implement the change order.

- For early activities (such as structural steel) with long lead procurement time, options for change (even VECPs) are limited.

- For a VECP that has time impact, consider using part of the gross savings to buy time (i.e., workaround or acceleration).

PCOs in the baseline

Most specifications make no provision to reflect PCOs until the resulting change orders are approved. This is typically long after the information is needed. At the Bremerton Naval Hospital Project (Santa Fe v. U.S., ASBCA No. 24578, ASBCA No. 25838, 94-2), the NavFac scheduling specifications allowed each PCO to be put in line with its impact point (at duration 0). That, at least, identifies the relative criticality of each PCO.

Figure 6-10 is a list of activity changes required to incorporate RFI 001 and the four PCOs into the baseline. (PCO numbers are now advanced by one each.) Figure 6-11 (pp. 140 to 146 and 148 to 154) shows the CPM output with the PCOs at zero duration. The output is the same as the original base (Fig. 6-10), except the float value for the PCOs (Fig. 6-11a, sorted by i-j) gives the relative criticality of the PCOs:

PCO #1 Duct bank   0

PCO #2 Sinkhole   0

PCO #3 Hot water   33

PCO #4 Bar joists   48

PCO #5 Air conditioning   68

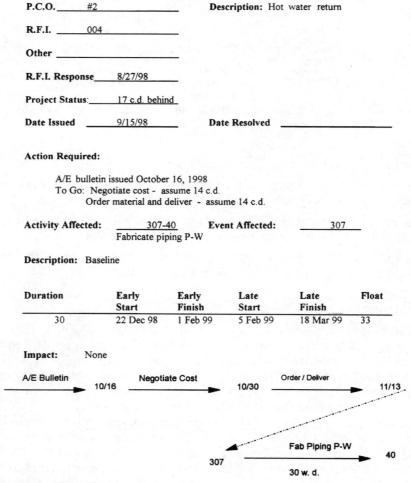

TIME IMPACT EVALUATION
(T.I.E.)
JOHN DOE PROJECT / WHATCO

P.C.O. _____#2_____   **Description:** Hot water return

R.F.I. _____004_____

Other _____

R.F.I. Response____8/27/98_____

Project Status:_____17 c.d. behind_

Date Issued _____9/15/98_____   **Date Resolved** _____

Action Required:

A/E bulletin issued October 16, 1998
To Go: Negotiate cost - assume 14 c.d.
         Order material and deliver - assume 14 c.d.

Activity Affected: _____307-40_____ **Event Affected:** _____307____
Fabricate piping P-W

Description: Baseline

Duration	Early Start	Early Finish	Late Start	Late Finish	Float
30	22 Dec 98	1 Feb 99	5 Feb 99	18 Mar 99	33

Impact: None

Figure 6-7 TIE for hot-water return.

TIME IMPACT EVALUATION
(T.I.E.)
JOHN DOE PROJECT / WHATCO

P.C.O. _____#3_____

R.F.I. _____

Other _____

R.F.I. Response_____

Project Status:_____

Date Issued   _____

Description: VECP - long-span joists for plant - warehouse - office vs. girders and secondary joists.

Date Resolved   ____9/30/98_____

Action Required:

Submit joists for approval

Activity Affected: _____0-216___   **Event Affected:** _____216___

Description: Submit structural steel

Duration	Early Start	Early Finish	Late Start	Late Finish	Float
20	1 Jul 98	28 Jul 98	7 Sep 98	2 Oct 98	48

Impact: 18 calendar days

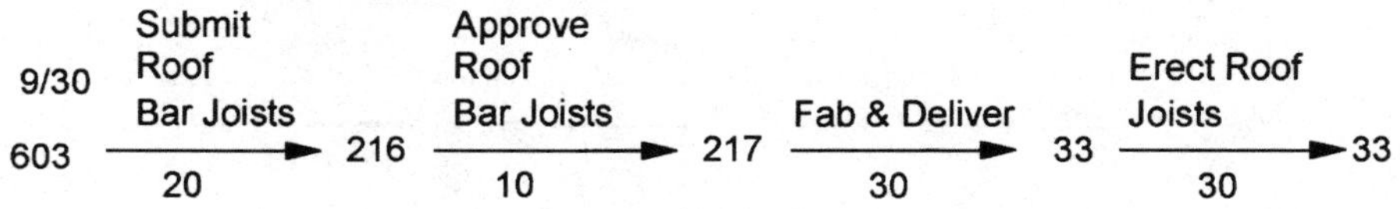

```
         Submit           Approve
9/30     Roof             Roof                                Erect Roof
         Bar Joists       Bar Joists       Fab & Deliver      Joists
603  ─────────────▶ 216 ────────────▶ 217 ───────────▶ 33 ──────────▶ 33
         20               10                30                 30
```

Roof joists replace part of structural steel

Workaround:

0 - 216: change to 10
216 - 217: change to 5
217 - 33 : change to 20

Should increase 33 - 34 to 10

Net improvement: 18

Figure 6-8   TIE for VECP #1 (PCO #3).

TIME IMPACT EVALUATION
(T.I.E.)
JOHN DOE PROJECT / WHATCO

P.C.O. _____#4_____   Description:

R.F.I. _____

Other _____Bulletin_____

R.F.I. Response_____

Project Status:_____

Date Issued _____9/03/98_____   Date Resolved _____10/01/98_____

Action Required:

Activity Affected: _____0-235_____   Event Affected: _____235 (LF)_____

Description: Submit packaged A/C

Duration	Early Start	Early Finish	Late Start	Late Finish	Float
30	1 Jul 98	11 Aug 98	5 Oct 98	13 Nov 98	68

Impact: No impact

T.I.E. DATE: October 8, 1998 **BY:** J. O'Brien

Fragnet:

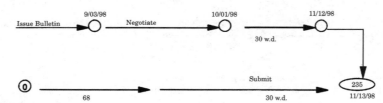

Evaluation:
LF   11 - 13 - 98
(10/01/98) + 42 c.d.
    11 - 12 - 98
        1 day float

Responsibility:
Owner - to get better reliability and operating characteristics.

Workaround:
None needed.

Figure 6-9   TIE for multiple chiller units.

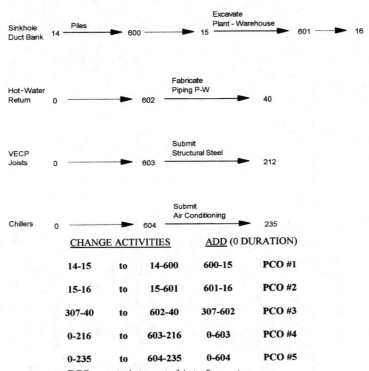

PCO Event Numbers

PCO Event	Relates to	PCO #	Revised PCO #
600	Duct Bank	–	#1
601	Sinkhole	#1	#2
602	Hot-Water Return	#2	#3
603	VECP - Joists	#3	#4
604	Chillers	#4	#5

CHANGE ACTIVITIES			ADD (0 DURATION)	
14-15	to	14-600	600-15	PCO #1
15-16	to	15-601	601-16	PCO #2
307-40	to	602-40	307-602	PCO #3
0-216	to	603-216	0-603	PCO #4
0-235	to	604-235	0-604	PCO #5

Figure 6-10 PCO events integrated into fragnets.

When a change order is approved, a time fragnet of it should be incorporated into the baseline. This can be as easy as adding the time factors to the in-line PCOs. This is done in Fig. 6-12 (p. 155). Figure 6-13 (pp. 156 to 162 and 164 to 170) is the CPM output with "no-earlier-than" dates locked into the PCO events. Figure 6-13a (i-j) shows the following float values:

PCO #1 Duct bank	8	
PCO #2 Sinkhole	5	
PCO #3 Hot water	33	
PCO #4 Bar joist	65	
PCO #5 Air conditioning	85	

(Text continues on p. 147)

O'BRIEN KREITZBERG INC.   FINEST HOUR   JOHN DOE 2

REPORT DATE 7APR97 RUN NO. 7   JOHN DOE   START DATE 1JUL98 FIN DATE 22APR99
            19:11
SCHEDULE REPORT BY I J   DATA DATE 1JUL98   PAGE NO. 1

PRED	SUCC	ORIG DUR	REM DUR	%	CODE	ACTIVITY DESCRIPTION	EARLY START	EARLY FINISH	LATE START	LATE FINISH	TOTAL FLOAT
0	1	3	3	0	1 1	CLEAR SITE	1JUL98	3JUL98	1JUL98	3JUL98	0
0	210	10	10	0	6 31	SUBMIT FOUNDATION REBAR	1JUL98	14JUL98	29JUL98	11AUG98	20
0	212	20	20	0	6 31	SUBMIT STRUCTURAL STEEL	1JUL98	28JUL98	27JUL98	21AUG98	18
0	214	20	20	0	6 31	SUBMIT CRANE	1JUL98	28JUL98	3AUG98	28AUG98	23
0	218	20	20	0	6 31	SUBMIT SIDING	1JUL98	28JUL98	1SEP98	28SEP98	44
0	220	20	20	0	6 31	SUBMIT PLANT ELECTRICAL LOAD CENTER	1JUL98	28JUL98	7JUL98	3AUG98	4
0	222	20	20	0	6 31	SUBMIT POWER PANELS-PLANT	1JUL98	28JUL98	13AUG98	9SEP98	31
0	224	20	20	0	6 31	SUBMIT EXTERIOR DOORS	1JUL98	28JUL98	21AUG98	17SEP98	37
0	225	30	30	0	6 31	SUBMIT PLANT EXECTRICAL FIXTURES	1JUL98	11AUG98	21AUG98	1OCT98	37
0	227	20	20	0	6 31	SUBMIT PLANT HEATING AND VENTILATING FANS	1JUL98	28JUL98	28AUG98	24SEP98	42
0	229	20	20	0	6 31	SUBMIT BOILER	1JUL98	28JUL98	16OCT98	12NOV98	77
0	231	20	20	0	6 31	SUBMIT OIL TANK	1JUL98	28JUL98	8DEC98	4JAN99	114
0	237	40	40	0	6 31	SUBMIT PRECAST	1JUL98	25AUG98	7SEP98	30OCT98	48
0	603	0	0	0	0	PCO #4	1JUL98	30JUN98	7SEP98	4SEP98	48
0	604	0	0	0	0	PCO #5	1JUL98	30JUN98	5OCT98	2OCT98	68
1	2	2	2	0	1 2	SURVEY AND LAYOUT	6JUL98	7JUL98	6JUL98	7JUL98	0
2	3	2	2	0	1 1	ROUGH GRADE	8JUL98	9JUL98	8JUL98	9JUL98	0
3	4	15	15	0	1 9	DRILL WELL	10JUL98	30JUL98	10JUL98	30JUL98	0
3	6	4	4	0	1 3	WATER TANK FOUNDATIONS	10JUL98	15JUL98	13JUL98	16JUL98	1
3	9	10	10	0	1 1	EXCAVATE FOR SEWER	10JUL98	23JUL98	16JUL98	29JUL98	4
3	10	1	1	0	1 1	EXCAVATE ELECTRICAL MANHOLES	10JUL98	10JUL98	29JUL98	29JUL98	13
3	12	6	6	0	1 4	OVERHEAD POLE LINE	10JUL98	17JUL98	3AUG98	10AUG98	16
4	5	2	2	0	1 9	INSTALL WELL PUMP	31JUL98	3AUG98	31JUL98	3AUG98	0
5	9	8	8	0	1 5	UNDERGROUND WATER PIPING	4AUG98	13AUG98	4AUG98	13AUG98	0

140

6	7	10	10	0	1	10	ERECT WATER TOWER	16JUL98	29JUL98	17JUL98	30JUL98	1
7	8	10	10	0	1	10	TANK PIPING AND VALVES	30JUL98	12AUG98	31JUL98	13AUG98	1
8	13	2	2	0	1	10	CONNECT WATER PIPING	14AUG98	17AUG98	14AUG98	17AUG98	0
9	11	5	5	0	1	5	INSTALL SEWER AND BACKFILL	24JUL98	30JUL98	5AUG98	5AUG98	4
10	11	5	5	0	1	4	INSTALL ELECTRICAL MANHOLES	13JUL98	17JUL98	30JUL98	5AUG98	13
11	12	3	3	0	1	4	ELECTRICAL DUCT BANK	31JUL98	4AUG98	6AUG98	10AUG98	4
12	13	5	5	0	1	4	PULL IN POWER FEEDER	5AUG98	11AUG98	11AUG98	17AUG98	4
13	14	1	1	0	1	2	BUILDING LAYOUT	18AUG98	18AUG98	18AUG98	18AUG98	0
14	23	3	3	0	2	1	EXCAVATE FOR OFFICE BUILDING	19AUG98	21AUG98	25NOV98	27NOV98	70
14	600	10	10	0	2	11	DRIVE AND POUR PILES	19AUG98	1SEP98	19AUG98	1SEP98	0
15	601	5	5	0	2	1	EXCAVATE FOR PLANT WAREHOUSE	2SEP98	8SEP98	2SEP98	8SEP98	0
16	17	5	5	0	2	3	POUR PILE CAPS PLANT-WAREHSE	9SEP98	15SEP98	9SEP98	15SEP98	0
17	18	10	10	0	2	3	FORM AND POUR GRADE BEAMS P-W	16SEP98	29SEP98	16SEP98	29SEP98	0
18	19	3	3	0	2	1	BACKFILL AND COMPACT P-W	30SEP98	2OCT98	30SEP98	2OCT98	0
18	21	5	5	0	2	3	FORM AND POUR RR LOAD DOCK P-W	30SEP98	6OCT98	12OCT98	16OCT98	8
18	22	5	5	0	2	3	FORM AND POUR TK LOAD DOCK P-W	30SEP98	6OCT98	12OCT98	16OCT98	8
19	20	5	5	0	2	5	UNDERSLAB PLUMBING P-W	5OCT98	9OCT98	5OCT98	9OCT98	0
20	22	5	5	0	2	4	UNDERSLAB CONDUIT P-W	12OCT98	16OCT98	12OCT98	16OCT98	0
21	22	0	0	0	0		RESTRAINT	7OCT98	6OCT98	19OCT98	16OCT98	8
22	29	10	10	0	2	3	FORM AND POUR SLABS P-W	19OCT98	30OCT98	19OCT98	30OCT98	0
23	24	4	4	0	2	3	SPREAD FOOTINGS OFFICE	24AUG98	27AUG98	30NOV98	3DEC98	70
24	25	6	6	0	2	3	FORM AND POUR GRADE BEAMS OFF	28AUG98	4SEP98	4DEC98	11DEC98	70
25	26	1	1	0	2	1	BACKFILL AND COMPACT OFFICE	7SEP98	7SEP98	14DEC98	14DEC98	70
26	27	3	3	0	2	5	UNDERSLAB PLUMBING OFFICE	8SEP98	10SEP98	15DEC98	17DEC98	70
27	28	3	3	0	2	4	UNDERSLAB CONDUIT OFFICE	11SEP98	15SEP98	18DEC98	22DEC98	70
28	99	3	3	0	2	3	FORM AND POUR OFFICE SLAB	16SEP98	18SEP98	23DEC98	25DEC98	70
29	30	10	10	0	3	6	ERECT STRUCT STEEL P-W	2NOV98	13NOV98	2NOV98	13NOV98	0
30	31	5	5	0	3	6	PLUMB STEEL AND BOLT P-W	16NOV98	20NOV98	16NOV98	20NOV98	0
31	32	5	5	0	3	6	ERECT CRANE WAY AND CRANE P-W	23NOV98	27NOV98	23NOV98	27NOV98	0
31	33	3	3	0	3	6	ERECT MONORAIL TRACK P-W	23NOV98	25NOV98	25NOV98	27NOV98	2
32	33	0	0	0	0		RESTRAINT	30NOV98	27NOV98	30NOV98	27NOV98	0
33	34	3	3	0	3	6	ERECT BAR JOISTS P-W	30NOV98	2DEC98	30NOV98	2DEC98	0

Figure 6-11a Relative criticality of PCOs—i-j sort.

141

PRED	SUCC	ORIG DUR	REM DUR	%	CODE	ACTIVITY DESCRIPTION	EARLY START	EARLY FINISH	LATE START	LATE FINISH	TOTAL FLOAT
34	35	3	3	0	7	ERECT ROOF PLANKS P-W	3DEC98	7DEC98	3DEC98	7DEC98	0
35	36	10	10	0	12	ERECT SIDING P-W	8DEC98	21DEC98	8DEC98	21DEC98	0
35	37	5	5	0	13	ROOFING P-W	8DEC98	14DEC98	15DEC98	21DEC98	5
36	37	0	0	0		RESTRAINT	22DEC98	21DEC98	22DEC98	21DEC98	0
37	80	10	10	0	15	PERIMETER FENCE	22DEC98	4JAN99	9APR99	22APR99	78
37	90	5	5	0	16	PAVE PARKING AREA	22DEC98	28DEC98	2APR99	8APR99	73
37	91	5	5	0	17	GRADE AND BALLAST RR SIDING	22DEC98	28DEC98	19MAR99	25MAR99	63
37	92	10	10	0	16	ACCESS ROAD	22DEC98	4JAN99	26MAR99	8APR99	68
37	93	20	20	0	4	AREA LIGHTING	22DEC98	18JAN99	26MAR99	22APR99	68
37	300	0	0	0		RESTRAINT	22DEC98	21DEC98	22DEC98	21DEC98	0
37	301	0	0	0		RESTRAINT	22DEC98	21DEC98	7JAN99	6JAN99	12
37	302	0	0	0		RESTRAINT	22DEC98	21DEC98	22JAN99	21JAN99	23
37	303	0	0	0		RESTRAINT	22DEC98	21DEC98	8JAN99	7JAN99	13
37	304	0	0	0		RESTRAINT	22DEC98	21DEC98	22JAN99	21JAN99	23
37	305	0	0	0		RESTRAINT	22DEC98	21DEC98	30MAR99	29MAR99	70
37	306	0	0	0		RESTRAINT	22DEC98	21DEC98	19FEB99	18FEB99	43
37	307	0	0	0		RESTRAINT	22DEC98	21DEC98	5FEB99	4FEB99	33
37	308	0	0	0		RESTRAINT	22DEC98	21DEC98	2APR99	1APR99	73
38	43	20	20	0	4	INSTALL POWER CONDUIT P-W	24DEC98	20JAN99	24DEC98	20JAN99	0
39	42	5	5	0	18	FRAME CEILINGS P-W	5JAN99	11JAN99	22JAN99	28JAN99	13
40	47	10	10	0	8	TEST PIPING SYSTEMS P-W	2FEB99	15FEB99	19MAR99	1APR99	33
41	47	5	5	0	8	PREOPERATIONAL BOILER CHECK	26JAN99	1FEB99	26MAR99	1APR99	43
42	44	10	10	0	19	DRYWELL PARTITIONS P-W	12JAN99	25JAN99	29JAN99	11FEB99	13

I	J	OD	RD	%	CAL	ACT	DESCRIPTION	EARLY START	EARLY FINISH	LATE START	LATE FINISH	FLOAT
43	49	15	15	0	3	4	INSTALL BRANCH CONDUIT P-W	21JAN99	10FEB99	21JAN99	10FEB99	0
44	45	0	0	0	3		RESTRAINT	26JAN99	25JAN99	18MAR99	17MAR99	37
44	46	0	0	0	3		RESTRAINT	26JAN99	25JAN99	12FEB99	11FEB99	13
44	48	10	10	0	3	20	CERAMIC TILE	26JAN99	8FEB99	19FEB99	4MAR99	18
44	58	10	10	0	3	21	HANG INTERIOR DOORS P-W	26JAN99	8FEB99	26MAR99	8APR99	43
45	51	5	5	0	3	4	ROOM OUTLETS P-W	11FEB99	17FEB99	18MAR99	24MAR99	25
46	52	25	25	0	3	8	INSTALL DUCTWORK P-W	26JAN99	1MAR99	12FEB99	18MAR99	13
47	58	5	5	0	3	8	LIGHTOFF BOILER AND TEST	16FEB99	22FEB99	2APR99	8APR99	33
48	53	5	5	0	3	22	PAINT ROOMS P-W	9FEB99	15FEB99	5MAR99	11MAR99	18
49	45	0	0	0	3		RESTRAINT	11FEB99	10FEB99	18MAR99	17MAR99	25
49	50	15	15	0	3	4	PULL WIRE P-W	11FEB99	3MAR99	11FEB99	3MAR99	0
50	54	5	5	0	3	4	INSTALL PANEL INTERNALS P-W	4MAR99	10MAR99	4MAR99	10MAR99	0
51	56	10	10	0	3	4	INSTALL ELECTRICAL FIXTURES	18FEB99	3MAR99	25MAR99	7APR99	25
52	58	15	15	0	3	8	INSULATE H&V SYSTEM P-W	2MAR99	22MAR99	19MAR99	8APR99	13
53	57	10	10	0	3	22	FLOOR TILE P-W	16FEB99	1MAR99	12MAR99	25MAR99	18
53	58	10	10	0	3	5	INSTALL PLUMBING FIXTURES P-W	16FEB99	1MAR99	26MAR99	8APR99	28
54	55	10	10	0	3	4	TERMINATE WIRES P-W	11MAR99	24MAR99	11MAR99	24MAR99	0
55	56	10	10	0	3	4	RINGOUT P-W	25MAR99	7APR99	25MAR99	7APR99	0
56	58	1	1	0	3	4	ENERGIZE POWER	8APR99	8APR99	8APR99	8APR99	0
57	59	10	10	0	3	24	INSTALL FURNISHING P-W	2MAR99	15MAR99	26MAR99	8APR99	18
58	80	5	5	0	5	27	ERECT FLAGPOLE	9APR99	15APR99	16APR99	22APR99	5
58	94	5	5	0	5	27	FINE GRADE	9APR99	15APR99	9APR99	15APR99	0
59	60	5	5	0	5	7	ERECT PRECAST ROOF OFFICE	28OCT98	3NOV98	4JAN99	8JAN99	48
60	61	10	10	0	4	14	EXTERIOR MASONRY OFFICE	4NOV98	17NOV98	11JAN99	22JAN99	48
60	98	0	0	0	4		RESTRAINT	4NOV98	3NOV98	5APR99	2APR99	108
61	62	5	5	0	4	21	EXTERIOR DOORS OFFICE	18NOV98	24NOV98	12FEB99	18FEB99	62
61	63	5	5	0	4	13	ROOFING OFFICE	18NOV98	24NOV98	12FEB99	18FEB99	62
61	64	10	10	0	4	8	INSTALL PIPING OFFICE	18NOV98	1DEC98	25JAN99	5FEB99	48
61	65	4	4	0	4	25	INSTALL ELEC BACKING BOXES	18NOV98	23NOV98	25JAN99	28JAN99	48
61	68	5	5	0	4		GLAZE OFFICE	18NOV98	24NOV98	12FEB99	18FEB99	62
61	77	15	15	0	4	8	DUCTWORK OFFICE	18NOV98	8DEC98	12MAR99	1APR99	32
62	63	0	0	0	0		RESTRAINT	25NOV98	24NOV98	19FEB99	18FEB99	-2
63	68	0	0	0	0		RESTRAINT	25NOV98	24NOV98	19FEB99	18FEB99	62

Figure 6-11a (Continued)

PRED	SUCC	ORIG DUR	REM DUR	%	CODE	ACTIVITY DESCRIPTION	EARLY START	EARLY FINISH	LATE START	LATE FINISH	TOTAL FLOAT
63	80	5	5	0	22	PAINT OFFICE EXTERIOR	25NOV98	1DEC98	16APR99	22APR99	102
64	67	4	4	0	8	TEST PIPING OFFICE	2DEC98	7DEC98	8FEB99	11FEB99	48
65	66	10	10	0	4	INSTALL CONDUIT OFFICE	24NOV98	7DEC98	29JAN99	11FEB99	48
66	67	0	0	0		RESTRAINT	8DEC98	7DEC98	12FEB99	11FEB99	48
66	74	10	10	0	4	PULL WIRE OFFICE	8DEC98	21DEC98	12MAR99	25MAR99	68
67	68	5	5	0	19	PARTITIONS OFFICE	8DEC98	14DEC98	12FEB99	18FEB99	48
68	69	5	5	0	19	DRYWALL	15DEC98	21DEC98	19FEB99	25FEB99	48
69	70	10	10	0	19	DRYWALL	22DEC98	4JAN99	26FEB99	11MAR99	48
69	73	10	10	0	20	CERAMIC TILE OFFICE	22DEC98	4JAN99	2APR99	15APR99	73
70	71	10	10	0	26	WOOD TRIM OFFICE	5JAN99	18JAN99	12MAR99	25MAR99	48
70	77	0	0	0		RESTRAINT	5JAN99	4JAN99	2APR99	1APR99	63
71	72	10	10	0	22	PAINT INTERIOR OFFICE	19JAN99	1FEB99	26MAR99	8APR99	48
71	80	5	5	0	21	HANG DOORS OFFICE	19JAN99	25JAN99	16APR99	22APR99	63
72	73	0	0	0		RESTRAINT	2FEB99	1FEB99	16APR99	15APR99	53
72	78	0	0	0		RESTRAINT	2FEB99	1FEB99	9APR99	8APR99	48
72	80	10	10	0	20	FLOOR TILE OFFICE	2FEB99	15FEB99	9APR99	22APR99	48
73	90	5	5	0	5	TOILET FIXTURES OFFICE	2FEB99	8FEB99	16APR99	22APR99	53
74	75	5	5	0	4	INSTALL PANEL INTERNALS OFFICE	22DEC98	28DEC98	26MAR99	1APR99	68
74	76	0	0	0		RESTRAINT	22DEC98	21DEC98	12APR99	9APR99	79
75	79	10	10	0	4	TERMINATE WIRES OFFICE	29DEC98	11JAN99	12APR99	15APR99	68
76	79	4	4	0	4	AIR CONDITIONING ELEC CONNECT	6JAN99	11JAN99	12APR99	15APR99	68
77	78	5	5	0	18	INSTALL CEILING GRID OFFICE	5JAN99	11JAN99	2APR99	8APR99	63
78	80	10	10	0	18	ACOUSTIC TILE OFFICE	2FEB99	15FEB99	9APR99	22APR99	48

Figure 6-11a (Continued) — activity schedule listing (rotated table)

I–J	OD	RD	%	Code A	Code B	Description	Early Start	Early Finish	Late Start	Late Finish	TF
79–80	5	5	0	4	4	RINGOUT ELECT	12JAN99	18JAN99	16APR99	22APR99	68
80–END	0	0	0			Project Finish	23APR99	22APR99	23APR99	22APR99	0
90–58	0	0	0			RESTRAINT	29DEC98	28DEC98	9APR99	8APR99	73
91–59	10	10	0	5	17	INSTALL RR SIDING	29DEC98	11JAN99	26MAR99	8APR99	63
92–58	0	0	0			RESTRAINT	5JAN99	4JAN99	9APR99	8APR99	68
93–80	0	0	0			RESTRAINT	19JAN99	18JAN99	23APR99	22APR99	68
94–80	5	5	0	5	27	SEED AND PLANT	16APR99	22APR99	16APR99	22APR99	0
98–76	5	5	0	4	8	INSTALL PACKAGE AIR CONDITONER	30DEC98	5JAN99	5APR99	9APR99	68
99–59	5	5	0	4	7	ERECT PRECAST STRUCT. OFFICE	21OCT98	27OCT98	28DEC98	1JAN99	48
210–211	10	10	0	6	32	APPROVE FONDATION REBAR	15JUL98	28JUL98	12AUG98	25AUG98	20
211–16	10	10	0	6	33	FAB&DEL FOUNDATION REBAR	29JUL98	11AUG98	26AUG98	8SEP98	20
212–213	10	10	0	6	32	APPROVE STRUCTURAL STEEL	29JUL98	11AUG98	24AUG98	4SEP98	18
213–29	40	40	0	6	33	FAB&DEL STRUCTURAL STEEL	12AUG98	6OCT98	7SEP98	30OCT98	18
214–215	10	10	0	6	32	APPROVE CRANE	29JUL98	11AUG98	31AUG98	11SEP98	23
215–31	50	50	0	6	33	FAB&DEL CRANE	12AUG98	20OCT98	14SEP98	20NOV98	23
216–217	10	10	0	6	32	APPROVE BAR JOISTS	29JUL98	11AUG98	5OCT98	16OCT98	48
217–33	30	30	0	6	33	FAB&DEL BAR JOISTS	12AUG98	22SEP98	19OCT98	27NOV98	48
218–219	10	10	0	6	32	APPROVE SIDING	29JUL98	11AUG98	29SEP98	12OCT98	44
219–35	40	40	0	6	33	FAB&DEL SIDING	12AUG98	6OCT98	13OCT98	7DEC98	44
220–221	10	10	0	6	32	APPROVE PLANT ELECTRICAL LOAD CENTER	29JUL98	11AUG98	4AUG98	17AUG98	4
221–300	90	90	0	6	33	FAB&DEL PLANT ELECTRICAL LOAD CENTER	12AUG98	15DEC98	18AUG98	21DEC98	4
222–223	10	10	0	6	32	APPROVE POWER PANELS-PLANT	29JUL98	11AUG98	10SEP98	23SEP98	31
223–301	75	75	0	6	33	FAB&DEL POWER PANELS-PLANT	12AUG98	24NOV98	24SEP98	6JAN99	31
224–225	10	10	0	6	32	APPROVE EXTERIOR DOORS	29JUL98	11AUG98	18SEP98	1OCT98	37
225–226	15	15	0	6	33	FAB&DEL EXTERIOR DOORS	12AUG98	1SEP98	19NOV98	9DEC98	71
225–302	80	80	0	6	32	APPROVE PLANT ELECTRICAL FIXTURES	12AUG98	1DEC98	2OCT98	21JAN99	37
226–51	75	75	0	6	33	FAB&DEL PLANT ELECTRICAL FIXTURES	2SEP98	15DEC98	10DEC98	24MAR99	71
227–223	10	10	0	6	32	APPROVE PLANT HEATING AND VENTILATING FANS	29JUL98	11AUG98	25SEP98	8OCT98	42
228–304	75	75	0	6	33	FAB&DEL PLANT HEATING AND VENTILATING FANS	12AUG98	24NOV98	9OCT98	21JAN99	42
229–230	10	10	0	6	32	APPROVE BOILER	29JUL98	11AUG98	13NOV98	26NOV98	77
230–306	60	60	0	6	33	FAB&DEL BOILER	12AUG98	3NOV98	27NOV98	18FEB99	77
231–232	10	10	0	6	32	APPROVE OIL TANK	29JUL98	11AUG98	5JAN99	18JAN99	114
232–305	50	50	0	6	33	FAB&DEL OIL TANK	12AUG98	20OCT98	19JAN99	29MAR99	::4

Figure 6-11a (Continued)

145

PRED	SUCC	ORIG DUR	REM DUR	%	CODE	ACTIVITY DESCRIPTION	EARLY START	EARLY FINISH	LATE START	LATE FINISH	TOTAL FLOAT
235	236	10	10	0	6 32	APPROVE PACKAGED A/C	12AUG98	25AUG98	16NOV98	27NOV98	68
236	98	90	90	0	6 33	FAB&DEL PACKAGED A/C	26AUG98	29DEC98	30NOV98	2APR99	68
237	238	10	10	0	6 32	APPROVE PRECAST	26AUG98	8SEP98	2NOV98	13NOV98	48
238	99	30	30	0	6 33	FAB&DEL PRECAST	9SEP98	20OCT98	16NOV98	25DEC98	48
3CO	39	2	2	0	3 4	SET ELECTRICAL LOAD CENTER P-W	22DEC98	23DEC98	22DEC98	23DEC98	0
301	43	10	10	0	3 4	POWER PANEL BACKFILL BOXES P-W	22DEC98	4JAN99	7JAN99	20JAN99	12
302	42	5	5	0	3 6	ERECT EXTERIOR DOORS P-W	22DEC98	28DEC98	22JAN99	28JAN99	23
303	39	10	10	0	3 14	MASONRY PARTITIONS P-W	22DEC98	4JAN99	8JAN99	21JAN99	13
304	46	15	15	0	3 8	INSTALL H AND V UNITS P-W	22DEC98	11JAN99	22JAN99	11FEB99	23
305	47	3	3	0	3 8	INSTALL FUEL TANK P-W	22DEC98	24DEC98	30MAR99	1APR99	70
306	41	25	25	0	3 8	ERECT BOILER AND AUXILIARY P-W	22DEC98	25JAN99	19FEB99	25MAR99	43
307	602	0	0	0		PCO #3	22DEC98	21DEC98	5FEB99	4FEB99	33
308	58	5	5	0	6	INSTALL MONORAIL WAREHOUSE	22DEC98	28DEC98	2APR99	8APR99	73
600	15	0	0	0		PCO #1	2SEP98	1SEP98	2SEP98	1SEP98	0
601	16	0	0	0		PCO #2	9SEP98	8SEP98	9SEP98	8SEP98	0
602	40	30	30	0	3 8	FABRICATE PIPING P-W	22DEC98	1FEB99	5FEB99	18MAR99	33
603	216	20	20	0	6 31	SUBMIT BAR JOISTS	1JUL98	28JUL98	7SEP98	2OCT98	48
604	235	30	30	0	6 31	SUBMIT PACKAGED A/C	1JUL98	11AUG98	5OCT98	13NOV98	68
START	0	0	0	0		Project Start	1JUL98	30JUN98	1JUL98	30JUN98	0

Figure 6-11a (Continued)

146

Figure 6-13, which is the CPM output for Fig. 6-12, is the same as the baseline except that the five change orders (Events 600 to 604) are included with no-earlier-than dates locked in. The integration of the five PCOs does impact the overall baseline. Where, in the individual fragnets, a delay of 17 calendar days was anticipated, it is indicated on the total float sort (Fig. 6-13*b*). The end of the project is extended from April 22, 1999, to May 17, 1999—25 calendar days. PCO #2 has five work days of float (one week). The "stretch out" is due to two things:

1. The submittal of the VECP joists does not start until September 30, 1998.

2. The procurement chain has not been reduced as suggested in the fragnet analysis.

Unfortunately, waiting to incorporate the change orders until they are formally approved may seriously impair the realism of the monthly updated schedule information. One approach to mitigate this delayed information is for the PM/CM to maintain a clone of the last approved baseline (or even the last baseline submitted for approval). On this clone, the PM/CM should overlay all expected PCOs and expected change orders.

Changing the Baseline

As the project proceeds, it is sensible to change the baseline periodically. In the John Doe example, it was appropriate to "reset the project clock" at the end of excavation of the plant and warehouse (Figs. 6-14, pp. 172 to 178, and 6-15, pp. 180 to 186). To reset the baseline, all activities, especially procurement, had to be reviewed and updated. (This is normally done monthly.) Of the total 187 activities, 39 were statused as complete and another 9 had durations reduced to reflect either progress or change of plan (joists).

PCOs #1 and #2 were critical and caused 17-calendar-days delay (from September 8, 1998, to September 25, 1998). The joist submittal process was expedited and now has 20 workdays of float. However, the seven-workday increase in Activity 33-34 Erect Bar Joists, which had been critical, continued as seven workdays more critical. The projected end date is May 19, 1999.

The project is really substantially complete at May 5, 1999 (Activity 56-58 Energize Power). This could easily be shown by resequencing Activity 58-94 Fine Grade and Activity 94-80 Seed and Plant.

O'BRIEN KREITZBERG INC.

FINEST HOUR

JOHN DOE 2

REPORT DATE 7APR97 RUN NO. 6
18:09

JOHN DOE

START DATE 1JUL98 FIN DATE 22APR99

SCHEDULE REPORT - SORT BY TF, ES

DATA DATE 1JUL98 PAGE NO. 1

PRED	SUCC	ORIG DUR	REM DUR	%	CODE	ACTIVITY DESCRIPTION	EARLY START	EARLY FINISH	LATE START	LATE FINISH	TOTAL FLOAT
0	1	3	3	0		CLEAR SITE	1JUL98	3JUL98	1JUL98	3JUL98	0
START	0	0	0	0		Project Start	1JUL98	30JUN98	1JUL98	30JUN98	0
1	2	2	2	0		SURVEY AND LAYOUT	6JUL98	7JUL98	6JUL98	7JUL98	0
2	3	2	2	0		ROUGH GRADE	8JUL98	9JUL98	8JUL98	9JUL98	0
3	4	15	15	0		DRILL WELL	10JUL98	30JUL98	10JUL98	30JUL98	0
4	5	2	2	0		INSTALL WELL PUMP	31JUL98	3AUG98	31JUL98	3AUG98	0
5	8	8	8	0		UNDERGROUND WATER PIPING	4AUG98	13AUG98	4AUG98	13AUG98	0
8	13	2	2	0		CONNECT WATER PIPING	14AUG98	17AUG98	14AUG98	17AUG98	0
13	14	1	1	0		BUILDING LAYOUT	18AUG98	18AUG98	18AUG98	18AUG98	0
14	600	10	10	0		DRIVE AND POUR PILES	19AUG98	1SEP98	19AUG98	1SEP98	0
15	601	5	5	0		EXCAVATE FOR PLANT WAREHOUSE	2SEP98	8SEP98	2SEP98	8SEP98	0
600	15	0	0	0		PCO #1	2SEP98	1SEP98	2SEP98	1SEP98	0
16	17	5	5	0		POUR PILE CAPS PLANT-WAREHSE	9SEP98	15SEP98	9SEP98	15SEP98	0
601	16	0	0	0		PCO #2	9SEP98	8SEP98	9SEP98	8SEP98	0
17	18	10	10	0		FORM AND POUR GRADE BEAMS P-W	16SEP98	29SEP98	16SEP98	29SEP98	0
18	19	3	3	0		BACKFILL AND COMPACT P-W	30SEP98	2OCT98	30SEP98	2OCT98	0
19	20	5	5	0		UNDERSLAB PLUMBING P-W	5OCT98	9OCT98	5OCT98	9OCT98	0
20	22	5	5	0		UNDERSLAB CONDUIT P-W	12OCT98	16OCT98	12OCT98	16OCT98	0
22	29	10	10	0		FORM AND POUR SLABS P-W	19OCT98	30OCT98	19OCT98	30OCT98	0
29	30	10	10	0		ERECT STRUCT STEEL P-W	2NOV98	13NOV98	2NOV98	13NOV98	0
30	31	5	5	0		PLUMB STEEL AND BOLT P-W	16NOV98	20NOV98	16NOV98	20NOV98	0
31	32	5	5	0		ERECT CRANE WAY AND CRANE P-W	23NOV98	27NOV98	23NOV98	27NOV98	0
32	33	0	0	0		RESTRAINT	30NOV98	27NOV98	30NOV98	27NOV98	0

Figure 6-11*b* Relative criticality of PCOs—TF sort.

i	j				Description					TF
33	34	3	3	0	ERECT BAR JOISTS P-W	30NOV98	2DEC98	30NOV98	2DEC98	0
34	35	3	3	0	ERECT ROOF PLANKS P-W	3DEC98	7DEC98	3DEC98	7DEC98	0
35	36	10	10	0	ERECT SIDING P-W	8DEC98	21DEC98	8DEC98	21DEC98	0
36	37	0	0	0	RESTRAINT	22DEC98	21DEC98	22DEC98	21DEC98	0
37	100	100	0	0	RESTRAINT	22DEC98	21DEC98	22DEC98	21DEC98	0
300	38	2	2	0	SET ELECTRICAL LOAD CENTER P-W	22DEC98	23DEC98	22DEC98	23DEC98	0
38	43	20	20	0	INSTALL POWER CONDUIT P-W	24DEC98	20JAN99	24DEC98	20JAN99	0
43	49	15	15	0	INSTALL BRANCH CONDUIT P-W	21JAN99	10FEB99	21JAN99	10FEB99	0
49	50	15	15	0	PULL WIRE P-W	11FEB99	3MAR99	11FEB99	3MAR99	0
50	54	5	5	0	INSTALL PANEL INTERNALS P-W	4MAR99	10MAR99	4MAR99	10MAR99	0
54	55	10	10	0	TERMINATE WIRES P-W	11MAR99	24MAR99	11MAR99	24MAR99	0
55	56	10	10	0	RINGOUT P-W	25MAR99	7APR99	25MAR99	7APR99	0
56	58	1	1	0	ENERGIZE POWER	8APR99	8APR99	8APR99	8APR99	0
58	94	5	5	0	FINE GRADE	9APR99	15APR99	9APR99	15APR99	0
94	30	5	5	0	SEED AND PLANT	16APR99	22APR99	16APR99	22APR99	0
90	END			0	Project Finish	23APR99	22APR99	23APR99	22APR99	0
3	6	4	4	0	WATER TANK FOUNDATIONS	10JUL98	15JUL98	13JUL98	16JUL98	1
6	7	10	10	0	ERECT WATER TOWER	16JUL98	29JUL98	17JUL98	30JUL98	1
7	8	10	10	0	TANK PIPING AND VALVES	30JUL98	12AUG98	31JUL98	13AUG98	1
31	33	3	3	0	ERECT MONORAIL TRACK P-W	23NOV98	25NOV98	25NOV98	27NOV98	2
0	220	20	20	0	SUBMIT PLANT ELECTRICAL LOAD CENTER	1JUL98	28JUL98	7JUL98	3AUG98	4
3	9	10	10	0	EXCAVATE FOR SEWER	10JUL98	23JUL98	16JUL98	29JUL98	4
9	11	5	5	0	INSTALL SEWER AND BACKFILL	24JUL98	30JUL98	30JUL98	5AUG98	4
220	221	10	10	0	APPROVE PLANT ELECTRICAL LOAD CENTER	29JUL98	11AUG98	4AUG98	17AUG98	4
11	12	3	3	0	ELECTRICAL DUCT BANK	31JUL98	4AUG98	6AUG98	10AUG98	4
12	13	5	5	0	PULL IN POWER FEEDER	5AUG98	11AUG98	11AUG98	17AUG98	4
221	300	90	90	0	FAB&DEL PLANT ELECTRICAL LOAD CENTER	12AUG98	15DEC98	18AUG98	21DEC98	4
35	37	5	5	0	ROOFING P-W	8DEC98	14DEC98	15DEC98	21DEC98	5
58	60	5	5	0	ERECT FLAGPOLE	9APR99	15APR99	16APR99	22APR99	5
19	21	5	5	0	FORM AND POUR RR LOAD DOCK P-W	30SEP98	6OCT98	12OCT98	16OCT98	9
19	22	5	5	0	FORM AND POUR TK LOAD DOCK P-W	30SEP98	6OCT98	12OCT98	16OCT98	9
21	22	0	0	0	RESTRAINT	7OCT98	6OCT98	19OCT98	16OCT98	9
47	301	0	0	0	RESTRAINT	22DEC98	21DEC98	7JAN99	6JAN99	12

PRED	SUCC	ORIG DUR	REM DUR	%	CODE	ACTIVITY DESCRIPTION	EARLY START	EARLY FINISH	LATE START	LATE FINISH	TOTAL FLOAT
301	43	10	10	0		POWER PANEL BACKFILL BOXES P-W	22DEC98	4JAN99	7JAN99	20JAN99	12
3	10	1	1	0		EXCAVATE ELECTRICAL MANHOLES	10JUL98	10JUL98	29JUL98	29JUL98	13
10	11	5	5	0		INSTALL ELECTRICAL MANHOLES	13JUL98	17JUL98	30JUL98	5AUG98	13
37	303	0	0	0		RESTRAINT	22DEC98	21DEC98	8JAN99	7JAN99	13
303	39	10	10	0		MASONRY PARTITIONS P-W	22DEC98	4JAN99	8JAN99	21JAN99	13
39	42	5	5	0		FRAME CEILINGS P-W	5JAN99	11JAN99	22JAN99	28JAN99	13
42	44	10	10	0		DRYWELL PARTITIONS P-W	12JAN99	25JAN99	29JAN99	11FEB99	13
44	46	0	0	0		RESTRAINT	26JAN99	25JAN99	12FEB99	11FEB99	13
46	52	25	25	0		INSTALL DUCTWORK P-W	26JAN99	1MAR99	12FEB99	18MAR99	13
52	58	15	15	0		INSULATE H&V SYSTEM P-W	2MAR99	22MAR99	19MAR99	8APR99	13
3	12	6	6	0		OVERHEAD POLE LINE	10JUL98	17JUL98	3AUG98	10AUG98	16
0	212	20	20	0		SUBMIT STRUCTURAL STEEL	1JUL98	28JUL98	27JUL98	21AUG98	18
212	213	10	10	0		APPROVE STRUCTURAL STEEL	29JUL98	11AUG98	24AUG98	4SEP98	18
213	29	40	40	0		FAB&DEL STRUCTURAL STEEL	12AUG98	6OCT98	7SEP98	30OCT98	18
44	48	10	10	0		CERAMIC TILE	26JAN99	8FEB99	19FEB99	4MAR99	18
48	53	5	5	0		PAINT ROOMS P-W	9FEB99	15FEB99	5MAR99	11MAR99	18
53	57	10	10	0		FLOOR TILE P-W	16FEB99	1MAR99	12MAR99	25MAR99	18
57	58	10	10	0		INSTALL FURNISHING P-W	2MAR99	15MAR99	26MAR99	8APR99	18
0	210	10	10	0		SUBMIT FOUNDATION REBAR	1JUL98	14JUL98	29JUL98	11AUG98	20
210	211	10	10	0		APPROVE FONDATION REBAR	15JUL98	28JUL98	12AUG98	25AUG98	20
211	16	10	10	0		FAB&DEL FOUNDATION REBAR	29JUL98	11AUG98	26AUG98	8SEP98	20
0	214	20	20	0		SUBMIT CRANE	1JUL98	28JUL98	3AUG98	28AUG98	23
214	215	10	10	0		APPROVE CRANE	29JUL98	11AUG98	31AUG98	11SEP98	23

I	J	Dur	Rem	%	Activity Description	Early Start	Early Finish	Late Start	Late Finish	TF
215	31	50	50	0	FAB&DEL CRANE	12AUG98	20OCT98	14SEP98	20NOV98	23
37	302	0	0	0	RESTRAINT	22DEC98	21DEC98	22JAN99	21JAN99	23
37	304	0	0	0	RESTRAINT	22DEC98	21DEC98	22JAN99	21JAN99	23
302	42	5	5	0	ERECT EXTERIOR DOORS P-W	22DEC98	28DEC98	22JAN99	28JAN99	23
304	46	15	15	0	INSTALL H AND V UNITS P-W	22DEC98	11JAN99	22JAN99	11FEB99	23
45	51	5	5	0	ROOM OUTLETS P-W	11FEB99	17FEB99	18MAR99	24MAR99	25
49	45	0	0	0	RESTRAINT	11FEB99	10FEB99	18MAR99	17MAR99	25
51	56	10	10	0	INSTALL ELECTRICAL FIXTURES	18FEB99	3MAR99	25MAR99	7APR99	25
53	58	10	10	0	INSTALL PLUMBING FIXTURES P-W	16FEB99	1MAR99	26MAR99	8APR99	28
0	222	20	20	0	SUBMIT POWER PANELS-PLANT	1JUL98	28JUL98	13AUG98	9SEP98	31
222	223	10	10	0	APPROVE POWER PANELS-PLANT	29JUL98	11AUG98	10SEP98	23SEP98	31
223	301	75	75	0	FAB&DEL POWER PANELS-PLANT	12AUG98	24NOV98	24SEP98	6JAN99	31
37	307	0	0	0	RESTRAINT	22DEC98	21DEC98	5FEB99	4FEB99	33
307	602	0	0	0	PCO #3	22DEC98	21DEC98	5FEB99	4FEB99	33
602	40	30	30	0	FABRICATE PIPING P-W	22DEC98	1FEB99	2FEB99	18MAR99	33
40	47	10	10	0	TEST PIPING SYSTEMS P-W	2FEB99	15FEB99	19MAR99	1APR99	33
47	58	5	5	0	LIGHTOFF BOILER AND TEST	16FEB99	22FEB99	2APR99	8APR99	33
0	224	20	20	0	SUBMIT EXTERIOR DOORS	1JUL98	28JUL98	21AUG98	17SEP98	37
0	225	30	30	0	SUBMIT PLANT EXECTRICAL FIXTURES	1JUL98	11AUG98	21AUG98	1OCT98	37
224	225	10	10	0	APPROVE EXTERIOR DOORS	29JUL98	11AUG98	18SEP98	1OCT98	37
225	302	80	80	0	FAB&DEL EXTERIOR DOORS	12AUG98	1DEC98	2OCT98	21JAN99	37
44	45	0	0	0	RESTRAINT	26JAN99	25JAN99	18MAR99	17MAR99	42
0	227	20	20	0	SUBMIT PLANT HEATING AND VENTILATING FANS	1JUL98	28JUL98	28AUG98	24SEP98	42
227	228	10	10	0	APPROVE PLANT HEATING AND VENTILATING FANS	29JUL98	11AUG98	25SEP98	8OCT98	42
228	304	75	75	0	FAB&DEL PLANT HEATING AND VENTILATING FANS	12AUG98	24NOV98	9OCT98	21JAN99	42
37	306	0	0	0	RESTRAINT	22DEC98	21DEC98	19FEB99	18FEB99	43
306	41	25	25	0	ERECT BOILER AND AUXILIARY P-W	22DEC98	25JAN99	19FEB99	25MAR99	43
41	47	5	5	0	PREOPERATIONAL BOILER CHECK	26JAN99	1FEB99	26MAR99	1APR99	43
44	58	10	10	0	HANG INTERIOR DOORS P-W	26JAN99	8FEB99	26MAR99	8APR99	43
0	213	20	20	0	SUBMIT SIDING	1JUL98	28JUL98	1SEP98	28SEP98	44
213	219	10	10	0	APPROVE SIDING	29JUL98	11AUG98	29SEP98	12OCT98	44
219	35	40	40	0	FAB&DEL SIDING	12AUG98	6OCT98	13OCT98	7DEC98	44
0	237	40	40	0	SUBMIT PRECAST	1JUL98	25AUG98	7SEP98	30OCT98	49

Figure 6-11b (Continued)

FINEST HOUR   JOHN DOE

REPORT DATE 7APR97   RUN NO. 6   JOHN DOE   START DATE 1JUL98   FIN DATE 22APR99

18:09

SCHEDULE REPORT - SORT BY TF, ES   DATA DATE 1JUL98   PAGE NO. 3

JOHN DOE 2

PRED	SUCC	ORIG DUR	REM DUR	CODE	%	ACTIVITY DESCRIPTION	EARLY START	EARLY FINISH	LATE START	LATE FINISH	TOTAL FLOAT
0	603	0	0		0	PCO #4	1JUL98	30JUN98	7SEP98	4SEP98	48
603	216	20	20		0	SUBMIT BAR JOISTS	1JUL98	28JUL98	7SEP98	2OCT98	48
216	217	10	10		0	APPROVE BAR JOISTS	29JUL98	11AUG98	5OCT98	16OCT98	48
217	33	30	30		0	FAB&DEL BAR JOISTS	12AUG98	22SEP98	19OCT98	27NOV98	48
237	238	10	10		0	APPROVE PRECAST	26AUG98	8SEP98	2NOV98	13NOV98	48
238	99	30	30		0	FAB&DEL PRECAST	9SEP98	20OCT98	16NOV98	25DEC98	48
99	59	5	5		0	ERECT PRECAST STRUCT. OFFICE	21OCT98	27OCT98	28DEC98	1JAN99	48
59	60	5	5		0	ERECT PRECAST ROOF OFFICE	28OCT98	3NOV98	4JAN99	8JAN99	48
60	61	10	10		0	EXTERIOR MASONRY OFFICE	4NOV98	17NOV98	11JAN99	22JAN99	48
61	64	10	10		0	INSTALL PIPING OFFICE	18NOV98	1DEC98	25JAN99	5FEB99	48
61	65	4	4		0	INSTALL ELEC BACKING BOXES	18NOV98	23NOV98	25JAN99	28JAN99	48
65	66	10	10		0	INSTALL CONDUIT OFFICE	24NOV98	7DEC98	29JAN99	11FEB99	48
64	67	4	4		0	TEST PIPING OFFICE	2DEC98	7DEC98	8FEB99	11FEB99	48
66	67	0	0		0	RESTRAINT	8DEC98	7DEC98	12FEB99	11FEB99	48
67	68	5	5		0	PARTITIONS OFFICE	8DEC98	14DEC98	12FEB99	18FEB99	48
68	69	5	5		0	DRYWALL	15DEC98	21DEC98	19FEB99	25FEB99	48
69	70	10	10		0	DRYWALL	22DEC98	4JAN99	26FEB99	11MAR99	48
70	71	10	10		0	WOOD TRIM OFFICE	5JAN99	18JAN99	12MAR99	25MAR99	48
71	72	10	10		0	PAINT INTERIOR OFFICE	19JAN99	1FEB99	26MAR99	8APR99	48
72	78	0	0		0	RESTRAINT	2FEB99	1FEB99	9APR99	8APR99	48
72	80	10	10		0	FLOOR TILE OFFICE	2FEB99	15FEB99	9APR99	22APR99	48
78	80	10	10		0	ACOUSTIC TILE OFFICE	2FEB99	15FEB99	9APR99	22APR99	48
72	73	0	0		0	RESTRAINT	2FEB99	1FEB99	16APR99	15APR99	53

PRED	SUCC	OD	RD	—	DESCRIPTION	ES	EF	LS	LF	TF
73	80	5	5	0	TOILET FIXTURES OFFICE	2FEB99	8FEB99	16APR99	22APR99	53
61	62	5	5	0	EXTERIOR DOORS OFFICE	18NOV98	24NOV98	12FEB99	18FEB99	62
61	63	5	5	0	ROOFING OFFICE	18NOV98	24NOV98	12FEB99	18FEB99	62
61	68	5	5	0	GLAZE OFFICE	18NOV98	24NOV98	12FEB99	18FEB99	62
62	63	0	0	0	RESTRAINT	25NOV98	24NOV98	19FEB99	18FEB99	62
63	68	0	0	0	RESTRAINT	25NOV98	24NOV98	19FEB99	18FEB99	62
37	91	5	5	0	GRADE AND BALLAST RR SIDING	22DEC98	28DEC98	19MAR99	25MAR99	63
91	58	10	10	0	INSTALL RR SIDING	29DEC98	11JAN99	26MAR99	8APR99	63
70	77	0	0	0	RESTRAINT	5JAN99	4JAN99	2APR99	1APR99	63
77	78	5	5	0	INSTALL CEILING GRID OFFICE	5JAN99	11JAN99	2APR99	8APR99	63
71	80	5	5	0	HANG DOORS OFFICE	19JAN99	25JAN99	16APR99	22APR99	63
0	604	0	0	0	PCO #5	30JUN98	30JUN98	5OCT98	2OCT98	68
604	235	30	30	0	SUBMIT PACKAGED A/C	1JUL98	11AUG98	5OCT98	13NOV98	68
235	236	10	10	0	APPROVE PACKAGED A/C	12AUG98	25AUG98	16NOV98	27NOV98	68
236	99	90	90	0	FAB&DEL PACKAGED A/C	26AUG98	29DEC98	30NOV98	2APR99	68
66	74	10	10	0	PULL WIRE OFFICE	8DEC98	21DEC98	12MAR99	25MAR99	68
37	92	10	10	0	ACCESS ROAD	22DEC98	4JAN99	26MAR99	8APR99	68
37	93	20	20	0	AREA LIGHTING	22DEC98	18JAN99	26MAR99	22APR99	68
74	75	5	5	0	INSTALL PANEL INTERNALS OFFICE	22DEC98	28DEC98	26MAR99	1APR99	68
75	79	10	10	0	TERMINATE WIRES OFFICE	29DEC98	11JAN99	2APR99	15APR99	68
98	76	5	5	0	INSTALL PACKAGE AIR CONDITONER	30DEC98	5JAN99	5APR99	9APR99	68
92	59	0	0	0	RESTRAINT	5JAN99	4JAN99	9APR99	8APR99	68
76	79	4	4	0	AIR CONDITIONING ELEC CONNECT	6JAN99	11JAN99	9APR99	15APR99	68
79	80	5	5	0	RINGOUT ELECT	12JAN99	18JAN99	16APR99	22APR99	68
93	80	0	0	0	RESTRAINT	19JAN99	18JAN99	23APR99	22APR99	68
14	23	3	3	0	EXCAVATE FOR OFFICE BUILDING	19AUG98	21AUG98	25NOV98	27NOV98	70
23	24	4	4	0	SPREAD FOOTINGS OFFICE	24AUG98	27AUG98	30NOV98	3DEC98	70
24	25	6	6	0	FORM AND POUR GRADE BEAMS OFF	28AUG98	4SEP98	4DEC98	11DEC98	70
25	26	1	1	0	BACKFILL AND COMPACT OFFICE	7SEP98	7SEP98	14DEC98	14DEC98	70
26	27	3	3	0	UNDERSLAB PLUMBING OFFICE	8SEP98	10SEP98	15DEC98	17DEC98	70
27	28	3	3	0	UNDERSLAB CONDUIT OFFICE	11SEP98	15SEP98	18DEC98	22DEC98	70
29	39	3	3	0	FORM AND POUR OFFICE SLAB	16SEP98	18SEP98	23DEC98	25DEC98	70
37	305	0	0	0	RESTRAINT	22DEC98	21DEC98	30MAR99	29MAR99	70

Figure 6-11b (Continued)

FINEST HOUR

JOHN DOE

JOHN DOE 2

START DATE 1JUL98   FIN DATE 22APR99

DATA DATE 1JUL98   PAGE NO.   4

PRED	SUCC	ORIG DUR	REM DUR	%	CODE	ACTIVITY DESCRIPTION	EARLY START	EARLY FINISH	LATE START	LATE FINISH	TOTAL FLOAT
305	47	3	3	0		INSTALL FUEL TANK P-W	22DEC98	24DEC98	30MAR99	1APR99	70
225	226	15	15	0		APPROVE PLANT ELECTRICAL FIXTURES	12AUG98	1SEP98	19NOV98	9DEC98	71
226	51	75	75	0		FAB&DEL PLANT ELECTRICAL FIXTURES	2SEP98	15DEC98	10DEC98	24MAR99	71
37	90	5	5	0		PAVE PARKING AREA	22DEC98	28DEC98	2APR99	8APR99	73
37	308	0	0	0		RESTRAINT	22DEC98	21DEC98	2APR99	1APR99	73
69	73	10	10	0		CERAMIC TILE OFFICE	22DEC98	4JAN99	2APR99	15APR99	73
308	58	5	5	0		INSTALL MONORAIL WAREHOUSE	22DEC98	28DEC98	2APR99	8APR99	73
90	58	0	0	0		RESTRAINT	29DEC98	28DEC98	9APR99	8APR99	73
0	229	20	20	0		SUBMIT BOILER	1JUL98	28JUL98	16OCT98	12NOV98	77
229	230	10	10	0		APPROVE BOILER	29JUL98	11AUG98	13NOV98	26NOV98	77
230	306	60	60	0		FAB&DEL BOILER	12AUG98	3NOV98	27NOV98	18FEB99	77
37	90	10	10	0		PERIMETER FENCE	22DEC98	4JAN99	9APR99	22APR99	78
74	76	0	0	0		RESTRAINT	22DEC98	21DEC98	12APR99	9APR99	79
61	77	15	15	0		DUCTWORK OFFICE	18NOV98	8DEC98	12MAR99	1APR99	82
63	80	5	5	0		PAINT OFFICE EXTERIOR	25NOV98	1DEC98	16APR99	22APR99	102
60	98	0	0	0		RESTRAINT	4NOV98	3NOV98	5APR99	2APR99	108
0	231	20	20	0		SUBMIT OIL TANK	1JUL98	28JUL98	8DEC98	4JAN99	114
231	232	10	10	0		APPROVE OIL TANK	29JUL98	11AUG98	5JAN99	18JAN99	114
232	305	50	50	0		FAB&DEL OIL TANK	12AUG98	20OCT98	19JAN99	29MAR99	114

Figure 6-11b (*Continued*)

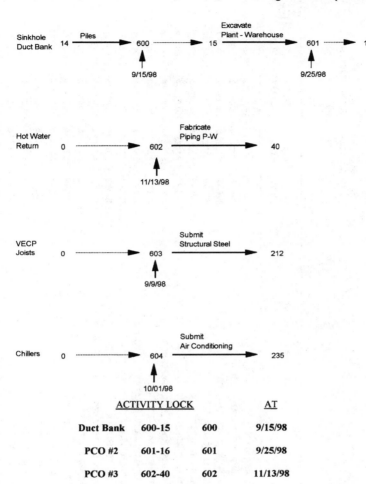

ACTIVITY LOCK			AT
Duct Bank	**600-15**	**600**	**9/15/98**
PCO #2	**601-16**	**601**	**9/25/98**
PCO #3	**602-40**	**602**	**11/13/98**
PCO #4	**603-216**	**603**	**9/30/98**
PCO #5	**604-235**	**604**	**10/01/98**

Figure 6-12 Change order integrated using PCO events.

O'BRIEN KREITZBERG INC.

FINEST HOUR

REPORT DATE 7APR97   RUN NO.   3
              18:22

JOHN DOE

JOHN DOE 3

START DATE 1JUL98   FIN DATE 17MAY99

SCHEDULE REPORT BY I J

DATA DATE 1JUL98   PAGE NO.   1

PRED	SUCC	ORIG DUR	REM DUR	%	CODE	ACTIVITY DESCRIPTION	EARLY START	EARLY FINISH	LATE START	LATE FINISH	TOTAL FLOAT
0	1	3	3	0	1 1	CLEAR SITE	1JUL98	3JUL98	24JUL98	28JUL98	17
0	210	10	10	0	6 31	SUBMIT FOUNDATION REBAR	1JUL98	14JUL98	21AUG98	3SEP98	37
0	212	20	20	0	6 31	SUBMIT STRUCTURAL STEEL	1JUL98	28JUL98	19AUG98	15SEP98	35
0	214	20	20	0	6 31	SUBMIT CRANE	1JUL98	28JUL98	26AUG98	22SEP98	40
0	218	20	20	0	6 31	SUBMIT SIDING	1JUL98	28JUL98	24SEP98	21OCT98	61
0	220	20	20	0	6 31	SUBMIT PLANT ELECTRICAL LOAD CENTER	1JUL98	28JUL98	30JUL98	26AUG98	21
0	222	20	20	0	6 31	SUBMIT POWER PANELS-PLANT	1JUL98	28JUL98	7SEP98	2OCT98	48
0	224	20	20	0	6 31	SUBMIT EXTERIOR DOORS	1JUL98	28JUL98	15SEP98	12OCT98	54
0	225	30	30	0	6 31	SUBMIT PLANT ELECTRICAL FIXTURES	1JUL98	11AUG98	15SEP98	26OCT98	54
0	227	20	20	0	6 31	SUBMIT PLANT HEATING AND VENTILATING FANS	1JUL98	28JUL98	22SEP98	19OCT98	59
0	229	20	20	0	6 31	SUBMIT BOILER	1JUL98	28JUL98	10NOV98	7DEC98	94
0	231	20	20	0	6 31	SUBMIT OIL TANK	1JUL98	28JUL98	31DEC98	27JAN99	131
0	237	40	40	0	6 31	SUBMIT PRECAST	1JUL98	25AUG98	30SEP98	24NOV98	65
0	603	0	0	0		PCO #4	1JUL98	30JUN98	30SEP98	29SEP98	65
0	604	0	0	0		PCO #5	1JUL98	30JUN98	28OCT98	27OCT98	85
1	2	2	2	0	1 2	SURVEY AND LAYOUT	6JUL98	7JUL98	29JUL98	30JUL98	17
2	3	2	2	0	1 1	ROUGH GRADE	8JUL98	9JUL98	31JUL98	3AUG98	17
3	4	15	15	0	1 9	DRILL WELL	10JUL98	30JUL98	4AUG98	24AUG98	17
3	6	4	4	0	1 3	WATER TANK FOUNDATIONS	10JUL98	15JUL98	5AUG98	10AUG98	18
3	9	10	10	0	1 1	EXCAVATE FOR SEWER	10JUL98	23JUL98	10AUG98	21AUG98	21
3	10	1	1	0	1 1	EXCAVATE ELECTRICAL MANHOLES	10JUL98	10JUL98	21AUG98	21AUG98	30
3	12	6	6	0	1 4	OVERHEAD POLE LINE	10JUL98	17JUL98	26AUG98	2SEP98	33
4	5	2	2	0	1 9	INSTALL WELL PUMP	31JUL98	3AUG98	25AUG98	26AUG98	17

I	J						DESCRIPTION					
5	8	8	8	0	1	5	UNDERGROUND WATER PIPING	4AUG98	13AUG98	27AUG98	7SEP98	17
6	7	10	10	0	1	10	ERECT WATER TOWER	16JUL98	29JUL98	11AUG98	24AUG98	18
7	8	10	10	0	1	10	TANK PIPING AND VALVES	30JUL98	12AUG98	25AUG98	7SEP98	18
8	13	2	2	0	1	10	CONNECT WATER PIPING	14AUG98	17AUG98	8SEP98	9SEP98	17
9	11	5	5	0	1	5	INSTALL SEWER AND BACKFILL	24JUL98	30JUL98	24AUG98	28AUG98	21
10	11	5	5	0	1	4	INSTALL ELECTRICAL MANHOLES	13JUL98	17JUL98	24AUG98	28AUG98	30
11	12	3	3	0	1	4	ELECTRICAL DUCT BANK	31JUL98	4AUG98	31AUG98	2SEP98	21
12	13	5	5	0	1	4	PULL IN POWER FEEDER	5AUG98	11AUG98	3SEP98	9SEP98	21
13	14	1	1	0	2	2	BUILDING LAYOUT	18AUG98	18AUG98	10SEP98	10SEP98	17
14	23	3	3	0	2	1	EXCAVATE FOR OFFICE BUILDING	19AUG98	21AUG98	18DEC98	22DEC98	87
14	600	10	10	0	2	11	DRIVE AND POUR PILES	19AUG98	1SEP98	11SEP98	24SEP98	17
15	601	5	5	0	2	1	EXCAVATE FOR PLANT WAREHOUSE	15SEP98	21SEP98	25SEP98	1OCT98	8
16	17	5	5	0	2	3	POUR PILE CAPS PLANT-WAREHSE	25SEP98	1OCT98	2OCT98	8OCT98	5
17	19	10	10	0	2	3	FORM AND POUR GRADE BEAMS P-W	2OCT98	15OCT98	9OCT98	22OCT98	5
19	19	3	3	0	2	1	BACKFILL AND COMPACT P-W	16OCT98	20OCT98	23OCT98	27OCT98	5
19	21	5	5	0	2	3	FORM AND POUR RR LOAD DOCK P-W	16OCT98	22OCT98	4NOV98	10NOV98	13
18	22	5	5	0	2	3	FORM AND POUR TK LOAD DOCK P-W	16OCT98	22OCT98	4NOV98	10NOV98	13
19	20	5	5	0	2	5	UNDERSLAB PLUMBING P-W	21OCT98	27OCT98	28OCT98	3NOV98	5
20	22	5	5	0	2	4	UNDERSLAB CONDUIT P-W	28OCT98	4NOV98	4NOV98	10NOV98	5
21	22	0	0	0	2	3	RESTRAINT	23OCT98	22OCT98	11NOV98	10NOV98	13
22	29	0	0	0	2	3	FORM AND POUR SLABS P-W	4NOV98	17NOV98	11NOV98	24NOV98	5
23	24	10	10	0	2	3	SPREAD FOOTINGS OFFICE	24AUG98	27AUG98	23DEC98	28DEC98	97
24	25	4	4	0	2	3	FORM AND POUR GRADE BEAMS OFF	28AUG98	4SEP98	29DEC98	5JAN99	87
25	26	6	6	0	2	1	BACKFILL AND COMPACT OFFICE	7SEP98	7SEP98	6JAN99	6JAN99	87
26	27	1	1	0	2	5	UNDERSLAB PLUMBING OFFICE	8SEP98	10SEP98	7JAN99	11JAN99	87
27	28	3	3	0	2	4	UNDERSLAB CONDUIT OFFICE	11SEP98	15SEP98	12JAN99	14JAN99	97
28	99	3	3	0	2	3	FORM AND POUR OFFICE SLAB	16SEP98	18SEP98	15JAN99	19JAN99	97
99	30	3	3	0	3	6	ERECT STRUCT STEEL P-W	18NOV98	25NOV98	8DEC98	8DEC98	5
30	31	10	10	0	3	6	PLUMB STEEL AND BOLT P-W	2DEC98	8DEC98	9DEC98	15DEC98	5
31	32	5	5	0	3	6	ERECT CRANE WAY AND CRANE P-W	9DEC98	15DEC98	16DEC98	22DEC98	5
31	33	5	5	0	3	6	ERECT MONORAIL TRACK P-W	9DEC98	11DEC98	18DEC98	22DEC98	7
32	33	3	3	0	0	0	RESTRAINT	16DEC98	15DEC98	23DEC98	22DEC98	5
33	34	0	0	0	0	6	ERECT BAR JOISTS P-W	23DEC98	25DEC98	23DEC98	25DEC98	0

Figure 6-13a CPM with dates locked into PCOs—i-j sort.

O'BRIEN KREITZBERG INC.

FINEST HOUR

JOHN DOE 3

REPORT DATE 7APR97 RUN NO. 3

JOHN DOE

START DATE 1JUL98 FIN DATE 17MAY99

18:22

SCHEDULE REPORT BY I J

DATA DATE 1JUL98 PAGE NO. 2

PRED	SUCC	ORIG DUR	REM DUR	%	CODE	ACTIVITY DESCRIPTION	EARLY START	EARLY FINISH	LATE START	LATE FINISH	TOTAL FLOAT
34	35	3	3	0	3	7	ERECT ROOF PLANKS P-W	28DEC98	30DEC98	28DEC98	30DEC98	0
35	36	10	10	0	3	12	ERECT SIDING P-W	31DEC98	13JAN99	31DEC98	13JAN99	0
35	37	5	5	0	3	13	ROOFING P-W	31DEC98	6JAN99	7JAN99	13JAN99	5
36	37	0	0	0			RESTRAINT	14JAN99	13JAN99	13JAN99		0
37	80	10	10	0	5	15	PERIMETER FENCE	14JAN99	27JAN99	4MAY99	17MAY99	78
37	90	5	5	0	5	16	PAVE PARKING AREA	14JAN99	20JAN99	27APR99	3MAY99	73
37	91	5	5	0	5	17	GRADE AND BALLAST RR SIDING	14JAN99	20JAN99	13APR99	19APR99	63
37	92	10	10	0	5	16	ACCESS ROAD	14JAN99	27JAN99	20APR99	3MAY99	68
37	93	20	20	0	5	4	AREA LIGHTING	14JAN99	10FEB99	20APR99	17MAY99	68
37	300	0	0	0			RESTRAINT	14JAN99	13JAN99	14JAN99	13JAN99	0
37	301	0	0	0			RESTRAINT	14JAN99	13JAN99	1FEB99	29JAN99	12
37	302	0	0	0			RESTRAINT	14JAN99	13JAN99	16FEB99	15FEB99	23
37	303	0	0	0			RESTRAINT	14JAN99	13JAN99	2FEB99	1FEB99	13
37	304	0	0	0			RESTRAINT	14JAN99	13JAN99	16FEB99	15FEB99	23
37	305	0	0	0			RESTRAINT	14JAN99	13JAN99	22APR99	21APR99	70
37	306	0	0	0			RESTRAINT	14JAN99	13JAN99	16MAR99	15MAR99	43
37	307	0	0	0			RESTRAINT	14JAN99	13JAN99	2MAR99	1MAR99	33
37	308	0	0	0			RESTRAINT	14JAN99	13JAN99	27APR99	26APR99	73
38	43	20	20	0	3	4	INSTALL POWER CONDUIT P-W	18JAN99	12FEB99	18JAN99	12FEB99	0
39	42	5	5	0	3	18	FRAME CEILINGS P-W	28JAN99	3FEB99	16FEB99	22FEB99	13
40	47	10	10	0	3	8	TEST PIPING SYSTEMS P-W	25FEB99	10MAR99	13APR99	26APR99	33
41	47	5	5	0	3	8	PREOPERATIONAL BOILER CHECK	18FEB99	24FEB99	20APR99	26APR99	43
42	44	10	10	0	3	19	DRYWELLL PARTITIONS P-W	4FEB99	17FEB99	23FEB99	8MAR99	13

I	J	OD	RD	%	ACT	DESCRIPTION	ES	EF	LS	LF	TF
43	19	15	15	0	4	INSTALL BRANCH CONDUIT P-W	15FEB99	5MAR99	15FEB99	5MAR99	0
44	45	0	0	0		RESTRAINT	18FEB99	17FEB99	12APR99	9APR99	37
44	46	0	0	0		RESTRAINT	18FEB99	17FEB99	9MAR99	8MAR99	13
44	48	10	10	0	20	CERAMIC TILE	18FEB99	3MAR99	16MAR99	29MAR99	18
44	58	10	10	0	21	HANG INTERIOR DOORS P-W	18FEB99	3MAR99	20APR99	3MAY99	43
45	51	5	5	0	4	ROOM OUTLETS P-W	8MAR99	12MAR99	12APR99	16APR99	25
46	52	25	25	0	8	INSTALL DUCTWORK P-W	18FEB99	24MAR99	9MAR99	12APR99	13
47	58	5	5	0	8	LIGHTOFF BOILER AND TEST	11MAR99	17MAR99	27APR99	3MAY99	33
48	53	5	5	0	22	PAINT ROOMS P-W	4MAR99	10MAR99	30MAR99	5APR99	18
49	45	0	0	0		RESTRAINT	8MAR99	5MAR99	12APR99	9APR99	25
49	50	15	15	0	4	PULL WIRE P-W	8MAR99	26MAR99	8MAR99	26MAR99	0
50	54	5	5	0	4	INSTALL PANEL INTERNALS P-W	29MAR99	2APR99	29MAR99	2APR99	0
51	56	10	10	0	4	INSTALL ELECTRICAL FIXTURES	15MAR99	26MAR99	19APR99	30APR99	25
52	58	15	15	0	8	INSULATE H&V SYSTEM P-W	25MAR99	14APR99	13APR99	3MAY99	13
53	57	10	10	0	22	FLOOR TILE P-W	11MAR99	24MAR99	6APR99	19APR99	18
53	58	10	10	0	5	INSTALL PLUMBING FIXTURES P-W	11MAR99	24MAR99	20APR99	3MAY99	28
54	55	10	10	0	4	TERMINATE WIRES P-W	5APR99	16APR99	5APR99	16APR99	0
55	56	10	10	0	4	RINGOUT P-W	19APR99	30APR99	19APR99	30APR99	0
56	59	1	1	0	4	ENERGIZE POWER	3MAY99	3MAY99	3MAY99	3MAY99	0
57	58	10	10	0	24	INSTALL FURNISHING P-W	25MAR99	7APR99	20APR99	3MAY99	18
58	80	5	5	0	27	ERECT FLAGPOLE	4MAY99	10MAY99	11MAY99	17MAY99	5
58	94	5	5	0	27	FINE GRADE	4MAY99	10MAY99	4MAY99	10MAY99	0
59	60	5	5	0	7	ERECT PRECAST ROOF OFFICE	28OCT98	3NOV98	27JAN99	2FEB99	65
60	61	10	10	0	14	EXTERIOR MASONRY OFFICE	4NOV98	17NOV98	3FEB99	16FEB99	65
60	98	0	0	0		RESTRAINT	4NOV98	3NOV98	28APR99	27APR99	125
61	62	5	5	0	21	EXTERIOR DOORS OFFICE	18NOV98	24NOV98	9MAR99	15MAR99	79
61	63	5	5	0	13	ROOFING OFFICE	18NOV98	24NOV98	9MAR99	15MAR99	79
61	64	10	10	0	8	INSTALL PIPING OFFICE	18NOV98	1DEC98	17FEB99	2MAR99	65
61	65	4	4	0	4	INSTALL ELEC BACKING BOXES	18NOV98	23NOV98	17FEB99	22FEB99	65
61	68	5	5	0	25	GLAZE OFFICE	18NOV98	24NOV98	9MAR99	15MAR99	79
61	77	15	15	0	8	DUCTWORK OFFICE	18NOV98	8DEC98	6APR99	26APR99	99
62	63	0	0	0		RESTRAINT	25NOV98	24NOV98	16MAR99	15MAR99	79
63	68	0	0	0		RESTRAINT	25NOV98	24NOV98	16MAR99	15MAR99	79

Figure 6-13a (Continued)

REPORT DATE 7APR97 RUN NO. 3   JOHN DOE   START DATE 1JUL98 FIN DATE 17MAY99
18:22
SCHEDULE REPORT BY I J   DATA DATE 1JUL98   PAGE NO. 3

PRED	SUCC	ORIG DUR	REM DUR	%	CODE	ACTIVITY DESCRIPTION	EARLY START	EARLY FINISH	LATE START	LATE FINISH	TOTAL FLOAT
63	80	5	5	0	22	PAINT OFFICE EXTERIOR	25NOV98	1DEC98	11MAY99	17MAY99	119
64	67	4	4	0	8	TEST PIPING OFFICE	2DEC98	7DEC98	3MAR99	8MAR99	65
65	66	10	10	0	4	INSTALL CONDUIT OFFICE	24NOV98	7DEC98	23FEB99	8MAR99	65
66	67	0	0	0		RESTRAINT	8DEC98	7DEC98	9MAR99	8MAR99	65
66	74	10	10	0	4	PULL WIRE OFFICE	8DEC98	21DEC98	6APR99	19APR99	85
67	68	5	5	0	19	PARTITIONS OFFICE	8DEC98	14DEC98	9MAR99	15MAR99	65
68	69	5	5	0	19	DRYWALL	15DEC98	21DEC98	16MAR99	22MAR99	65
69	70	10	10	0	19	DRYWALL	22DEC98	4JAN99	23MAR99	5APR99	65
69	73	10	10	0	20	CERAMIC TILE OFFICE	22DEC98	4JAN99	27APR99	10MAY99	90
70	71	10	10	0	26	WOOD TRIM OFFICE	5JAN99	18JAN99	6APR99	19APR99	65
70	77	0	0	0		RESTRAINT	5JAN99	4JAN99	27APR99	26APR99	80
71	72	10	10	0	22	PAINT INTERIOR OFFICE	19JAN99	1FEB99	20APR99	3MAY99	65
71	80	5	5	0	21	HANG DOORS OFFICE	19JAN99	25JAN99	11MAY99	17MAY99	80
72	73	0	0	0		RESTRAINT	2FEB99	1FEB99	11MAY99	10MAY99	70
72	78	0	0	0		RESTRAINT	2FEB99	1FEB99	4MAY99	3MAY99	65
72	80	10	10	0	20	FLOOR TILE OFFICE	2FEB99	15FEB99	4MAY99	17MAY99	65
73	80	5	5	0	5	TOILET FIXTURES OFFICE	2FEB99	8FEB99	11MAY99	17MAY99	70
74	75	5	5	0	5	INSTALL PANEL INTERNALS OFFICE	22DEC98	28DEC98	20APR99	26APR99	85
74	76	0	0	0		RESTRAINT	22DEC98	21DEC98	5MAY99	4MAY99	96
75	79	10	10	0	4	TERMINATE WIRES OFFICE	29DEC98	11JAN99	27APR99	10MAY99	85
76	79	4	-4	0	4	AIR CONDITIONING ELEC CONNECT	8APR99	13APR99	5MAY99	10MAY99	19
77	78	5	5	0	18	INSTALL CEILING GRID OFFICE	5JAN99	11JAN99	27APR99	3MAY99	80
78	80	10	10	0	18	ACOUSTIC TILE OFFICE	2FEB99	15FEB99	4MAY99	17MAY99	65

160

PRED	SUCC	OD	RD	%			DESCRIPTION	ES	EF	LS	LF	TF
79	80	5	5	0	4	4	RINGOUT ELECT	14APR99	20APR99	11MAY99	17MAY99	19
80	END	0	0	0			Project Finish	18MAY99	17MAY99	18MAY99	17MAY99	0
90	58	0	0	0			RESTRAINT	21JAN99	20JAN99	4MAY99	3MAY99	73
91	58	10	10	0	5	17	INSTALL RR SIDING	21JAN99	3FEB99	20APR99	3MAY99	63
92	58	0	0	0			RESTRAINT	28JAN99	27JAN99	4MAY99	3MAY99	68
93	80	0	0	0			RESTRAINT	11FEB99	10FEB99	18MAY99	17MAY99	68
94	80	5	5	0	5	27	SEED AND PLANT	11MAY99	17MAY99	11MAY99	17MAY99	0
98	76	5	5	0	4	8	INSTALL PACKAGE AIR CONDITONER	1APR99	7APR99	28APR99	4MAY99	19
99	59	5	5	0	4	7	ERECT PRECAST STRUCT. OFFICE	21OCT98	27OCT98	20JAN99	26JAN99	65
210	211	10	10	0	6	32	APPROVE FONDATION REBAR	15JUL98	28JUL98	4SEP98	17SEP98	37
211	16	10	10	0	6	33	FAB&DEL FOUNDATION REBAR	29JUL98	11AUG98	18SEP98	1OCT98	37
212	213	10	10	0	6	32	APPROVE STRUCTURAL STEEL	29JUL98	11AUG98	16SEP98	29SEP98	35
213	29	40	40	0	6	33	FAB&DEL STRUCTURAL STEEL	12AUG98	6OCT98	30SEP98	24NOV98	35
214	215	10	10	0	6	32	APPROVE CRANE	29JUL98	11AUG98	23SEP98	6OCT98	40
215	31	50	50	0	6	33	FAB&DEL CRANE	12AUG98	20OCT98	7OCT98	15DEC98	40
216	217	10	10	0	6	32	APPROVE BAR JOISTS	28OCT98	10NOV98	28OCT98	10NOV98	0
217	33	30	30	0	6	33	FAB&DEL BAR JOISTS	11NOV98	22DEC98	11NOV98	22DEC98	0
218	219	10	10	0	6	32	APPROVE SIDING	29JUL98	11AUG98	22OCT98	4NOV98	61
219	35	40	40	0	6	33	FAB&DEL SIDING	12AUG98	6OCT98	5NOV98	30DEC98	61
220	221	10	10	0	6	32	APPROVE PLANT ELECTRICAL LOAD CENTER	29JUL98	11AUG98	27AUG98	9SEP98	21
221	300	90	90	0	6	33	FAB&DEL PLANT ELECTRICAL LOAD CENTER	12AUG98	15DEC98	10SEP98	13JAN99	21
222	223	10	10	0	6	32	APPROVE POWER PANELS-PLANT	29JUL98	11AUG98	5OCT98	16OCT98	48
223	301	75	75	0	6	33	FAB&DEL POWER PANELS-PLANT	12AUG98	24NOV98	19OCT98	29JAN99	48
224	225	10	10	0	6	32	APPROVE EXTERIOR DOORS	29JUL98	11AUG98	13OCT98	26OCT98	54
225	226	15	15	0	6	32	APPROVE PLANT ELECTRICAL FIXTURES	12AUG98	1SEP98	14DEC98	1JAN99	98
225	302	80	80	0	6	33	FAB&DEL EXTERIOR DOORS	12AUG98	1DEC98	27OCT98	15FEB99	54
226	51	75	75	0	6	33	FAB&DEL PLANT ELECTRICAL FIXTURES	2SEP98	15DEC98	4JAN99	16APR99	88
227	228	10	10	0	6	32	APPROVE PLANT HEATING AND VENTILATING FANS	29JUL98	11AUG98	20OCT98	2NOV98	59
228	304	75	75	0	6	33	FAB&DEL PLANT HEATING AND VENTILATING FANS	12AUG98	24NOV98	3NOV98	15FEB99	59
229	230	10	10	0	6	32	APPROVE BOILER	29JUL98	11AUG98	8DEC98	21DEC98	94
230	306	60	60	0	6	33	FAB&DEL BOILER	12AUG98	3NOV98	22DEC98	15MAR99	94
231	232	10	10	0	6	32	APPROVE OIL TANK	29JUL98	11AUG98	28JAN99	10FEB99	131
232	305	50	50	0	6	33	FAB&DEL OIL TANK	12AUG98	20OCT98	11FEB99	21APR99	131

Figure 6-13a (Continued)

O'BRIEN KREITZBERG INC.   FINEST HOUR   JOHN DOE 3

REPORT DATE 7APR97 RUN NO. 3   JOHN DOE   START DATE 1JUL98 FIN DATE 17MAY99
     18:22

SCHEDULE REPORT BY I J   DATA DATE 1JUL98 PAGE NO. 4

PRED	SUCC	ORIG DUR	REM DUR		CODE		ACTIVITY DESCRIPTION	EARLY START	EARLY FINISH	LATE START	LATE FINISH	TOTAL FLOAT
235	236	10	10	0	6	32	APPROVE PACKAGED A/C	12NOV98	25NOV98	9DEC98	22DEC98	19
236	99	90	90	0	6	33	FAB&DEL PACKAGED A/C	26NOV98	31MAR99	23DEC98	27APR99	19
237	238	10	10	0	6	32	APPROVE PRECAST	26AUG98	8SEP98	25NOV98	8DEC98	65
238	99	30	30	0	6	33	FAB&DEL PRECAST	9SEP98	20OCT98	9DEC98	19JAN99	65
300	38	2	2	0	3	4	SET ELECTRICAL LOAD CENTER P-W	14JAN99	15JAN99	14JAN99	15JAN99	0
301	43	10	10	0	3	4	POWER PANEL BACKFILL BOXES P-W	14JAN99	27JAN99	1FEB99	12FEB99	12
302	42	5	5	0	3	6	ERECT EXTERIOR DOORS P-W	14JAN99	20JAN99	16FEB99	22FEB99	23
303	39	10	10	0	3	14	MASONRY PARTITIONS P-W	14JAN99	27JAN99	2FEB99	15FEB99	13
304	46	15	15	0	3	8	INSTALL H AND V UNITS P-W	14JAN99	3FEB99	16FEB99	8MAR99	23
305	47	3	3	0	3	8	INSTALL FUEL TANK P-W	14JAN99	18JAN99	22APR99	26APR99	70
306	41	25	25	0	3	8	ERECT BOILER AND AUXILIARY P-W	14JAN99	17FEB99	16MAR99	19APR99	43
307	602	0	0	0		3	PCO #3	14JAN99	13JAN99	2MAR99	1MAR99	33
308	58	5	5	0	3	6	INSTALL MONORAIL WAREHOUSE	14JAN99	20JAN99	27APR99	3MAY99	73
600	15	0	0	0		3	PCO #1	15SEP98*	14SEP98	25SEP98	24SEP98	8
601	16	0	0	0		3	PCO #2	25SEP98*	24SEP98	2OCT98	1OCT98	5
602	40	30	30	0	3	8	FABRICATE PIPING P-W	14JAN99*	24FEB99	2MAR99	12APR99	33
603	216	20	20	0	6	31	SUBMIT BAR JOISTS	30SEP98*	27OCT98	30SEP98	27OCT98	0
604	235	30	30	0	6	31	SUBMIT PACKAGED A/C	1OCT98*	11NOV98	28OCT98	8DEC98	19
START	0	0	0	0			Project Start	1JUL98	30JUN98	24JUL98	23JUL98	17

Figure 6-13a (*Continued*)

O'BRIEN KREITZBERG INC.   FINEST HOUR   JOHN DOE 3

REPORT DATE 7APR97 RUN NO. 2   START DATE 1JUL98 FIN DATE 17MAY99
            18:20   JOHN DOE
SCHEDULE REPORT - SORT BY TF, ES   DATA DATE 1JUL98 PAGE NO. 1

PRED	SUCC	ORIG DUR	REM DUR	%	CODE	ACTIVITY DESCRIPTION	EARLY START	EARLY FINISH	LATE START	LATE FINISH	TOTAL FLOAT
603	216	20	20	0		SUBMIT BAR JOISTS	30SEP98*	27OCT98	30SEP98	27OCT98	0
216	217	10	10	0		APPROVE BAR JOISTS	28OCT98	10NOV98	28OCT98	10NOV98	0
217	33	30	30	0		FAB&DEL BAR JOISTS	11NOV98	22DEC98	11NOV98	22DEC98	0
33	34	3	3	0		ERECT BAR JOISTS P-W	23DEC98	25DEC98	23DEC98	25DEC98	0
34	35	3	3	0		ERECT ROOF PLANKS P-W	28DEC98	30DEC98	28DEC98	30DEC98	0
35	36	10	10	0		ERECT SIDING P-W	31DEC98	13JAN99	31DEC98	13JAN99	0
36	37	0	0	0		RESTRAINT	14JAN99	13JAN99	14JAN99	13JAN99	0
37	300	0	0	0		RESTRAINT	14JAN99	13JAN99	14JAN99	13JAN99	0
300	38	2	2	0		SET ELECTRICAL LOAD CENTER P-W	14JAN99	15JAN99	14JAN99	15JAN99	0
38	43	20	20	0		INSTALL POWER CONDUIT P-W	18JAN99	12FEB99	18JAN99	12FEB99	0
43	49	15	15	0		INSTALL BRANCH CONDUIT P-W	15FEB99	5MAR99	15FEB99	5MAR99	0
49	50	15	15	0		PULL WIRE P-W	8MAR99	26MAR99	8MAR99	26MAR99	0
50	54	5	5	0		INSTALL PANEL INTERNALS P-W	29MAR99	2APR99	29MAR99	2APR99	0
54	55	10	10	0		TERMINATE WIRES P-W	5APR99	16APR99	5APR99	16APR99	0
55	56	10	10	0		RINGOUT P-W	19APR99	30APR99	19APR99	30APR99	0
56	58	1	1	0		ENERGIZE POWER	3MAY99	3MAY99	3MAY99	3MAY99	0
58	94	5	5	0		FINE GRADE	4MAY99	10MAY99	4MAY99	10MAY99	0
94	80	5	5	0		SEED AND PLANT	11MAY99	17MAY99	11MAY99	17MAY99	0
80	END	0	0	0		Project Finish	18MAY99	17MAY99	18MAY99	17MAY99	0
16	17	5	5	0		POUR PILE CAPS PLANT-WAREHSE	25SEP98	1OCT98	2OCT98	8OCT98	5
601	16	0	0	0		PCO #2	25SEP98*	24SEP98	2OCT98	1OCT98	5
17	18	10	10	0		FORM AND POUR GRADE BEAMS P-W	2OCT98	15OCT98	9OCT98	22OCT98	5
18	19	3	3	0		BACKFILL AND COMPACT P-W	16OCT98	20OCT98	23OCT98	27OCT98	5

i	j				Activity					
19	20	5	5	0	UNDERSLAB PLUMBING P-W	21OCT98	27OCT98	28OCT98	3NOV98	5
20	22	5	5	0	UNDERSLAB CONDUIT P-W	28OCT98	3NOV98	4NOV98	10NOV98	5
22	29	10	10	0	FORM AND POUR SLABS P-W	4NOV98	17NOV98	11NOV98	24NOV98	5
29	30	10	10	0	ERECT STRUCT STEEL P-W	18NOV98	1DEC98	25NOV98	8DEC98	5
30	31	5	5	0	PLUMB STEEL AND BOLT P-W	2DEC98	8DEC98	9DEC98	15DEC98	5
31	32	5	5	0	ERECT CRANE WAY AND CRANE P-W	9DEC98	15DEC98	16DEC98	22DEC98	5
32	33	0	0	0	RESTRAINT	16DEC98	15DEC98	23DEC98	22DEC98	5
35	37	5	5	0	ROOFING P-W	31DEC98	6JAN99	7JAN99	13JAN99	5
58	80	5	5	0	ERECT FLAGPOLE	4MAY99	10MAY99	11MAY99	17MAY99	5
31	33	3	3	0	ERECT MONORAIL TRACK P-W	9DEC98	11DEC98	18DEC98	22DEC98	7
15	601	5	5	0	EXCAVATE FOR PLANT WAREHOUSE	15SEP98	21SEP98	25SEP98	1OCT98	8
600	15	0	0	0	PCO #1	15SEP98*	14SEP98	25SEP98	24SEP98	8
37	301	0	0	0	RESTRAINT	14JAN99	13JAN99	1FEB99	29JAN99	12
301	43	10	10	0	POWER PANEL BACKFILL BOXES P-W	14JAN99	27JAN99	1FEB99	12FEB99	12
18	21	5	5	0	FORM AND POUR RR LOAD DOCK P-W	16OCT98	22OCT98	4NOV98	10NOV98	13
18	22	5	5	0	FORM AND POUR TK LOAD DOCK P-W	16OCT98	22OCT98	4NOV98	10NOV98	13
21	22	0	0	0	RESTRAINT	23OCT98	22OCT98	11NOV98	10NOV98	13
37	303	0	0	0	RESTRAINT	14JAN99	13JAN99	2FEB99	1FEB99	13
303	39	10	10	0	MASONRY PARTITIONS P-W	14JAN99	27JAN99	2FEB99	15FEB99	13
39	42	5	5	0	FRAME CEILINGS P-W	28JAN99	3FEB99	16FEB99	22FEB99	13
42	44	10	10	0	DRYWELL PARTITIONS P-W	4FEB99	17FEB99	23FEB99	8MAR99	13
44	46	0	0	0	RESTRAINT	18FEB99	17FEB99	9MAR99	9MAR99	13
46	52	25	25	0	INSTALL DUCTWORK P-W	18FEB99	24MAR99	9MAR99	12APR99	13
52	58	15	15	0	INSULATE H&V SYSTEM P-W	25MAR99	14APR99	13APR99	3MAY99	13
0	1	3	3	0	CLEAR SITE	1JUL98	3JUL98	24JUL98	28JUL98	17
START	0	0	0	0	Project Start	1JUL98	30JUN98	24JUL98	23JUL98	17
1	2	2	2	0	SURVEY AND LAYOUT	6JUL98	7JUL98	29JUL98	30JUL98	17
2	3	2	2	0	ROUGH GRADE	8JUL98	9JUL98	31JUL98	3AUG98	17
3	4	15	15	0	DRILL WELL	10JUL98	30JUL98	4AUG98	24AUG98	17
4	5	2	2	0	INSTALL WELL PUMP	31JUL98	3AUG98	25AUG98	26AUG98	17
5	8	8	8	0	UNDERGROUND WATER PIPING	4AUG98	13AUG98	27AUG98	7SEP98	17
8	13	2	2	0	CONNECT WATER PIPING	14AUG98	17AUG98	8SEP98	9SEP98	17
13	14	1	1	0	BUILDING LAYOUT	18AUG98	18AUG98	10SEP98	10SEP98	17

Figure 6-13b CPM with dates locked into PCOs—TF sort.

O'BRIEN KREITZBERG INC.

FINEST HOUR

JOHN DOE 3

REPORT DATE 7APR97 RUN NO. 2   JOHN DOE   START DATE 1JUL98 FIN DATE 17MAY99
            18:20

SCHEDULE REPORT - SORT BY TF, ES   DATA DATE 1JUL98 PAGE NO. 2

PRED	SUCC	ORIG DUR	REM DUR	%	CODE	ACTIVITY DESCRIPTION	EARLY START	EARLY FINISH	LATE START	LATE FINISH	TOTAL FLOAT
14	600	10	10	0		DRIVE AND POUR PILES	19AUG98	1SEP98	11SEP98	24SEP98	17
3	6	4	4	0		WATER TANK FOUNDATIONS	10JUL98	15JUL98	5AUG98	10AUG98	18
6	7	10	10	0		ERECT WATER TOWER	16JUL98	29JUL98	11AUG98	24AUG98	18
7	8	10	10	0		TANK PIPING AND VALVES	30JUL98	12AUG98	25AUG98	7SEP98	18
44	48	10	10	0		CERAMIC TILE	18FEB99	3MAR99	16MAR99	29MAR99	18
48	53	5	5	0		PAINT ROOMS P-W	4MAR99	10MAR99	30MAR99	5APR99	18
53	57	10	10	0		FLOOR TILE P-W	11MAR99	24MAR99	6APR99	19APR99	18
57	58	10	10	0		INSTALL FURNISHING P-W	25MAR99	7APR99	20APR99	3MAY99	18
604	235	30	30	0		SUBMIT PACKAGED A/C	1OCT98*	11NOV98	28OCT98	8DEC98	19
235	236	10	10	0		APPROVE PACKAGED A/C	12NOV98	25NOV98	9DEC98	22DEC98	19
236	98	90	90	0		FAB&DEL PACKAGED A/C	26NOV98	31MAR99	23DEC98	27APR99	19
98	76	5	5	0		INSTALL PACKAGE AIR CONDITONER	1APR99	7APR99	28APR99	4MAY99	19
76	79	4	4	0		AIR CONDITIONING ELEC CONNECT	8APR99	13APR99	5MAY99	10MAY99	19
79	90	5	5	0		RINGOUT ELECT	14APR99	20APR99	11MAY99	17MAY99	19
0	220	20	20	0		SUBMIT PLANT ELECTRICAL LOAD CENTER	1JUL98	28JUL98	30JUL98	26AUG98	21
3	9	10	10	0		EXCAVATE FOR SEWER	10JUL98	23JUL98	10AUG98	21AUG98	21
9	11	5	5	0		INSTALL SEWER AND BACKFILL	24JUL98	30JUL98	24AUG98	28AUG98	21
220	221	10	10	0		APPROVE PLANT ELECTRICAL LOAD CENTER	29JUL98	11AUG98	27AUG98	9SEP98	21
11	12	3	3	0		ELECTRICAL DUCT BANK	31JUL98	4AUG98	31AUG98	2SEP98	21
12	13	5	5	0		PULL IN POWER FEEDER	5AUG98	11AUG98	3SEP98	9SEP98	21
221	300	90	90	0		FAB&DEL PLANT ELECTRICAL LOAD CENTER	12AUG98	15DEC98	10SEP98	13JAN99	21
37	302	0	0	0		RESTRAINT	14JAN99	13JAN99	16FEB99	15FEB99	23
37	304	0	0	0		RESTRAINT	14JAN99	13JAN99	16FEB99	15FEB99	23

302	42	5	5	0	ERECT EXTERIOR DOORS P-W	14JAN99	20JAN99	16FEB99	22FEB99	23
304	46	15	15	0	INSTALL H AND V UNITS P-W	14JAN99	3FEB99	16FEB99	8MAR99	23
45	51	5	5	0	ROOM OUTLETS P-W	8MAR99	12MAR99	12APR99	16APR99	25
49	45	0	0	0	RESTRAINT	8MAR99	5MAR99	12APR99	9APR99	25
51	56	10	10	0	INSTALL ELECTRICAL FIXTURES	15MAR99	26MAR99	19APR99	30APR99	25
53	58	10	10	0	INSTALL PLUMBING FIXTURES P-W	11MAR99	24MAR99	20APR99	3MAY99	28
3	10	1	1	0	EXCAVATE ELECTRICAL MANHOLES	10JUL98	10JUL98	21AUG98	21AUG98	30
10	11	5	5	0	INSTALL ELECTRICAL MANHOLES	13JUL98	17JUL98	24AUG98	28AUG98	30
3	12	6	6	0	OVERHEAD POLE LINE	10JUL98	17JUL98	26AUG98	2SEP98	33
37	307	0	0	0	RESTRAINT	14JAN99	13JAN99	2MAR99	1MAR99	33
307	602	0	0	0	PCO #3	14JAN99	13JAN99	2MAR99	1MAR99	33
602	40	30	30	0	FABRICATE PIPING P-W	14JAN99*	24FEB99	2MAR99	12APR99	33
40	47	10	10	0	TEST PIPING SYSTEMS P-W	25FEB99	10MAR99	13APR99	26APR99	33
47	58	5	5	0	LIGHTOFF BOILER AND TEST	11MAR99	17MAR99	27APR99	3MAY99	33
0	212	20	20	0	SUBMIT STRUCTURAL STEEL	1JUL98	28JUL98	19AUG98	15SEP98	35
212	213	10	10	0	APPROVE STRUCTURAL STEEL	29JUL98	11AUG98	16SEP98	29SEP98	35
213	29	40	40	0	FAB&DEL STRUCTURAL STEEL	12AUG98	6OCT98	30SEP98	24NOV98	35
0	210	10	10	0	SUBMIT FOUNDATION REBAR	1JUL98	14JUL98	21AUG98	3SEP98	37
210	211	10	10	0	APPROVE FONDATION REBAR	15JUL98	28JUL98	4SEP98	17SEP98	37
211	16	10	10	0	FAB&DEL FOUNDATION REBAR	29JUL98	11AUG98	18SEP98	1OCT98	37
44	45	0	0	0	RESTRAINT	18FEB99	17FEB99	12APR99	9APR99	37
0	214	20	20	0	SUBMIT CRANE	1JUL98	28JUL98	26AUG98	22SEP98	40
214	215	10	10	0	APPROVE CRANE	29JUL98	11AUG98	23SEP98	6OCT98	40
215	31	50	50	0	FAB&DEL CRANE	12AUG98	20OCT98	7OCT98	15DEC98	40
37	306	0	0	0	RESTRAINT	14JAN99	13JAN99	16MAR99	15MAR99	43
306	41	25	25	0	ERECT BOILER AND AUXILIARY P-W	14JAN99	17FEB99	16MAR99	19APR99	43
41	47	5	5	0	PREOPERATIONAL BOILER CHECK	18FEB99	24FEB99	20APR99	26APR99	43
44	58	10	10	0	HANG INTERIOR DOORS P-W	18FEB99	3MAR99	20APR99	3MAY99	43
0	222	20	20	0	SUBMIT POWER PANELS-PLANT	1JUL98	28JUL98	7SEP98	2OCT98	49
222	223	10	10	0	APPROVE POWER PANELS-PLANT	29JUL98	11AUG98	5OCT98	16OCT98	-8
223	301	75	75	0	FAB&DEL POWER PANELS-PLANT	12AUG98	24NOV98	19OCT98	29JAN99	3
0	224	20	20	0	SUBMIT EXTERIOR DOORS	1JUL98	28JUL98	15SEP98	12OCT98	54
0	225	30	30	0	SUBMIT PLANT EXECTRICAL FIXTURES	1JUL98	11AUG98	15SEP98	26OCT98	54

Figure 6-13b (*Continued*)

PRED	SUCC	ORIG DUR	REM DUR	%	CODE	ACTIVITY DESCRIPTION	EARLY START	EARLY FINISH	LATE START	LATE FINISH	TOTAL FLOAT
224	225	10	10	0		APPROVE EXTERIOR DOORS	29JUL98	11AUG98	13OCT98	26OCT98	54
225	302	80	80	0		FAB&DEL EXTERIOR DOORS	12AUG98	1DEC98	27OCT98	15FEB99	54
0	227	20	20	0		SUBMIT PLANT HEATING AND VENTILATING FANS	1JUL98	28JUL98	22SEP98	19OCT98	59
227	228	10	10	0		APPROVE PLANT HEATING AND VENTILATING FANS	29JUL98	11AUG98	20OCT98	2NOV98	59
228	304	75	75	0		FAB&DEL PLANT HEATING AND VENTILATING FANS	12AUG98	24NOV98	3NOV98	15FEB99	59
0	218	20	20	0		SUBMIT SIDING	1JUL98	28JUL98	24SEP98	21OCT98	61
218	219	10	10	0		APPROVE SIDING	29JUL98	11AUG98	22OCT98	4NOV98	61
219	35	40	40	0		FAB&DEL SIDING	12AUG98	6OCT98	5NOV98	30DEC98	61
37	91	5	5	0		GRADE AND BALLAST RR SIDING	14JAN99	20JAN99	13APR99	19APR99	63
91	58	10	10	0		INSTALL RR SIDING	21JAN99	3FEB99	20APR99	3MAY99	63
0	237	40	40	0		SUBMIT PRECAST	1JUL98	25AUG98	30SEP98	24NOV98	65
0	603	0	0	0		PCO #4	1JUL98	30JUN98	30SEP98	29SEP98	65
237	238	10	10	0		APPROVE PRECAST	26AUG98	8SEP98	25NOV98	8DEC98	65
238	90	30	30	0		FAB&DEL PRECAST	9SEP98	20OCT98	9DEC98	19JAN99	65
90	59	5	5	0		ERECT PRECAST STRUCT. OFFICE	21OCT98	27OCT98	20JAN99	26JAN99	65
59	60	5	5	0		ERECT PRECAST ROOF OFFICE	28OCT98	3NOV98	27JAN99	2FEB99	65
60	61	10	10	0		EXTERIOR MASONRY OFFICE	4NOV98	17NOV98	3FEB99	16FEB99	65
61	64	10	10	0		INSTALL PIPING OFFICE	18NOV98	1DEC98	17FEB99	2MAR99	65
61	65	4	4	0		INSTALL ELEC BACKING BOXES	18NOV98	23NOV98	17FEB99	22FEB99	65
65	66	10	10	0		INSTALL CONDUIT OFFICE	24NOV98	7DEC98	23FEB99	8MAR99	65
64	67	4	4	0		TEST PIPING OFFICE	2DEC98	7DEC98	3MAR99	8MAR99	65
66	67	0	0	0		RESTRAINT	8DEC98	7DEC98	9MAR99	8MAR99	65
67	68	5	5	0		PARTITIONS OFFICE	8DEC98	14DEC98	9MAR99	15MAR99	65

Figure 6-13b (Continued)

						Early Start	Early Finish	Late Start	Late Finish	
68	69	5	5	0	DRYWALL	15DEC98	21DEC98	16MAR99	22MAR99	65
69	70	10	10	0	DRYWALL	22DEC98	4JAN99	23MAR99	5APR99	65
70	71	10	10	0	WOOD TRIM OFFICE	5JAN99	18JAN99	6APR99	19APR99	65
71	72	10	10	0	PAINT INTERIOR OFFICE	19JAN99	1FEB99	20APR99	3MAY99	65
72	78	0	0	0	RESTRAINT	2FEB99	1FEB99	4MAY99	3MAY99	65
72	80	10	10	0	FLOOR TILE OFFICE	2FEB99	15FEB99	4MAY99	17MAY99	65
78	80	10	10	0	ACOUSTIC TILE OFFICE	2FEB99	15FEB99	4MAY99	17MAY99	68
37	92	10	10	0	ACCESS ROAD	14JAN99	27JAN99	20APR99	3MAY99	68
37	93	20	20	0	AREA LIGHTING	14JAN99	10FEB99	20APR99	17MAY99	68
92	53	0	0	0	RESTRAINT	28JAN99	27JAN99	4MAY99	3MAY99	68
73	80	0	0	0	RESTRAINT	11FEB99	10FEB99	18MAY99	17MAY99	70
37	305	0	0	0	RESTRAINT	14JAN99	13JAN99	22APR99	21APR99	70
305	47	3	3	0	INSTALL FUEL TANK P-W	14JAN99	18JAN99	22APR99	26APR99	70
72	73	0	0	0	RESTRAINT	2FEB99	1FEB99	11MAY99	10MAY99	70
73	80	5	5	0	TOILET FIXTURES OFFICE	2FEB99	8FEB99	11MAY99	17MAY99	73
37	90	5	5	0	PAVE PARKING AREA	14JAN99	20JAN99	27APR99	3MAY99	73
308	308	0	0	0	RESTRAINT	14JAN99	13JAN99	27APR99	26APR99	73
308	58	5	5	0	INSTALL MONORAIL WAREHOUSE	14JAN99	20JAN99	27APR99	3MAY99	73
90	58	0	0	0	RESTRAINT	21JAN99	20JAN99	27APR99	3MAY99	78
37	80	10	10	0	PERIMETER FENCE	14JAN99	27JAN99	4MAY99	17MAY99	79
61	62	5	5	0	EXTERIOR DOORS OFFICE	18NOV98	24NOV98	9MAR99	15MAR99	79
61	63	5	5	0	ROOFING OFFICE	18NOV98	24NOV98	9MAR99	15MAR99	79
61	68	5	5	0	GLAZE OFFICE	18NOV98	24NOV98	9MAR99	15MAR99	79
62	63	0	0	0	RESTRAINT	25NOV98	24NOV98	16MAR99	15MAR99	79
63	68	0	0	0	RESTRAINT	25NOV98	24NOV98	16MAR99	15MAR99	79
70	77	0	0	0	RESTRAINT	5JAN99	4JAN99	27APR99	26APR99	80
77	78	5	5	0	INSTALL CEILING GRID OFFICE	5JAN99	11JAN99	27APR99	3MAY99	80
71	80	5	5	0	HANG DOORS OFFICE	19JAN99	25JAN99	11MAY99	17MAY99	80
0	604	0	0	0	PCO #5	1JUL98	30JUN98	28OCT98	27OCT98	85
66	71	10	10	0	PULL WIRE OFFICE	8DEC98	21DEC98	6APR99	19APR99	85
74	75	5	5	0	INSTALL PANEL INTERNALS OFFICE	22DEC98	28DEC98	20APR99	26APR99	85
75	79	10	10	0	TERMINATE WIRES OFFICE	29DEC98	11JAN99	27APR99	10MAY99	85
14	23	3	3	0	EXCAVATE FOR OFFICE BUILDING	19AUG98	21AUG98	18DEC98	22DEC98	37

O'BRIEN KREITZBERG INC.

FINEST HOUR

JOHN DOE 3

REPORT DATE 7APR97 RUN NO. 2   JOHN DOE   START DATE 1JUL98 FIN DATE 17MAY99
18:20
SCHEDULE REPORT - SORT BY TF, ES   DATA DATE 1JUL98 PAGE NO. 4

PRED	SUCC	ORIG DUR	REM DUR	%	CODE	ACTIVITY DESCRIPTION	EARLY START	EARLY FINISH	LATE START	LATE FINISH	TOTAL FLOAT
23	24	4	4	0		SPREAD FOOTINGS OFFICE	24AUG98	27AUG98	23DEC98	28DEC98	87
24	25	6	6	0		FORM AND POUR GRADE BEAMS OFF	28AUG98	4SEP98	29DEC98	5JAN99	87
25	26	1	1	0		BACKFILL AND COMPACT OFFICE	7SEP98	7SEP98	6JAN99	6JAN99	87
26	27	3	3	0		UNDERSLAB PLUMBING OFFICE	8SEP98	10SEP98	7JAN99	11JAN99	87
27	28	3	3	0		UNDERSLAB CONDUIT OFFICE	11SEP98	15SEP98	12JAN99	14JAN99	87
28	99	3	3	0		FORM AND POUR OFFICE SLAB	16SEP98	18SEP98	15JAN99	19JAN99	87
225	226	15	15	0		APPROVE PLANT ELECTRICAL FIXTURES	12AUG98	1SEP98	14DEC98	1JAN99	88
226	51	75	75	0		FAB&DEL PLANT ELECTRICAL FIXTURES	2SEP98	15DEC98	4JAN99	16APR99	88
69	73	10	10	0		CERAMIC TILE OFFICE	22DEC98	4JAN99	27APR99	10MAY99	90
0	229	20	20	0		SUBMIT BOILER	1JUL98	28JUL98	10NOV98	7DEC98	94
229	230	10	10	0		APPROVE BOILER	29JUL98	11AUG98	8DEC98	21DEC98	94
230	306	60	60	0		FAB&DEL BOILER	12AUG98	3NOV98	22DEC98	15MAR99	94
74	76	0	0	0		RESTRAINT	22DEC98	21DEC98	5MAY99	4MAY99	96
61	77	15	15	0		DUCTWORK OFFICE	18NOV98	8DEC98	6APR99	26APR99	99
63	80	5	5	0		PAINT OFFICE EXTERIOR	25NOV98	1DEC98	11MAY99	17MAY99	119
60	98	0	0	0		RESTRAINT	4NOV98	3NOV98	28APR99	27APR99	125
0	231	20	20	0		SUBMIT OIL TANK	1JUL98	28JUL98	31DEC98	27JAN99	131
231	232	10	10	0		APPROVE OIL TANK	29JUL98	11AUG98	28JAN99	10FEB99	131
232	305	50	50	0		FAB&DEL OIL TANK	12AUG98	20OCT98	11FEB99	21APR99	131

Figure 6-13b (*Continued*)

O'BRIEN KREITZBERG INC.

FINEST HOUR

JOHN DOE 3

REPORT DATE 8APR97 RUN NO. 12
18:53

JOHN DOE

START DATE 1JUL98   FIN DATE 19MAY99

SCHEDULE REPORT BY I J

DATA DATE 25SEP98   PAGE NO.   1

PRED	SUCC	ORIG DUR	REM DUR	%	CODE	ACTIVITY DESCRIPTION	EARLY START	EARLY FINISH	LATE START	LATE FINISH	TOTAL FLOAT
0	1	3	0	100	1	CLEAR SITE					
0	210	10	0	100	31	SUBMIT FOUNDATION REBAR					
0	212	20	0	100	31	SUBMIT STRUCTURAL STEEL					
0	214	20	0	100	31	SUBMIT CRANE					
0	218	20	0	100	31	SUBMIT SIDING					
0	220	20	0	100	31	SUBMIT PLANT ELECTRICAL LOAD CENTER					
0	222	20	0	100	31	SUBMIT POWER PANELS-PLANT					
0	224	20	0	100	31	SUBMIT EXTERIOR DOORS					
0	225	30	0	100	31	SUBMIT PLANT ELECTRICAL FIXTURES					
0	227	20	0	100	31	SUBMIT PLANT HEATING AND VENTILATING FANS					
0	229	20	20	0	31	SUBMIT BOILER	25SEP98	22OCT98	12NOV98	9DEC98	34
0	231	20	20	0	31	SUBMIT OIL TANK	25SEP98	22OCT98	4JAN99	29JAN99	71
0	237	40	40	0	31	SUBMIT PRECAST	25SEP98	19NOV98	2OCT98	26NOV98	5
0	603	0	0	0		PCO #4	25SEP98	24SEP98	28OCT98	27OCT98	23
0	604	0	0	0		PCO #5	25SEP98	24SEP98	30OCT98	29OCT98	25
1	2	2	0	100	2	SURVEY AND LAYOUT					
2	3	2	0	100	1	ROUGH GRADE					
3	4	15	0	100	9	DRILL WELL					
3	6	4	0	100	3	WATER TANK FOUNDATIONS					
3	9	10	0	100	1	EXCAVATE FOR SEWER					
3	10	1	0	100	1	EXCAVATE ELECTRICAL MANHOLES					
3	12	6	0	100	4	OVERHEAD POLE LINE					
4	5	2	0	100	9	INSTALL WELL PUMP					

172

i	j						description					
5	8	8	0	100	1	5	UNDERGROUND WATER PIPING					
6	7	10	0	100	1	10	ERECT WATER TOWER					
7	8	10	0	100	1	10	TANK PIPING AND VALVES					
8	13	2	0	100	1	10	CONNECT WATER PIPING					
9	11	5	0	100	1	5	INSTALL SEWER AND BACKFILL					
10	11	5	0	100	1	4	INSTALL ELECTRICAL MANHOLES					
11	12	3	0	100	1	4	ELECTRICAL DUCT BANK					
12	13	5	0	100	1	4	PULL IN POWER FEEDER					
13	14	1	0	100	2	2	BUILDING LAYOUT					
14	23	3	3	0	2	1	EXCAVATE FOR OFFICE BUILDING	25SEP98	29SEP98	22DEC98	24DEC98	62
14	600	10	0	100	2	11	DRIVE AND POUR PILES					
15	601	5	0	100	2	1	EXCAVATE FOR PLANT WAREHOUSE					
16	17	5	5	0	2	3	POUR PILE CAPS PLANT-WAREHSE	25SEP98	1OCT98	25SEP98	1OCT98	0
17	18	10	10	0	2	3	FORM AND POUR GRADE BEAMS P-W	2OCT98	15OCT98	2OCT98	15OCT98	0
18	19	3	3	0	2	1	BACKFILL AND COMPACT P-W	16OCT98	20OCT98	16OCT98	20OCT98	0
18	21	5	5	0	2	3	FORM AND POUR RR LOAD DOCK P-W	16OCT98	22OCT98	28OCT98	3NOV98	8
18	22	5	5	0	2	3	FORM AND POUR TK LOAD DOCK P-W	16OCT98	22OCT98	28OCT98	3NOV98	8
19	20	5	5	0	2	5	UNDERSLAB PLUMBING P-W	21OCT98	27OCT98	21OCT98	27OCT98	0
20	22	5	5	0	2	4	UNDERSLAB CONDUIT P-W	28OCT98	3NOV98	27OCT98	3NOV98	0
21	22	0	0	0	2	3	RESTRAINT	23OCT98	3NOV98	28OCT98	3NOV98	8
22	29	10	10	0	2	3	FORM AND POUR SLABS P-W	4NOV98	17NOV98	4NOV98	17NOV98	0
23	24	4	4	0	2	6	SPREAD FOOTINGS OFFICE	30SEP98	5OCT98	25DEC98	30DEC98	62
24	25	6	6	0	2	1	FORM AND POUR GRADE BEAMS OFF	6OCT98	13OCT98	31DEC98	7JAN99	62
25	26	1	1	0	2	5	BACKFILL AND COMPACT OFFICE	14OCT98	14OCT98	8JAN99	8JAN99	62
26	27	3	3	0	2	4	UNDERSLAB PLUMBING OFFICE	15OCT98	19OCT98	11JAN99	13JAN99	62
27	28	3	3	0	2	3	UNDERSLAB CONDUIT OFFICE	20OCT98	22OCT98	14JAN99	18JAN99	62
28	99	3	3	0	2	6	FORM AND POUR OFFICE SLAB	23OCT98	27OCT98	19JAN99	21JAN99	62
99	30	10	10	0	3	6	ERECT STRUCT STEEL P-W	18NOV98	1DEC98	18NOV98	1DEC98	0
30	31	5	5	0	3	6	PLUMB STEEL AND BOLT P-W	2DEC98	8DEC98	2DEC98	8DEC98	0
31	32	5	5	0	3	6	ERECT CRANE WAY AND CRANE P-W	9DEC98	15DEC98	9DEC98	15DEC98	0
31	33	3	3	0	3	6	ERECT MONORAIL TRACK P-W	9DEC98	11DEC98	11DEC98	15DEC98	2
32	33	0	0	0	3	0	RESTRAINT	16DEC98	15DEC98	16DEC98	15DEC98	0
33	34	10	10	0	3	3	N603 ERECT BAR JOISTS P-W	16DEC98	29DEC98	16DEC98	29DEC98	0

Figure 6-14 Reset baseline—i-j sort.

O'BRIEN KREITZBERG INC.

FINEST HOUR

JOHN DOE 3

REPORT DATE 8APR97 RUN NO. 12
18:53

SCHEDULE REPORT BY I J

JOHN DOE

START DATE 1JUL98 FIN DATE 19MAY99

DATA DATE 25SEP98 PAGE NO. 2

PRED	SUCC	ORIG DUR	REM DUR	/%	CODE	ACTIVITY DESCRIPTION	EARLY START	EARLY FINISH	LATE START	LATE FINISH	TOTAL FLOAT
34	35	3	3	0	3	7	ERECT ROOF PLANKS P-W	30DEC98	1JAN99	30DEC98	1JAN99	0
35	36	10	10	0	3	12	ERECT SIDING P-W	4JAN99	15JAN99	4JAN99	15JAN99	0
35	37	5	5	0	3	13	ROOFING P-W	4JAN99	8JAN99	11JAN99	15JAN99	5
36	37	0	0	0			RESTRAINT	18JAN99	15JAN99	18JAN99	15JAN99	0
37	80	10	10	0	5	15	PERIMETER FENCE	18JAN99	29JAN99	6MAY99	19MAY99	78
37	90	5	5	0	5	16	PAVE PARKING AREA	18JAN99	22JAN99	29APR99	5MAY99	73
37	91	5	5	0	5	17	GRADE AND BALLAST RR SIDING	18JAN99	22JAN99	15APR99	21APR99	63
37	92	10	10	0	5	16	ACCESS ROAD	18JAN99	29JAN99	22APR99	5MAY99	68
37	93	20	20	0	5	4	AREA LIGHTING	18JAN99	12FEB99	22APR99	19MAY99	68
37	300	0	0	0			RESTRAINT	18JAN99	15JAN99	18JAN99	15JAN99	0
37	301	0	0	0			RESTRAINT	18JAN99	15JAN99	3FEB99	2FEB99	12
37	302	0	0	0			RESTRAINT	18JAN99	15JAN99	18FEB99	17FEB99	23
37	303	0	0	0			RESTRAINT	18JAN99	15JAN99	4FEB99	3FEB99	13
37	304	0	0	0			RESTRAINT	18JAN99	15JAN99	18FEB99	17FEB99	23
37	305	0	0	0			RESTRAINT	18JAN99	15JAN99	26APR99	23APR99	70
37	306	0	0	0			RESTRAINT	18JAN99	15JAN99	18MAR99	17MAR99	43
37	307	0	0	0			RESTRAINT	18JAN99	15JAN99	4MAR99	3MAR99	33
37	308	0	0	0			RESTRAINT	18JAN99	15JAN99	29APR99	28APR99	73
38	43	20	20	0	3	4	INSTALL POWER CONDUIT P-W	20JAN99	16FEB99	20JAN99	16FEB99	0
39	42	5	5	0	3	18	FRAME CEILINGS P-W	1FEB99	5FEB99	18FEB99	24FEB99	13
40	47	10	10	0	3	8	TEST PIPING SYSTEMS P-W	1MAR99	12MAR99	15APR99	28APR99	33
41	47	5	5	0	3	8	PREOPERATIONAL BOILER CHECK	5MAR99	11MAR99	22APR99	28APR99	34
42	44	10	10	0	3	19	DRYWELL1 PARTITIONS P-W	8FEB99	19FEB99	25FEB99	10MAR99	13

I	J						Activity Description	Early Start	Early Finish	Late Start	Late Finish	Total Float
43	49	15	15	0	3	4	INSTALL BRANCH CONDUIT P-W	17FEB99	9MAR99	17FEB99	9MAR99	0
44	45	0	0	0	0		RESTRAINT	22FEB99	19FEB99	14APR99	13APR99	37
44	46	0	0	0	0		RESTRAINT	22FEB99	19FEB99	11FEB99	10MAR99	13
44	48	10	10	0	3	20	CERAMIC TILE	22FEB99	5MAR99	18MAR99	31MAR99	18
44	58	10	10	0	3	21	HANG INTERIOR DOORS P-W	22FEB99	5MAR99	22APR99	5MAY99	43
45	51	5	5	0	3	4	ROOM OUTLETS P-W	10MAR99	16MAR99	14APR99	20APR99	25
46	52	25	25	0	3	8	INSTALL DUCTWORK P-W	22FEB99	26MAR99	11APR99	14APR99	13
47	58	5	5	0	3	8	LIGHTOFF BOILER AND TEST	15MAR99	19MAR99	29APR99	29APR99	33
48	53	5	5	0	3	22	PAINT ROOMS P-W	8MAR99	12MAR99	1APR99	7APR99	18
49	45	0	0	0	0		RESTRAINT	10MAR99	9MAR99	14APR99	13APR99	25
49	50	15	15	0	3	4	PULL WIRE P-W	10MAR99	30MAR99	10MAR99	30MAR99	0
50	54	5	5	0	3	4	INSTALL PANEL INTERNALS P-W	31MAR99	6APR99	31MAR99	6APR99	0
51	56	10	10	0	3	4	INSTALL ELECTRICAL FIXTURES	17MAR99	30MAR99	21APR99	4MAY99	25
52	58	15	15	0	3	8	INSULATE H&V SYSTEM P-W	29MAR99	16APR99	15APR99	5MAY99	13
53	57	10	10	0	3	22	FLOOR TILE P-W	15MAR99	26MAR99	8APR99	21APR99	18
53	58	10	10	0	3	5	INSTALL PLUMBING FIXTURES P-W	15MAR99	26MAR99	22APR99	5MAY99	28
54	55	10	10	0	3	4	TERMINATE WIRES P-W	7APR99	20APR99	7APR99	20APR99	0
55	56	10	10	0	3	4	RINGOUT P-W	21APR99	4MAY99	21APR99	4MAY99	0
56	58	1	1	0	0	4	ENERGIZE POWER	5MAY99	5MAY99	5MAY99	5MAY99	0
57	58	10	10	0	3	24	INSTALL FURNISHING P-W	29MAR99	9APR99	22APR99	5MAY99	18
58	80	5	5	0	5	27	ERECT FLAGPOLE	6MAY99	12MAY99	13MAY99	19MAY99	5
58	94	5	5	0	5	27	FINE GRADE	6MAY99	12MAY99	6MAY99	12MAY99	0
59	60	5	5	0	4	7	ERECT PRECAST ROOF OFFICE	22JAN99	28JAN99	29JAN99	4FEB99	5
60	61	10	10	0	4	14	EXTERIOR MASONRY OFFICE	29JAN99	29JAN99	5FEB99	18FEB99	5
60	98	0	0	0	0		RESTRAINT	29JAN99	28JAN99	30APR99	29APR99	65
61	62	5	5	0	4	21	EXTERIOR DOORS OFFICE	12FEB99	18FEB99	11MAR99	17MAR99	19
61	63	5	5	0	4	13	ROOFING OFFICE	12FEB99	18FEB99	11MAR99	17MAR99	19
61	64	10	10	0	4	8	INSTALL PIPING OFFICE	12FEB99	25FEB99	19FEB99	4MAR99	5
61	65	4	4	0	4	4	INSTALL ELEC BACKING BOXES	12FEB99	17FEB99	19FEB99	24FEB99	5
61	68	5	5	0	4	25	GLAZE OFFICE	12FEB99	18FEB99	11MAR99	17MAR99	19
61	77	15	15	0	4	8	DUCTWORK OFFICE	12FEB99	4MAR99	8APR99	28APR99	39
62	63	0	0	0	0		RESTRAINT	19FEB99	18FEB99	18MAR99	17MAR99	19
63	68	0	0	0	0		RESTRAINT	19FEB99	18FEB99	18MAR99	17MAR99	19

Figure 6-14 (Continued)

O'BRIEN KREITZBERG INC.　　　　　　　FINEST HOUR

JOHN DOE 3

REPORT DATE 8APR97 RUN NO. 12　　JOHN DOE　　　　　START DATE 1JUL98 FIN DATE 19MAY99

18:53

SCHEDULE REPORT BY I J　　　　　　　　　　　　　　DATA DATE 25SEP98 PAGE NO. 3

PRED	SUCC	ORIG DUR	REM DUR	%	CAL	CODE	ACTIVITY DESCRIPTION	EARLY START	EARLY FINISH	LATE START	LATE FINISH	TOTAL FLOAT
63	80	5	5	0	4	22	PAINT OFFICE EXTERIOR	19FEB99	25FEB99	13MAY99	19MAY99	59
64	67	4	4	0	4	8	TEST PIPING OFFICE	26FEB99	3MAR99	5MAR99	10MAR99	5
65	66	10	10	0	4	4	INSTALL CONDUIT OFFICE	18FEB99	3MAR99	25FEB99	10MAR99	5
66	67	0	0	0	4		RESTRAINT	4MAR99	3MAR99	11MAR99	10MAR99	5
66	74	10	10	0	4	4	PULL WIRE OFFICE	4MAR99	17MAR99	8APR99	21APR99	25
67	68	5	5	0	4	19	PARTITIONS OFFICE	4MAR99	10MAR99	11MAR99	17MAR99	5
68	69	5	5	0	4	19	DRYWALL	11MAR99	17MAR99	18MAR99	24MAR99	5
69	70	10	10	0	4	19	DRYWALL	18MAR99	31MAR99	25MAR99	7APR99	5
69	73	10	10	0	4	20	CERAMIC TILE OFFICE	18MAR99	31MAR99	29APR99	12MAY99	30
70	71	10	10	0	4	26	WOOD TRIM OFFICE	1APR99	14APR99	8APR99	21APR99	5
70	77	0	0	0	4		RESTRAINT	1APR99	31MAR99	29APR99	28APR99	20
71	72	10	10	0	4	22	PAINT INTERIOR OFFICE	15APR99	28APR99	22APR99	5MAY99	5
71	80	5	5	0	4	21	HANG DOORS OFFICE	15APR99	21APR99	13MAY99	19MAY99	20
72	73	0	0	0	4		RESTRAINT	29APR99	28APR99	13MAY99	12MAY99	10
72	78	0	0	0	4		RESTRAINT	29APR99	28APR99	6MAY99	5MAY99	5
72	80	10	10	0	4	20	FLOOR TILE OFFICE	29APR99	12MAY99	6MAY99	19MAY99	5
73	80	5	5	0	4	5	TOILET FIXTURES OFFICE	29APR99	5MAY99	13MAY99	19MAY99	10
74	75	5	5	0	5	4	INSTALL PANEL INTERNALS OFFICE	18MAR99	24MAR99	22APR99	28APR99	25
74	76	0	0	0	5		RESTRAINT	18MAR99	17MAR99	7MAY99	6MAY99	36
75	79	10	10	0	4	4	TERMINATE WIRES OFFICE	25MAR99	7APR99	29APR99	12MAY99	25
76	79	4	4	0	4	4	AIR CONDITIONING ELEC CONNECT	8APR99	13APR99	7MAY99	12MAY99	21
77	78	5	5	0	4	18	INSTALL CEILING GRID OFFICE	1APR99	7APR99	29APR99	5MAY99	20
78	80	10	10	0	4	18	ACOUSTIC TILE OFFICE	29APR99	12MAY99	6MAY99	19MAY99	5

176

79	80	5	5	0	4	4	RINGOUT ELECT	14APR99	20APR99	13MAY99	19MAY99	21
80	END	0	0	0			Project Finish	20MAY99	19MAY99	20MAY99	19MAY99	0
90	58	0	0	0			RESTRAINT	25JAN99	22JAN99	6MAY99	5MAY99	73
91	58	10	10	0	5	17	INSTALL RR SIDING	25JAN99	5FEB99	22APR99	5MAY99	63
92	58	0	0	0			RESTRAINT	1FEB99	29JAN99	6MAY99	5MAY99	68
93	80	0	0	0			RESTRAINT	15FEB99	12FEB99	20MAY99	19MAY99	68
94	80	5	5	0	5	27	SEED AND PLANT	13MAY99	19MAY99	13MAY99	19MAY99	0
98	76	5	5	0	4	8	INSTALL PACKAGE AIR CONDITONER	1APR99	7APR99	30APR99	6MAY99	21
99	59	5	5	0	4	7	ERECT PRECAST STRUCT. OFFICE	15JAN99	21JAN99	22JAN99	28JAN99	5
210	211	10	0	100	6	32	APPROVE FONDATION REBAR					
211	16	10	0	100	6	33	FAB&DEL FOUNDATION REBAR					
212	213	10	0	100	6	32	APPROVE STRUCTURAL STEEL					
213	29	40	20	50	6	33	FAB&DEL STRUCTURAL STEEL	25SEP98	22OCT98	21OCT98	17NOV98	18
214	215	10	0	100	6	32	APPROVE CRANE					
215	31	50	25	50	6	33	FAB&DEL CRANE	25SEP98	29OCT98	4NOV98	8DEC98	28
216	217	10	5	50	6	32	APPROVE BAR JOISTS	14OCT98	20OCT98	11NOV98	17NOV98	20
217	33	30	20	33	6	33	FAB&DEL BAR JOISTS	21OCT98	17NOV98	18NOV98	15DEC98	20
218	219	10	0	100	6	32	APPROVE SIDING					
219	35	40	40	0	6	33	FAB&DEL SIDING	25SEP98	19NOV98	9NOV98	1JAN99	31
220	221	10	0	100	6	32	APPROVE PLANT ELECTRICAL LOAD CENTER					
221	300	90	60	33	6	N3	FAB&DEL PLANT ELECTRICAL LOAD CENTER	25SEP98	17DEC98	26OCT98	15JAN99	21
222	223	10	0	100	6	32	APPROVE POWER PANELS-PLANT					
223	301	75	60	20	6	33	FAB&DEL POWER PANELS-PLANT	25SEP98	17DEC98	11NOV98	2FEB99	33
224	225	10	0	100	6	32	APPROVE EXTERIOR DOORS					
225	226	15	15	0	6	32	APPROVE PLANT ELECTRICAL FIXTURES	25SEP98	15OCT98	16DEC98	5JAN99	58
225	302	80	60	25	6	33	FAB&DEL EXTERIOR DOORS	25SEP98	17DEC98	26NOV98	17FEB99	44
226	51	75	75	0	6	33	FAB&DEL PLANT ELECTRICAL FIXTURES	16OCT98	28JAN99	6JAN99	20APR99	58
227	228	10	0	100	6	32	APPROVE PLANT HEATING AND VENTILATING FANS					
228	304	75	60	20	6	33	FAB&DEL PLANT HEATING AND VENTILATING FANS	25SEP98	17DEC98	26NOV98	17FEB99	44
229	230	10	10	0	6	32	APPROVE BOILER	23OCT98	5NOV98	10DEC98	23DEC98	34
230	306	60	60	0	6	33	FAB&DEL BOILER	6NOV98	28JAN99	24DEC98	17MAR99	34
231	232	10	10	0	6	32	APPROVE OIL TANK	23OCT98	5NOV98	1FEB99	12FEB99	71
232	305	50	50	0	6	33	FAB&DEL OIL TANK	6NOV98	14JAN99	15FEB99	23APR99	71

Figure 6-14 *(Continued)*

REPORT DATE 8APR97 RUN NO. 12   JOHN DOE   START DATE 1JUL98 FIN DATE 19MAY99

18:53

SCHEDULE REPORT BY I J   DATA DATE 25SEP98 PAGE NO. 4

PRED	SUCC	ORIG DUR	REM DUR	%	CODE	ACTIVITY DESCRIPTION	EARLY START	EARLY FINISH	LATE START	LATE FINISH	TOTAL FLOAT
235	236	10	10	0	6 32	APPROVE PACKAGED A/C	12NOV98	25NOV98	11DEC98	24DEC98	21
236	98	90	90	0	6 33	FAB&DEL PACKAGED A/C	26NOV98	31MAR99	25DEC98	29APR99	21
237	238	10	10	0	6 32	APPROVE PRECAST	20NOV98	3DEC98	27NOV98	10DEC98	5
238	99	30	30	0	6 33	FAB&DEL PRECAST	4DEC98	14JAN99	11DEC98	21JAN99	5
300	38	2	2	0	3 4	SET ELECTRICAL LOAD CENTER P-W	18JAN99	19JAN99	18JAN99	19JAN99	0
301	43	10	10	0	3 4	POWER PANEL BACKFILL BOXES P-W	18JAN99	29JAN99	3FEB99	16FEB99	12
302	42	5	5	0	3 6	ERECT EXTERIOR DOORS P-W	18JAN99	22JAN99	18FEB99	24FEB99	23
303	39	10	10	0	3 14	MASONRY PARTITIONS P-W	18JAN99	29JAN99	4FEB99	17FEB99	13
304	46	15	15	0	3 8	INSTALL H AND V UNITS P-W	18JAN99	5FEB99	18FEB99	10MAR99	23
305	47	3	3	0	3 8	INSTALL FUEL TANK P-W	18JAN99	20JAN99	26APR99	28APR99	70
306	41	25	25	0	3 8	ERECT BOILER AND AUXILIARY P-W	29JAN99	4MAR99	18MAR99	21APR99	34
307	602	0	0	0	3	PCO #3	18JAN99	15JAN99	4MAR99	3MAR99	33
308	58	5	5	0	3 6	INSTALL MONORAIL WAREHOUSE	18JAN99	22JAN99	29APR99	5MAY99	73
600	15	0	0	0		PCO #1	25SEP98*	24SEP98	25SEP98	24SEP98	0
601	16	0	0	0		PCO #2	25SEP98*	24SEP98	25SEP98	24SEP98	0
602	40	30	30	0	3 8	FABRICATE PIPING P-W	18JAN99*	26FEB99	4MAR99	14APR99	33
603	216	20	10	50	6 31	SUBMIT BAR JOISTS	30SEP98*	13OCT98	28OCT98	10NOV98	20
604	235	30	30	0	6 31	SUBMIT PACKAGED A/C	1OCT98*	11NOV98	30OCT98	10DEC98	21
START	0	0	0	0		Project Start	25SEP98	24SEP98	25SEP98	24SEP98	0

Figure 6-14 (Continued)

O'BRIEN KREITZBERG INC.

FINEST HOUR

JOHN DOE 3

REPORT DATE 8APR97 RUN NO. 11   JOHN DOE

18:51

START DATE 1JUL98 FIN DATE 19MAY99

SCHEDULE REPORT - SORT BY TF, ES

DATA DATE 25SEP98 PAGE NO. 1

PRED	SUCC	ORIG DUR	REM DUR	%	CODE	ACTIVITY DESCRIPTION	EARLY START	EARLY FINISH	LATE START	LATE FINISH	TOTAL FLOAT
0	1	3	0	100		CLEAR SITE					
0	210	10	0	100		SUBMIT FOUNDATION REBAR					
0	212	20	0	100		SUBMIT STRUCTURAL STEEL					
0	214	20	0	100		SUBMIT CRANE					
0	218	20	0	100		SUBMIT SIDING					
0	220	20	0	100		SUBMIT PLANT ELECTRICAL LOAD CENTER					
0	222	20	0	100		SUBMIT POWER PANELS-PLANT					
0	224	20	0	100		SUBMIT EXTERIOR DOORS					
0	225	30	0	100		SUBMIT PLANT EXECTRICAL FIXTURES					
0	227	20	0	100		SUBMIT PLANT HEATING AND VENTILATING FANS					
1	2	2	0	100		SURVEY AND LAYOUT					
2	3	2	0	100		ROUGH GRADE					
3	4	15	0	100		DRILL WELL					
3	6	4	0	100		WATER TANK FOUNDATIONS					
3	9	10	0	100		EXCAVATE FOR SEWER					
3	10	1	0	100		EXCAVATE ELECTRICAL MANHOLES					
3	12	6	0	100		OVERHEAD POLE LINE					
4	5	2	0	100		INSTALL WELL PUMP					
5	8	8	0	100		UNDERGROUND WATER PIPING					
6	7	10	0	100		ERECT WATER TOWER					
7	8	10	0	100		TANK PIPING AND VALVES					
8	13	2	0	100		CONNECT WATER PIPING					
9	11	5	0	100		INSTALL SEWER AND BACKFILL					

I	J	OD	RD	%	Description	Date 1	Date 2	TF
10	11	5	0	100	INSTALL ELECTRICAL MANHOLES			
11	12	3	0	100	ELECTRICAL DUCT BANK			
12	13	5	0	100	PULL IN POWER FEEDER			
13	14	1	0	100	BUILDING LAYOUT			
14	600	10	0	100	DRIVE AND POUR PILES			
15	601	5	0	100	EXCAVATE FOR PLANT WAREHOUSE			
210	211	10	0	100	APPROVE FONDATION REBAR			
211	16	10	0	100	FAB&DEL FOUNDATION REBAR			
212	213	10	0	100	APPROVE STRUCTURAL STEEL			
214	215	10	0	100	APPROVE CRANE			
218	219	10	0	100	APPROVE SIDING			
220	221	10	0	100	APPROVE PLANT ELECTRICAL LOAD CENTER			
222	223	10	0	100	APPROVE POWER PANELS-PLANT			
224	225	10	0	100	APPROVE EXTERIOR DOORS			
227	228	10	0	100	APPROVE PLANT HEATING AND VENTILATING FANS			
16	17	5	5	0	POUR PILE CAPS PLANT-WAREHSE	25SEP98	1OCT98	0
600	15	0	0	0	PCO #1	25SEP98*	24SEP98	0
601	16	0	0	0	PCO #2	25SEP98*	24SEP98	0
START	0	0	0	0	Project Start	25SEP98	24SEP98	0
17	18	10	10	0	FORM AND POUR GRADE BEAMS P-W	2OCT98	15OCT98	0
18	19	3	3	0	BACKFILL AND COMPACT P-W	16OCT98	20OCT98	0
19	20	5	5	0	UNDERSLAB PLUMBING P-W	21OCT98	27OCT98	0
20	22	5	5	0	UNDERSLAB CONDUIT P-W	28OCT98	3NOV98	0
22	29	10	10	0	FORM AND POUR SLABS P-W	4NOV98	17NOV98	0
29	30	10	10	0	ERECT STRUCT STEEL P-W	18NOV98	1DEC98	0
30	31	5	5	0	PLUMB STEEL AND BOLT P-W	2DEC98	8DEC98	0
31	32	5	5	0	ERECT CRANE WAY AND CRANE P-W	9DEC98	15DEC98	0
32	33	0	0	0	RESTRAINT	16DEC98	15DEC98	0
33	34	10	10	0	ERECT BAR JOISTS P-W	16DEC98	29DEC98	0
34	35	3	3	0	ERECT ROOF PLANKS P-W	30DEC98	1JAN99	0
35	36	10	10	0	ERECT SIDING P-W	4JAN99	15JAN99	0
36	37	0	0	0	RESTRAINT	18JAN99	15JAN99	0
37	300	0	0	0	RESTRAINT	18JAN99	15JAN99	0

Figure 6-15 Reset baseline—TF sort.

O'BRIEN KREITZBERG INC.   FINEST HOUR   JOHN DOE 3

REPORT DATE 8APR97 RUN NO. 11   JOHN DOE   START DATE 1JUL98 FIN DATE 19MAY99
            18:51
SCHEDULE REPORT - SORT BY TF, ES   DATA DATE 25SEP98 PAGE NO. 2

PRED	SUCC	ORIG DUR	REM DUR	%	CODE	ACTIVITY DESCRIPTION	EARLY START	EARLY FINISH	LATE START	LATE FINISH	TOTAL FLOAT
300	38	2	2	0		SET ELECTRICAL LOAD CENTER P-W	18JAN99	19JAN99	18JAN99	19JAN99	0
38	43	20	20	0		INSTALL POWER CONDUIT P-W	20JAN99	16FEB99	20JAN99	16FEB99	0
43	49	15	15	0		INSTALL BRANCH CONDUIT P-W	17FEB99	9MAR99	17FEB99	9MAR99	0
49	50	15	15	0		PULL WIRE P-W	10MAR99	30MAR99	10MAR99	30MAR99	0
50	54	5	5	0		INSTALL PANEL INTERNALS P-W	31MAR99	6APR99	31MAR99	6APR99	0
54	55	10	10	0		TERMINATE WIRES P-W	7APR99	20APR99	7APR99	20APR99	0
55	56	10	10	0		RINGOUT P-W	21APR99	4MAY99	21APR99	4MAY99	0
56	58	1	1	0		ENERGIZE POWER	5MAY99	5MAY99	5MAY99	5MAY99	0
58	94	5	5	0		FINE GRADE	6MAY99	12MAY99	6MAY99	12MAY99	0
94	80	5	5	0		SEED AND PLANT	13MAY99	19MAY99	13MAY99	19MAY99	0
80	END	0	0	0		Project Finish	20MAY99	19MAY99	20MAY99	19MAY99	0
31	33	3	3	0		ERECT MONORAIL TRACK P-W	9DEC98	11DEC98	11DEC98	15DEC98	2
0	237	40	40	0		SUBMIT PRECAST	25SEP98	19NOV98	20OCT98	26NOV98	5
237	238	10	10	0		APPROVE PRECAST	20NOV98	3DEC98	27NOV98	10DEC98	5
238	99	30	30	0		FAB&DEL PRECAST	4DEC98	14JAN99	11DEC98	21JAN99	5
35	37	5	5	0		ROOFING P-W	4JAN99	8JAN99	11JAN99	15JAN99	5
99	59	5	5	0		ERECT PRECAST STRUCT. OFFICE	15JAN99	21JAN99	22JAN99	28JAN99	5
59	60	5	5	0		ERECT PRECAST ROOF OFFICE	22JAN99	28JAN99	29JAN99	4FEB99	5
60	61	10	10	0		EXTERIOR MASONRY OFFICE	29JAN99	11FEB99	5FEB99	18FEB99	5
61	64	10	10	0		INSTALL PIPING OFFICE	12FEB99	25FEB99	19FEB99	4MAR99	5
61	65	4	4	0		INSTALL ELEC BACKING BOXES	12FEB99	17FEB99	19FEB99	24FEB99	5
65	66	10	10	0		INSTALL CONDUIT OFFICE	18FEB99	3MAR99	25FEB99	10MAR99	5
64	67	4	4	0		TEST PIPING OFFICE	26FEB99	3MAR99	5MAR99	10MAR99	5

66	67	0	0	0	RESTRAINT	4MAR99	3MAR99	11MAR99	10MAR99	5
67	68	5	5	0	PARTITIONS OFFICE	4MAR99	10MAR99	11MAR99	17MAR99	5
68	69	5	5	0	DRYWALL	11MAR99	17MAR99	18MAR99	24MAR99	5
69	70	10	10	0	DRYWALL	18MAR99	31MAR99	25MAR99	7APR99	5
70	71	10	10	0	WOOD TRIM OFFICE	1APR99	14APR99	8APR99	21APR99	5
71	72	10	10	0	PAINT INTERIOR OFFICE	15APR99	28APR99	22APR99	5MAY99	5
72	78	0	0	0	RESTRAINT	29APR99	28APR99	6MAY99	5MAY99	5
72	80	10	10	0	FLOOR TILE OFFICE	29APR99	12MAY99	6MAY99	19MAY99	5
78	80	10	10	0	ACOUSTIC TILE OFFICE	29APR99	12MAY99	6MAY99	19MAY99	5
58	80	5	5	0	ERECT FLAGPOLE	6MAY99	12MAY99	13MAY99	19MAY99	5
18	21	5	5	0	FORM AND POUR RR LOAD DOCK P-W	16OCT98	22OCT98	28OCT98	3NOV98	8
18	22	5	5	0	FORM AND POUR TK LOAD DOCK P-W	16OCT98	22OCT98	28OCT98	3NOV98	8
21	22	0	0	0	RESTRAINT	23OCT98	22OCT98	4NOV98	3NOV98	8
72	73	0	0	0	RESTRAINT	29APR99	28APR99	13MAY99	12MAY99	10
73	80	5	5	0	TOILET FIXTURES OFFICE	29APR99	5MAY99	13MAY99	19MAY99	10
37	301	0	0	0	RESTRAINT	18JAN99	15JAN99	3FEB99	2FEB99	12
301	43	10	10	0	POWER PANEL BACKFILL BOXES P-W	18JAN99	29JAN99	3FEB99	16FEB99	12
37	303	0	0	0	RESTRAINT	18JAN99	15JAN99	4FEB99	3FEB99	13
303	39	10	10	0	MASONRY PARTITIONS P-W	18JAN99	29JAN99	4FEB99	17FEB99	13
39	42	5	5	0	FRAME CEILINGS P-W	1FEB99	5FEB99	18FEB99	24FEB99	13
42	44	10	10	0	DRYWELLL PARTITIONS P-W	8FEB99	19FEB99	25FEB99	10MAR99	13
44	46	0	0	0	RESTRAINT	22FEB99	19FEB99	11MAR99	10MAR99	13
46	52	25	25	0	INSTALL DUCTWORK P-W	22FEB99	26MAR99	11MAR99	14APR99	13
52	58	15	15	0	INSULATE H&V SYSTEM P-W	29MAR99	16APR99	15APR99	5MAY99	13
213	29	40	20	50	FAB&DEL STRUCTURAL STEEL	25SEP98	22OCT98	21OCT98	17NOV98	18
44	48	10	10	0	CERAMIC TILE	22FEB99	5MAR99	18MAR99	31MAR99	18
48	53	5	5	0	PAINT ROOMS P-W	8MAR99	12MAR99	1APR99	7APR99	18
53	57	10	10	0	FLOOR TILE P-W	15MAR99	26MAR99	8APR99	21APR99	18
57	58	10	10	0	INSTALL FURNISHING P-W	29MAR99	9APR99	22APR99	5MAY99	18
61	62	5	5	0	EXTERIOR DOORS OFFICE	12FEB99	18FEB99	11MAR99	17MAR99	19
61	63	5	5	0	ROOFING OFFICE	12FEB99	18FEB99	11MAR99	17MAR99	19
61	68	5	5	0	GLAZE OFFICE	12FEB99	18FEB99	11MAR99	17MAR99	19
62	63	0	0	0	RESTRAINT	19FEB99	18FEB99	18MAR99	17MAR99	19

Figure 6-15 (*Continued*)

REPORT DATE 8APR97 RUN NO. 11   JOHN DOE   START DATE 1JUL98 FIN DATE 19MAY99
  18:51

SCHEDULE REPORT - SORT BY TF, ES   DATA DATE 25SEP98 PAGE NO. 3

PRED	SUCC	ORIG DUR	REM DUR	%	CODE	ACTIVITY DESCRIPTION	EARLY START	EARLY FINISH	LATE START	LATE FINISH	TOTAL FLOAT
63	68	0	0	0		RESTRAINT	19FEB99	18FEB99	18MAR99	17MAR99	19
603	216	20	10	50		SUBMIT BAR JOISTS	30SEP98*	13OCT98	28OCT98	10NOV98	20
216	217	10	5	50		APPROVE BAR JOISTS	14OCT98	20OCT98	11NOV98	17NOV98	20
217	33	30	20	33		FAB&DEL BAR JOISTS	21OCT98	17NOV98	18NOV98	15DEC98	20
70	77	0	0	0		RESTRAINT	1APR99	31MAR99	29APR99	28APR99	20
77	78	5	5	0		INSTALL CEILING GRID OFFICE	1APR99	7APR99	29APR99	5MAY99	20
71	80	5	5	0		HANG DOORS OFFICE	15APR99	21APR99	13MAY99	19MAY99	20
221	300	90	60	33		FAB&DEL PLANT ELECTRICAL LOAD CENTER	25SEP98	17DEC98	26OCT98	15JAN99	21
604	235	30	30	0		SUBMIT PACKAGED A/C	1OCT98*	11NOV98	30OCT98	10DEC98	21
235	236	10	10	0		APPROVE PACKAGED A/C	12NOV98	25NOV98	11DEC98	24DEC98	21
236	98	90	90	0		FAB&DEL PACKAGED A/C	26NOV98	31MAR99	25DEC98	29APR99	21
98	76	5	5	0		INSTALL PACKAGE AIR CONDITONER	1APR99	7APR99	30APR99	6MAY99	21
76	79	4	4	0		AIR CONDITIONING ELEC CONNECT	8APR99	13APR99	7MAY99	12MAY99	21
79	80	5	5	0		RINGOUT ELECT	14APR99	20APR99	13MAY99	19MAY99	21
0	603	0	0	0		PCO #4	25SEP98	24SEP98	28OCT98	27OCT98	23
37	302	0	0	0		RESTRAINT	18JAN99	15JAN99	18FEB99	17FEB99	23
37	304	0	0	0		RESTRAINT	18JAN99	15JAN99	18FEB99	17FEB99	23
302	42	5	5	0		ERECT EXTERIOR DOORS P-W	18JAN99	22JAN99	18FEB99	24FEB99	23
304	46	15	15	0		INSTALL H AND V UNITS P-W	18JAN99	5FEB99	18FEB99	10MAR99	23
0	604	0	0	0		PCO #5	25SEP98	24SEP98	30OCT98	29OCT98	25
66	74	10	10	0		PULL WIRE OFFICE	4MAR99	17MAR99	8APR99	21APR99	25
45	51	5	5	0		ROOM OUTLETS P-W	10MAR99	16MAR99	14APR99	20APR99	25
49	45	0	0	0		RESTRAINT	10MAR99	9MAR99	14APR99	13APR99	25

i	j	%	%	%	Activity Description					TF
51	56	10	10	0	INSTALL ELECTRICAL FIXTURES	17MAR99	30MAR99	21APR99	4MAY99	25
74	75	5	5	0	INSTALL PANEL INTERNALS OFFICE	18MAR99	24MAR99	22APR99	28APR99	25
75	79	10	10	0	TERMINATE WIRES OFFICE	25MAR99	7APR99	29APR99	12MAY99	25
215	31	50	25	50	FAB&DEL CRANE	25SEP98	29OCT98	4NOV98	8DEC98	28
53	58	10	10	0	INSTALL PLUMBING FIXTURES P-W	15MAR99	26MAR99	22APR99	5MAY99	28
69	73	10	10	0	CERAMIC TILE OFFICE	18MAR99	31MAR99	29APR99	12MAY99	30
219	35	40	40	0	FAB&DEL SIDING	25SEP98	19NOV98	9NOV98	1JAN99	31
223	301	75	60	20	FAB&DEL POWER PANELS-PLANT	25SEP98	17DEC98	11NOV98	2FEB99	33
37	307	0	0	0	RESTRAINT	18JAN99	15JAN99	4MAR99	3MAR99	33
307	602	0	0	0	PCO #3	18JAN99	15JAN99	4MAR99	3MAR99	33
602	40	30	30	0	FABRICATE PIPING P-W	18JAN99*	26FEB99	4MAR99	14APR99	33
40	47	10	10	0	TEST PIPING SYSTEMS P-W	1MAR99	12MAR99	15APR99	28APR99	33
47	58	5	5	0	LIGHTOFF BOILER AND TEST	15MAR99	19APR99	29APR99	5MAY99	33
0	229	20	20	0	SUBMIT BOILER	25SEP98	22OCT98	12NOV98	9DEC98	34
229	230	10	10	0	APPROVE BOILER	23OCT98	5NOV98	10DEC98	23DEC98	34
230	306	60	60	0	FAB&DEL BOILER	6NOV98	28JAN99	24DEC98	17MAR99	34
306	41	25	25	0	ERECT BOILER AND AUXILIARY P-W	29JAN99	4MAR99	18MAR99	21APR99	34
41	47	5	5	0	PREOPERATIONAL BOILER CHECK	5MAR99	11MAR99	22APR99	28APR99	34
74	76	0	0	0	RESTRAINT	18MAR99	17MAR99	7MAY99	6MAY99	36
44	45	0	0	0	RESTRAINT	22FEB99	19FEB99	14APR99	13APR99	37
61	77	15	15	0	DUCTWORK OFFICE	12FEB99	4MAR99	8APR99	28APR99	39
37	306	0	0	0	RESTRAINT	18JAN99	15JAN99	18MAR99	17MAR99	43
44	59	10	10	0	HANG INTERIOR DOORS P-W	22FEB99	5MAR99	22APR99	5MAY99	43
225	302	80	60	25	FAB&DEL EXTERIOR DOORS	25SEP98	17DEC98	26NOV98	17FEB99	44
228	304	75	60	20	FAB&DEL PLANT HEATING AND VENTILATING FANS	25SEP98	17DEC98	26NOV98	17FEB99	44
225	226	15	15	0	APPROVE PLANT ELECTRICAL FIXTURES	16OCT98	15OCT98	16DEC98	5JAN99	58
226	51	75	75	0	FAB&DEL PLANT ELECTRICAL FIXTURES	19FEB99	28JAN99	6JAN99	20APR99	58
63	80	5	5	0	PAINT OFFICE EXTERIOR	25SEP98	25FEB99	13MAY99	19MAY99	59
14	23	3	3	0	EXCAVATE FOR OFFICE BUILDING	30SEP98	29SEP98	22DEC98	24DEC98	62
23	24	4	4	0	SPREAD FOOTINGS OFFICE	6OCT98	5OCT98	25DEC98	30DEC98	62
24	25	6	6	0	FORM AND POUR GRADE BEAMS OFF	13OCT98	13OCT98	31DEC98	7JAN99	62
25	26	1	1	0	BACKFILL AND COMPACT OFFICE	14OCT98	14OCT98	8JAN99	8JAN99	62
26	27	3	3	0	UNDERSLAB PLUMBING OFFICE	15OCT98	19OCT98	11JAN99	13JAN99	62

Figure 6-15 (*Continued*)

REPORT DATE 8APR97 RUN NO. 11   JOHN DOE   START DATE 1JUL98 FIN DATE 19MAY99
18:51
SCHEDULE REPORT - SORT BY TF, ES   DATA DATE 25SEP98 PAGE NO. 4

PRED	SUCC	ORIG DUR	REM DUR	%	CODE	ACTIVITY DESCRIPTION	EARLY START	EARLY FINISH	LATE START	LATE FINISH	TOTAL FLOAT
27	28	3	3	0		UNDERSLAB CONDUIT OFFICE	2OCT98	22OCT98	14JAN99	18JAN99	62
28	99	3	3	0		FORM AND POUR OFFICE SLAB	23OCT98	27OCT98	19JAN99	21JAN99	62
37	91	5	5	0		GRADE AND BALLAST RR SIDING	18JAN99	22JAN99	15APR99	21APR99	63
91	58	10	10	0		INSTALL RR SIDING	25JAN99	5FEB99	22APR99	5MAY99	63
60	98	0	0	0		RESTRAINT	29JAN99	28JAN99	30APR99	29APR99	65
37	92	10	10	0		ACCESS ROAD	18JAN99	29JAN99	22APR99	5MAY99	68
37	93	20	20	0		AREA LIGHTING	18JAN99	12FEB99	22APR99	19MAY99	68
92	58	0	0	0		RESTRAINT	1FEB99	29JAN99	6MAY99	5MAY99	68
93	80	0	0	0		RESTRAINT	15FEB99	12FEB99	20MAY99	19MAY99	68
37	305	0	0	0		RESTRAINT	18JAN99	15JAN99	26APR99	23APR99	70
305	47	3	3	0		INSTALL FUEL TANK P-W	18JAN99	20JAN99	26APR99	28APR99	70
0	231	20	20	0		SUBMIT OIL TANK	25SEP98	22OCT98	4JAN99	29JAN99	71
231	232	10	10	0		APPROVE OIL TANK	23OCT98	5NOV98	1FEB99	12FEB99	71
232	305	50	50	0		FAB&DEL OIL TANK	6NOV98	14JAN99	15FEB99	23APR99	71
37	90	5	5	0		PAVE PARKING AREA	18JAN99	22JAN99	29APR99	5MAY99	73
37	308	0	0	0		RESTRAINT	18JAN99	15JAN99	29APR99	28APR99	73
308	58	5	5	0		INSTALL MONORAIL WAREHOUSE	18JAN99	22JAN99	29APR99	5MAY99	73
90	58	0	0	0		RESTRAINT	25JAN99	22JAN99	6MAY99	5MAY99	73
37	80	10	10	0		PERIMETER FENCE	18JAN99	29JAN99	6MAY99	19MAY99	78

Figure 6-15 *(Continued)*

7

The Cost of Time

Methods to determine the time impact of change orders were discussed in Chap. 6. Once determined, what do you do with these days? Through exchange of information (and perhaps an audit of same), a cost model of the base contract can be constructed. The purpose of the cost model is to make a reasonable estimate of the cost of spending an additional calendar day on the project.

General Conditions

General conditions are costs that are essentially fixed and cannot be reduced or deleted on a day-for-day basis. Here are some examples:

Trailer complex. Usually includes offices, parts warehousing, tool sheds and boxes, washroom and toilet facilities, heat and light utilities for the complex, security personnel, compound fencing, telephone system, base operating costs, and so on.

Management. Includes superintendent, secretaries, purchasing agent, master mechanic, and so on.

Light equipment. Cars and pickup trucks assigned to site management personnel not used directly in placing the construction.

Fixed plant. Batch plant, aggregate storage, and so on, specifically dedicated to this project.

Fixed equipment. Installed equipment such as tower cranes or shoring not related to specific work activities.

Other Costs

Time-related costs other than general conditions include the following:

Home office. Usually applied as a percentage (3 or 5 percent is typical) based upon an audit of last year's annual statement and/or audit.

Special facilities. If a project-specific source area (such as a borrow pit with equipment and trucks) or disposal landfill (approved leased land area plus equipment) is idled, this cost per day should be included as the cost of time.

Heavy equipment. If heavy equipment is on the project—even if it's not involved with the delayed critical activity—and its use is delayed, the cost per day (based upon either the weekly or monthly rates) can be added to the cost of time.

Reality Check

The "fixed" cost of a lost day will usually be 20 percent+ of the contract cost. In the John Doe Project, it can be expected that (in order of magnitude) the cost per day would be based upon the calendar days in the base contract (April 22, 1999 – July 1, 1998 = 296 c.d.):

$$\text{Estimated general conditions} = 20\% \ (\$5,000,000) = \$1,000,000$$

$$\text{Fixed cost/c.d.} = \$1,000,000 \div 296 = \$3378/\text{c.d.}$$

On that basis, the cost of the time applicable to the first two change orders (duct interference and sinkhole) would be as follows:

$$17 \text{ c.d.} \ \times \ \$3378 = \$57,426$$

The first change (duct interference) is paid for directly by the owner. The second change (PCO #2, sinkhole) has a direct cost of $8525. Note how much more significant the impact of time is versus the abstract cost of the work done. This is not unusual.

Adding Indirect Costs to the Change Order

There are some who advocate reaching into the indirect cost pool to try to capture costs that can be changed from indirect to direct. In this mode, costs such as postage, phone calls, secretarial costs, photographs, engineering, estimating, and mileage would be changed from indirect to direct. This approach is usually penny-wise and pound-foolish. The reason that overhead pools are an accepted accounting approach recognizes that there are support costs that cannot readily be assigned to specific items such as a change order. If such assignment were required, experience predicts that less than half those overhead costs could be recovered.

An attempt to recover some overhead items from change orders is a slippery slope. To the extent that a contractor purports to claim certain categories of overhead cost as direct, be aware that those categories should then be removed from the overhead pool. For instance, if this approach can increase a direct change order estimate by 5 percent, experience suggests that the overhead pool should be reduced by 10 percent or more. In the John Doe example for PCO #2 (sink-hole), this approach would increase the change order estimate #2 as follows:

$$\text{Assume } 20\% \text{ OH\&P} = 15\% \text{ OH} + 5\% \text{ profit}$$

$$\$1163 + 5\% (\$58) = \$1221$$

or an increase of \$58. However, the decrease in overhead pools would produce the following decrease in daily rate:

$$\text{Daily rate (decreased)} = \$3378 - 10\% (\$338) = \$3040$$

$$\text{Impact on PCO \#2} = \$3040/\text{c.d.} \times 17 \text{ c.d.} = \$51,680$$

or a decrease of \$5746.

Phased Cost of Time

Using an average daily cost of time is a convenient approach (see Fig. 7-1), but a phased approach may be more reasonable. In the phased approach, the costs are spread on different bases. For instance:

- Costs that average over time
- Costs that start with a lag and finish before the finish
- Special costs that vary over time

Table 7-1 and Fig. 7-2 illustrate this breakdown for the John Doe Project. Under the phased scenario shown in Fig. 7-2, the 17 added days before October

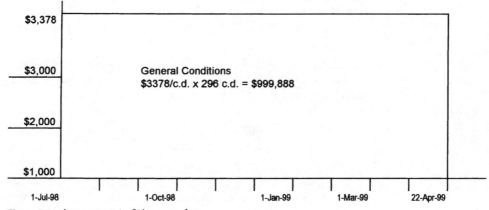

Figure 7-1 Average cost of time per day.

TABLE 7-1   Calculation of Overhead per Calendar Day and Time Period

Time period	Calendar days (c.d.)	General overhead per c.d.			Total per c.d.	Overhead period
		Type 1	Type 2	Type 3		
July 1–Sept. 1, 1998	62	$1,500	—	—	$1,500	$ 93,000
Sept.–Oct. 1, 1998	30	$1,500	$2,000	—	$3,500	$105,000
Oct. 1, 1998–Feb. 1, 1999	123	$1,500	$2,000	$1,576	$5,076	$624,348
Feb. 1–Mar. 1, 1999	28	$1,500	$2,000	—	$3,500	$ 98,000
Mar. 1–Apr. 22, 1999	53	$1,500	—	—	$1,500	$ 79,500
Total	296					$999,848

1, 1998, would be valued at $1500 × 17 days, or $25,500. This is about 44 percent of the straight average cost. It also compares with the bell-shaped average curve of probability (i.e., slow start building up to high average, followed by matching decay slope to finish). Even in this scenario the cost of time is more than twice the abstract cost of the change order.

Special Costs

The cost of time may also include special costs relating to the specific PCO. This could include factors such as the following:

Equipment costs.   If the delay related to the PCO requires certain on-site equipment to go into standby mode (idle), the cost of that equipment can be included as a part of the cost of time. The cost would probably be calculated at the long-term (monthly) rate apportioned to the length of delay impact calculated. The monthly cost of equipment should be from a currently published guide such as *The Rental Rate Blue Book for Construction Equipment* or the Associated General Contractors *Contractor's Equipment Manual.*

Reduced productivity.   If the PCO imposes special conditions that will result in reduced productivity, such as overmanning, shift work, stacking of trades, concurrent operations, joint occupancy, extended overtime, acceleration, and so on.

Escalation.   If the PCO pushes any work or material acquisition into a period of cost escalation, these costs can be included as part of the PCO cost.

Home office expense (HOE).   This expense is often incorporated as part of the contractor's profit line or a contingency line in the bid. The typical range is 3 to 5 percent of the contract value. HOE is usually accounted on a company basis and therefore not allocated to each project. If the contract

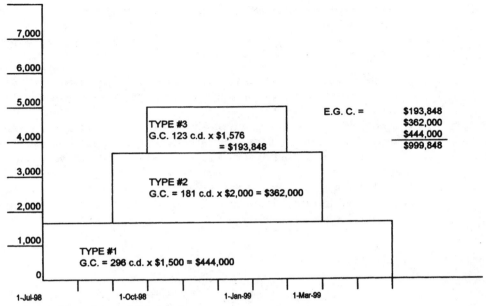

Figure 7-2 Phased cost of time per day.

limits all overhead (including HOE) and profit to a maximum of 20 percent, the total will probably exceed the maximum by more than the HOE amount.

One method of allocating HOE to a project was developed in the Eichleay case (Eichleay Corp. ASBCA No. 5183, 60-2 BCA). The Eichleay decision used revenues on the projects versus company revenues in the same period as a ratio for allocating HOE to the instant contract. The following formula was used:

(A)   Home office overhead (HOE) $= \dfrac{\dfrac{\text{contract billings}}{\text{(HOE contract period)}}}{\text{billings for contract period}}$
         (allocable to contract)

(B)   Daily contract HOE $= \dfrac{\text{allocable HOE (A)}}{\text{days of performance}^*}$

Delay HOE = daily HOE (B) × days of delay[*]

In the initial John Doe Project delay, assume the following:

John Doe billings (1998)	=	$ 3,050,000
XYZ total billings (1998)	=	25,546,000
XYZ home office overhead (1998)	=	1,021,840

---

[*]Days should be in consistent units, usually calendar days.

The HOE overhead for the John Doe Project in 1998:

$$\text{HOE/c.d.} = \frac{\$\ 3,050,000}{\$25,546,000}\ (\$1,021,840) = \$122,000$$

XYZ worked John Doe from July 1, 1998, through December 31, 1998, or 184 calendar days. The 17 c.d. delay through excavation had a HOE delay cost of $17 \times \$663 = \$11,271$.

The Eichleay case was actually a suspension case, and a number of state courts have indicated their preference for more specific methods of allocation, if available. If the delay period is one in which the contractor's overall revenues are decreasing, the Eichleay approach tends to allocate an unreasonably large portion of the expenses to the specific claim and, conversely, if the contractor is in an expanding business atmosphere, this approach tends to undervalue the cost of the problem job.

Forward Pricing

Developing change order costs before they occur has long been called *forward pricing*. More recently, the term has come to refer specifically to anticipating special costs, in particular, loss of productivity (disruption).

The Department of Metropolitan Services (Metro) was responsible for a $735 million program at Seattle's new wastewater treatment plant, West Point/Renton Waste Water Treatment Plants (WWTPs).* Artifacts located on the site had the potential to delay the liquid streams contract (held by M. A. Mortenson).

After six months of disputes about change orders and concern for changes to come, Metro and Mortenson agreed to forward pricing. Metro had to accelerate the program to complete it on time. Factor formulas were developed to agree how to settle future change orders (i.e., price impacts).

Engineering News-Record (ENR) reported that "Metro officials and their contractors believe partnering and forward pricing helped save a troubled project and bring it to completion on time and $16.5 million below budget." ENR went on to quote Brian E. Kasen, former construction manager for Metro: "It gave us a tool a lot of projects wish they had." Finally, *ENR* quoted Mortenson Vice President Steve Halverson: "On a heavily accelerated job, the impact costs at [West Point] were something like 0.8 of 1%."

From a macroview, the results (at West Point/Renton) WWTPs have to be viewed as a success. In the face of unmanageable delays (artifacts, etc.), the program was brought in on time and under budget. Who could argue with that? The answer is the Metropolitan King County Council, which launched a microview audit of the program. The audit, done in hindsight (as all audits are),

---

Engineering News-Record, June 10, 1996, p. 10.

offers some cautions on areas of change order management that require attention. Audit comments included the following:

> [Metro and/or Mortenson] "failed to document $20 million worth of payments."

> "The change order impact factors were arbitrary and the formulas assumed impacts have occurred when, in fact, there were no actual costs associated with those impacts."

The use of forward pricing by Metro at West Point/Renton highlights the risk of this approach, particularly in public projects. The parties tried to predict the cost of problems yet to be met. Even though a good overall result was achieved, a microview audit made it appear that the contractors had been given a windfall of $20 million.

Owners (public and private) typically use the liquidated damages clause to penalize contractors for delay. If earlier completion has advantage for the owner, a bonus clause can be effective. In this case, contractors can invest their own resources (overtime, increased equipment, etc.) to win the bonuses.

Time Extensions due to Change Orders

The methodology for making a request for time extension, even if it's part of a change order request, is described in the specifications. It includes the following:

- *Reasons for time extension.* Include change of scope, force majeure (weather, strikes, floods, storms, and other factors beyond the contractor's control), and direction (by bulletin or change order), and so on.

- *Time impact evaluation (TIE).* TIE must demonstrate impact and the amount of impact to the critical path.

- *Notice.* Timely notice of the intent to request a time extension must be made in writing. Failure to make timely notice of the need for a time extension (usually within 30 calendar days of the onset of a delay event) will be deemed sufficient cause to deny an extension of time.

Split Evaluation

Although the methodology does not allude to it, common practice is to bifurcate the evaluation of cost and time for a change order. The evaluation of time lags behind—with the owner hoping it will "go away" (it won't) and the contractor waiting to see how much time is needed.

Owner Reluctance

Owners are usually reluctant to give time extensions, regardless of the justification. Often this flows from real constraints: an EPA consent order with a cost per day if exceeded, a revenue-producing project with a need to go into revenue service, or a functional date (Olympics, school-year opening, etc.).

In addition, the concept of liquidated costs (i.e., cost to contractor for each day over the contract date) seems to be patently obvious to the least-experienced owner. The same owner is not only reluctant to give time but, worse yet, is appalled at the idea of paying for it.

Owner reluctance example—sludge processing facility

At a $20 million new sludge processing facility in the Northeast, the owner and A/E were working against a consent date that left them 21 months to design the facility and 24 to construct it. (See Case A in Fig. 8-1.) In this case, the bid/award is included in the design phase.

After one year, design was now estimated to be complete at month 27, which was 6 months late. (See Case B in Fig. 8-1.) To offset this, the A/E recommended a fast-track approach. A separate structural steel package would be prepared, at additional cost, and bid/awarded separately. The steel would be at the job site the same time as Case A. A separate parallel foundations package would have foundations ready when the steel was delivered.

The owner refused Case B because of the additional design money needed and because of the additional coordination responsibility to the owner. To appease the owner, Case C was implemented. (See Case C in Fig. 8-1.) Construct time was cut from 24 months to 18 months. Design was given 27 months, without bid/award, rather than 21 months.

After award, the contractor ran into unforeseen conditions, steel delivery delays, and a major design problem in the precast exterior design. When the contractor advised the owner of the serious nature of the early discovered problems, heated disagreements arose. At an early job meeting, the owner's representative announced that there would be no time extensions, regardless of reason.

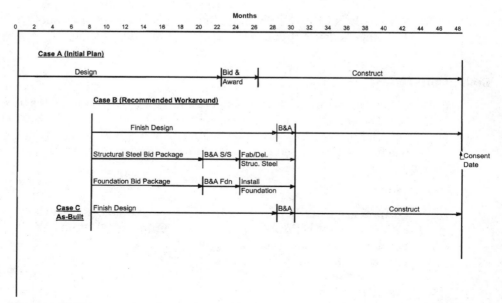

Figure 8-1 Sludge processing plant.

The contractor continued to lobby for time extensions. Concerned about $10,000-per-calendar-day liquidated damages, the contractor tried to mitigate the problem, in particular trying to winterize the building (with Visqueen) so the crew could work double shifts and overtime to work on piping and electrical work. Because of the very cold weather, worker productivity fell off.

The contractor substantially overspent its budget, and when the owner withheld $3 million in liquidated damages, the contractor went bankrupt. The matter went to a long arbitration, during which the owner's stonewall approach was thoroughly aired. What also came to light was the fact that the owner's situation consent date, the owner continued as before: barging sludge to another plant for treatment at a cost considerably less than the $300,000/month liquidated damages being withheld.

Weather Delays

Weather delays is an area that the typical specification addresses only obliquely. The specification writer, setting out to protect the owner (and A/E), writes a weather specification that makes it difficult for the contractor to claim weather days. The typical specification includes language such as, "Only weather more severe than normal shall be the basis for a time extension." This is designed to give relief for only unusual weather on the premise that the contractor should or will include a weather delays contingency—which the contractor won't.

It also leaves interpretation of what is meant by "severe." This usually involves use of last year's NOAA records, or even the past five years' NOAA average, as a baseline—neither of which is relevant to this project.

A number of state departments of transportation have experimented with the concept of naming the average number of effective workdays to be expected in each calendar month: for instance, in Pennsylvania, four in January, four in February, and so on. Thus, in the example, the only time extension in the case of January and February would be for workdays less than four. Unfortunately, this approach still dances around the issue.

The equitable approach would be to allow a one-day extension for each day actually lost to weather when the job is in the open and/or when temperature precludes work inside. Careful contemporaneous records should be kept by all parties (Fig. 8-2).

Assuming that the fieldwork is on a critical path, the project would be extended day-for-day. If long lead items were critical, a time impact evaluation would show a day-for-day use of float until the float is expended. In most cases, the result would be a day-for-day extension of the end date at no cost to the owner. (While not compensable to the contractor, each day of weather extension is an effective offset to liquidated damages.)

Getting the schedule back to or near the contract date is advantageous to all parties. Accordingly, particularly when a partnering approach has been agreed to, this is an opportunity for the parties to seek workarounds and/or limited acceleration (at cost to the owner) to get back on schedule.

PLANNING & CONSTRUCTION DIVISION
DAILY INSPECTION REPORT

PROJECT: _____ REPORT NO:_____

CONTRACTORS:

G.C. _____ DATE: _____

E.C. _____ WEATHER: _____

M.C._____ TEMPERATURE: _____ 9 am _____1 pm

WORK FORCE	EQUIPMENT	SUBCONTRACTORS WORKING

MATERIAL DELIVERED: _____

_____

WORK PERFORMED/REMARKS: _____

_____

_____

_____

_____

_____

_____

_____

_____

_____

_____

_____

_____

_____

_____

_____

_____

_____

_____

_____

VISITORS: _____

PREPARED BY:_____

10/85 — 3357—F-112

Figure 8-2 Daily inspection report.

Weather delay example—hospital project

In a pre-partnering-era hospital project, it was clear that the contractor had front-end-loaded the schedule. That is, early activity durations in the first year were comfortable—with a view toward avoiding hassling from the owner's PM/CM team. (The trade-off was that activity durations in years two and three were very tight.) At about month six, the contractor was lagging behind even with the comfortable schedule. To pick up some more slack, the contractor put in a PCO for a 55-day weather delay with justification. The owner's staff recommended approval, and the no-cost time extension went through. At the end of year one, the contractor tried to ask for compensable time extensions. The contractor's own soft schedule plus the 55-day no-cost cushion defeated all of these requests.

Constructive Acceleration

The stonewall approach in the previous sludge processing example resulted in constructive acceleration. This occurs when an owner refuses a justified time extension while holding the contract completion date (see Ehte, Inc. v. S. S. Mullen, Inc., 469 F. 2d 1127, 9th Cir.; 1972).

To prove constructive acceleration, the contractor must prove the following:

- One (or more) excusable delays were entitlement to a time extension.
- On the basis of the excusable delays, timely notice was given and an equitable time extension was requested.
- The owner refused or ignored the contractor's request for a time extension.
- The owner, despite the request for equitable time extension, continued to require that the contractor complete the work by the contract completion date.

Constructive acceleration at Camp Drum, N.Y.

In early 1990, the U.S. Army Corps of Engineers (Corps) was involved in the construction of a complete home base for the 10th Mountain Division. The division had a scheduled operational schedule that could not slip and the Corps so informed the contractor. When RFIs disclosed changes that had to be implemented, the contractor knew that the end date for construction must be maintained. The Corps included acceleration costs in the form of premium time and any other identifiable costs in the change orders. At Fort Drum, there were no directives to accelerate; however, in making it clear that the contract end dates could not slip, the Corps was admitting constructive acceleration and was forthright in paying the costs. The project was completed on time.

Contractor failure to request time extension

It is said that the law hates a volunteer. When a contract provides certain rights (for instance, the right to request a time extension), and the contractor elects not to make such a request, in effect, the contractor has volunteered not to ask for the time extension. (See Straus Group of Old Bridge v. Quaker Construction Management in Chap. 6.)

In Marriott Corp. v. Dasta Construction Co. (26 F. 3d 1057, 11th Cir.; 1994), Marriott acted as developer/manager for a large resort complex in Florida. Dasta and another contractor submitted a bid for the exterior skin and drywall on a 28-story part of the complex. The bid drawings were 50 percent complete because the project was on a fast track.

To be successful, a fast-track project must stay on budget. As part of the pre-bid information, Dasta reviewed the CPM plan. After making its bid, Dasta received a copy of the new CPM plan, which showed the work proceeding in an efficient sequence with normal working hours.

After the contract was signed, Dasta arrived at the site and found that the concrete work was five months behind schedule. This concerned Dasta because much of its work was to be placed directly on the concrete. Dasta was delayed by frequent, significant modifications in the plans and specifications. These changes prevented Dasta from proceeding in the orderly and efficient sequence shown in the CPM plan. Dasta also questioned the adequacy of the vertical transportation and safety measures provided by Marriott. When the concrete contractor failed to correct its work, Dasta took on that work. In addition, Dasta accepted other change order work.

Marriott issued an acceleration plan in the form of the CPM. Dasta bid on additional work based on that plan. As the accelerated completion date neared, there were delays in the progress schedule, and Dasta had a cash flow problem. Marriott told Dasta that it wanted the project as soon as possible and that it would make a reasonable settlement. Dasta borrowed $1.5 million, based upon Marriott's assurances. This sum, however, did not cover all the suppliers and subcontractors. Marriott covered the balance, debiting the payments against the balance owed Dasta. Initially, Dasta presented a claim to Marriott for additional time spent on the project (delay) and an inefficiency claim for costs due to owner interference (i.e., failure of other contractors to perform, owner-manager failure to provide vertical transportation, etc.). Marriott rejected these claims and a suit ensued.

The case ensued under Florida law, which often upholds no-damages-for-delay clauses. The court found that the no-damages-for-delay clause barred most of the defendant's claims. Further, Dasta was unable to file for additional delay claims because it had, in a timely manner, failed to request extensions of time even on those delays that it claimed were caused by the owner's interference with its (Dasta's) performance.

Voluntary acceleration

If a contractor is asked to do something that the contractor suspects might be out of scope, and the contractor proceeds to do it without either questioning whether it is in scope or giving notice that it might not be, the contractor probably cannot recover the value of the work if it is, indeed, out of scope. Similarly, if a contractor decides to perform ahead of schedule without directive to do so, the contractor would not have a claim for acceleration against any other party to the construction contract. This is deemed *voluntary acceleration*. (See McNutt Const. Co., ENGBCA No. 4724, 85-3 B.C.A.; 1985.)

Mitigating Time Impact of Change Orders

The change order process in a small project—say, a custom-designed residence—can be a one- or two-day affair. Those involved would be owner, architect, and builder, along with, perhaps, the approval of the mortgage bank. In large, complex, and/or heavy construction, the approval process is usually time-consuming even for simple changes. (The small change at the Camden County Municipal Utilities Authority in Chap. 5 was finalized 14 months after the initial incident.)

Time impact cannot be assessed until (1) a change (PCO) is identified, (2) the scope is defined and estimated, and (3) the change order process is initiated. For this to be expeditious requires a project management system and change management procedures to be in place.

In many large projects, the change order process is viewed as mere "paperwork." In the dynamic (read: "macho") world of construction in the field, paperwork is viewed with disdain and is allowed to fall behind while "real work" is being put to be in place.

If the change order process was just a matter of record keeping, there might be some empathy for this viewpoint. However, the change order process is an indispensable part of construction management: the ability to pay for work-in-place. The PM/CM cannot approve progress payments for change order work-in-place without an approved change order to pay against. However, no matter how much the PM/CM wants to approve payment, it can't unless the money is "in the bank" (i.e., the account for the specific change order has been funded through the change order process).

In major projects where the change order process paperwork has been allowed to flounder, the very real result is subcontractors running into serious cash flow problems because, at the urging of the PM/CM or general contractor, they laid it on the line and got work in place. In responding to this pressure for progress, they were told (or thought it was implied) that they would be "taken care of."

Ideally, the best course is for the PM/CM to go into a major program recognizing that change and the change process must be managed. However, in situations where the lack of change order management has lead to chaos, what should be done? In any number of such situations, the owner has brought in an outside team with the charter to catch up on the paperwork. In the more successful rescue or recovery efforts, the charter includes the right to issue unilateral change orders in the interim to facilitate some fast payments for work-in-place.

9

Change Management and Claims Procedures

Each major program should have a set of project management procedures. As PM/CM for the Southeastern Pennsylvania Transportation Authority (SEPTA), O'Brien-Kreitzberg & Associates (OKA) authored the project management procedures manual for the RailWorks program. (RailWorks entered its construction phase in 1990, which was successfully completed in 1993.) Section IV of that manual, "Change Management and Claims," functioned particularly well.

RailWorks

When SEPTA assumed operation of the former Pennsylvania and Reading rail lines in 1983, it inherited a network of bridges, track, and overhead power lines (catenary) that had already been in service for many years. Decades of deferred maintenance and virtually no dedicated capital funding had resulted in a usable but deteriorating rail system.

The commuter tunnel, completed in October 1984, connected the once-separate rail lines; it allowed all regional rail lines to access the three center-city Philadelphia rail stations. Several months after the tunnel's completion, an engineering inspection study indicated a need for renovation of many of the system's bridges, some of which had stood for nearly 100 years.

The 4-mile stretch of track north of the new tunnel was the renovation property. The stretch consisted of track and catenary system and 25 rail bridges—a total of 16 track miles—forming the main line, or throat, of the old Reading line. Six SEPTA regional rail lines feed into this central corridor.

The completed project was budgeted at about $300 million. The project, named SEPTA RailWorks, entailed more infrastructure rehabilitation of this regional rail corridor. The major components of the work included the renovation of bridges, complete replacement of 21 bridges, replacement of all track, a new catenary system, and replacement of related equipment, including switches and signals. All of the bridges span active highway crossings in a congested urban setting.

Throughout the project, the contractors were faced with a local population living in close proximity to the construction operations. (Two of the bridges were actually preassembled in the yards of neighboring residents.) The residents continued to use the streets in spite of the barricades and all attempts to close the streets. The entire area was laced with sanitary and storm sewers from the last century. Their existing condition was documented both before the work began and again after completion to minimize undue change orders.

The RailWorks program required intensive preparation and planning by the owner (SEPTA), the PM/CM (OKA), and all contractors. Most of the actual fieldwork (demolition and new facilities) was put in place in two summer "windows" as follows:

Window	Shutdown	Return to revenue service	Successful
I	April 15, 1992	October 1, 1992	Yes
II	May 1, 1993	September 5, 1993	Yes

There was great pressure to succeed on time, since liquidated damages were $70,000 per calendar day.

It was apparent from the outset that the essence of the project was the management and coordination of six prime contractors (in each phase) and their 74 subcontractors, combined with the sharply limited time available for the accomplishment of the work in a highly restricted and congested work area. Using the OKA generic procedures book as a base, a project-specific procedures book was developed before the first shutdown. In this same early time frame, a very successful partnering program was implemented. SEPTA, the construction management team, and the initial six prime contractors were involved.

To make the transition among the completion of the general contractor's work, the corresponding start of electrical work, and the integrated system tests as seamless as possible, several mandatory milestones were included in the contract specifications. A special CPM coordination schedule was developed by extracting relevant detailed window schedule information from each of the contractors. (The window periods were those when the rail system was shut down.) This process permitted an accurate weekly monitoring of the work wherever interdependent operations by more than one contractor was necessary to achieve early access.

The RailWorks construction contracts were packaged to give access to local contractors. This was successful, and the initial bids totaled $140 million—more than $50 million under budget. Change orders were under 10 percent, giving the opportunity to put out successful RFPs for "supplemental change orders." These changes, within the context of the program but technically "extra work," gave SEPTA the opportunity to build more value into the line.

For instance, one supplemental change allowed the mainline trains to go through a little-used local stop at 80 mph rather than 45 mph.

The differentiation between regular change orders and supplemental change orders gave the SEPTA Board a handle on the soundness of the basic design (versus supplements).

RailWorks Procedures Manual

Figure 9-1 is Section IV, "Change Management and Claims," of the RailWorks procedures manual. Figure 9-2 shows the flowchart for the contract change procedure, as described in the manual. Figures 9-3 through 9-17 are various forms from the RailWorks project.

Figure 9-1   Section IV of RailWorks procedures manual.

Section IV   Change Management and Claims

1. Purpose and Objectives
   A. This section outlines the basic requirements for identifying, evaluating, approving and processing construction contract changes relating to the Mainline Bridge, Station & System Improvement Project. Change Management is a team effort by SEPTA, the Construction Manager, the Design Engineer and the Contractors to execute necessary contract modifications which may be required to fulfill the overall project objectives.
   B. The Mainline Bridge Improvement Project is unique in several ways:
      1. The vast majority of field construction is to be accomplished in two relatively short "windows" during which all commuter rail service will be suspended.
      2. Much of the work involves replacement and rehabilitation of existing structures and facilities.
      3. Multiple construction contractors will be working in a limited area at the same time.
      4. These project conditions lead to the following conclusions:
         a. It is reasonable to expect a substantial number of Change Orders to be generated based upon changed conditions, additional work, etc.
         b. Changes to one construction contract are likely to have a "ripple effect" through one or more other contracts.
         c. Due to the limited time available for construction, disputes delaying the progress of the work must be avoided.
         d. Prompt resolution of problems encountered during the "window" outages is essential to maintain progress and successful project completion.
      5. The Construction Manager has been charged with the formulation of specific Change Order procedures for use during this project. These procedures are intended to be responsive to the needs of the project while retaining SEPTA's existing procedures.

2. Change Order Procedures
   A. Changes not involving Cost or Time

(Continued)

Figure 9-1 *(Continued)*

1. The Construction Manager, principally through the Resident Engineer will have the authority to direct and approve minor changes in the work not involving the value of the contract or the time of performance. Such changes do not require a formal contract modification.

   a. SEPTA project personnel also have this authority. All such changes are to be coordinated through the CM Resident Engineer.

   b. These changes are limited to those that are consistent with the intent of the contract plans and specifications and do not adversely affect the performance, quality or maintainability of the project.

B. Changes in Project Cost and Time

1. All changes in the Work involving modification of project cost or time of performance shall be authorized by Change Order, executed under applicable conditions of the construction contract documents. No Change Order, modification or amendment to a contract will be binding unless executed in writing by SEPTA, with the concurrence of the appropriate funding agencies if required.

2. MBIP Change Order procedures *when time of initiation or performance is not a critical issue* are as follows:

   1. *Initiation / Notification*

Description:	Any party involved in the project may initiate a revision to the contract; SEPTA, the Design Engineer, the CM or the Contractor. All potential changes to the contract will be channelled through the CM.
Responsibility:	OKA is responsible for the receipt and processing of all potential changes to the contract. Change notifications can be directed to either the Project Office or the Contract Field Office.
Action Req'd:	Log and file in Document Control System.
Distribution:	DPM Construction, DPM Controls, Resident Engineer, SEPTA Project Engineer.

   2. *Preliminary Analysis*

Description:	Evaluation of contract documents to establish contractor entitlement to a contract revision and additional compensation. Evaluation of "time criticality" of resolution and performance.
Responsibility:	The OKA Resident Engineer for the specific contract has primary responsibility.
Action Req'd:	Completion of Preliminary Entitlement Evaluation form. Classification of change, if any. If this evaluation indicates no entitlement, Resident Engineer will respond to Contractor. If this evaluation indicates entitlement and no cost proposal has been received, Resident Engineer will request same from Contractor.
Distribution:	DPM Controls, DPM Construction, SEPTA Project Engineer, SEPTA Contract Administrator, Contractor (if appropriate).

   2a. *Contractor Response*

Description:	Contractor accepts determination of no entitlement.
Responsibility:	Contractor.
Action Req'd:	If the Contractor accepts the Resident Engineer's determination that no entitlement exists, that fact will be documented. If no confirmation is offered by the Contractor, the Resident Engineer will confirm same by letter or "Speed Memo."
Distribution:	DPM Construction, DPM Controls, SEPTA Project Engineer.

   2b. *Contractor Response*

Figure 9-1 (*Continued*)

Description:	Contractor disputes determination of no entitlement.
Responsibility:	Contractor.
Action Req'd:	Documentation of Contractor's objection to RE's evaluation.
Distribution:	DPM Construction, DPM Controls, Resident Engineer.

2c. *Entitlement Dispute*

Description:	Review of Contractor dispute involving contract change entitlement.
Responsibility:	PM, DPM Construction, DPM Controls, SEPTA Contract Administrator.
Action Req'd:	CM will review issue in detail. If CM reverses initial determination, Contractor will be notified and Step 3 of this procedure will be initiated. If CM upholds initial determination, Contractor will be notified and matter referred to SEPTA Contract Administrator.
Distribution:	PM, DPM Construction, DPM Controls, Resident Engineer, Contractor, SEPTA Contract Administrator, SEPTA Project Manager, SEPTA Project Engineer.

3. *Scope Definition*

Description:	Review of RE's preliminary analysis. Define issues & work involved. Initiate detailed evaluation.
Responsibility:	DPM Controls.
Action Req'd:	Notify contractor to prepare cost proposal (if req'd). Notify design engineer (in all cases), and obtain design input (if req'd). Initiate OKA review.
Distribution:	Appropriate OKA staff, Design Engineer, DPM Construction, SEPTA Project Engineer.

3a. *Supplemental Design*

Description:	Additional design necessitated by change request.
Responsibility:	Design Engineer.
Action Req'd:	Upon receipt of direction by SEPTA, the Design Engineer will prepare supplementary designs required by the proposed contract change. Upon completion, the Design Engineer will furnish the appropriate number of copies to OKA.
Distribution:	DPM Construction, DPM Controls, Resident Engineer, SEPTA Project Engineer, Contractor (if appropriate).

3b. *Concurrent Cost Estimate*

Description:	Preparation of cost estimate for proposed change by the Design Engineer.
Responsibility:	Design Engineer.
Action Req'd:	Upon receipt of direction by SEPTA, the Design Engineer will prepare cost estimates of the proposed contract change. Upon completion, the Design Engineer will furnish the appropriate number of copies to OKA.
Distribution:	DPM Construction, DPM Controls, Resident Engineer, SEPTA Project Engineer.

3c. *Contractor Proposal Preparation*

Description:	Contractor cost proposal: for revised contract work.
Responsibility:	Contractor.

(*Continued*)

Figure 9-1 *(Continued)*

Action Req'd:	Preparation of cost proposal, including any requests for time extension, for revisions to contract.
Distribution:	Contractor Internal.

3d. *Contractor Submission of Cost Proposal*

Description:	Submission of Contractor cost and/or time extension request.
Responsibility:	Contractor.
Action Req'd:	Analysis by appropriate CM staff upon completion of independent estimate.
Distribution:	DPM Construction, DPM Controls, appropriate Estimating and Scheduling, SEPTA Project Engineer.

4. *Time and Cost Analysis*

Description:	Preparation of independent estimate of cost and/or time effects of proposed contract modification.
Responsibility:	Appropriate CM estimators and/or schedulers.
Action Req'd:	Preparation of independent cost estimate or time impact analysis in accord with detailed procedures for same.
Distribution:	DPM Construction, DPM Controls.

5. *Review of Contractor Proposal*

Description:	Review of Contractor change proposal.
Responsibility:	DPM Construction, DPM Controls, appropriate OKA Estimating and/or Scheduling.
Action Req'd:	Review contractor proposal for completeness and reasonableness. Review should occur *after* completion of independent analysis.
Distribution:	Upon receipt of proposal: DPM Controls only. After completion of independent analysis: DPM Construction, Resident Engineer, Estimating and Scheduling, SEPTA Project Engineer.

6. *Summary & Recommendations*

Description:	Preparation of recommendation for action.
Responsibility:	DPM Construction, DPM Controls, Resident Engineer.
Action Req'd:	Review all appropriate documentation. Prepare final recommendation and forward to SEPTA. Completion of Contract Change Summary.
Distribution:	SEPTA Project Manager, SEPTA Contract Administrator, SEPTA Project Engineer, CM Project Manager.

7. *SEPTA Evaluation*

Description:	Review of OKA Contract Change Summary.
Responsibility:	SEPTA Project Manager, SEPTA Project Engineer, SEPTA Contract Administrator.
Action Req'd:	Accept or reject OKA evaluation. Determine if negotiation with Contractor is required.
Distribution:	SEPTA internal, CM Project Manager.

8. *Negotiation (if read)*

Description:	Negotiate proposed change to contract with Contractor.
Responsibility:	SEPTA Contract Administrator will take the lead in all negotiations with support from the CM as required.
Action Req'd:	Formulation of negotiation strategy, engage in negotiations with Contractor. Evaluate actions necessary resulting from

Figure 9-1 *(Continued)*

unsuccessful negotiations (9a).

Distribution: Results of negotiation to PM, SEPTA Project Manager, SEPTA Resident Engineer.

9. *SEPTA Staff Summary*

Description: Documentation and sign-off sheet for SEPTA Board approval of contract change.

Responsibility: SEPTA Project Engineer with assistance from the CM.

Action Req'd: Writeup of proposed contract change in accord with standard SEPTA procedures.

Distribution: SEPTA Project Manager, SEPTA Contract Administrator.

9a. *Proceed Directive (T&M)*

Description: Directive for Contractor to proceed with disputed work.

Responsibility: SEPTA Project manager.

Action Req'd: If agreement cannot be reached via negotiation, the Contractor may be directed to proceed with the work. Documentation of actual expenses ("Time and Material") will be used as the basis of compensation.

Distribution: PM, DPM Construction, DPM Controls, Resident Engineer, Contractor, SEPTA Contract Administrator, SEPTA Project Engineer.

10. *SEPTA Board Approval / Change Order Execution*

Description: Ratification and execution of contract modification.

Responsibility: SEPTA.

Action Req'd: Concurrence of SEPTA Board with proposed contract modification, execution of contract modification.

Distribution: Contractor, PM, DPM Construction, DPM Controls, Resident Engineer, SEPTA Project Manager, SEPTA Project Engineer.

3. This procedure is summarized in the flow chart [Fig. 9-2]. Numbered boxes correspond to individual steps outlined above.

4. During performance of the Mainline Bridge Improvement Project, this general procedure will be employed for all contract changes where time of initiation or execution is not "critical."

5. Detailed procedures for execution of changes when time of initiation or execution is "critical" are under development at the time of this writing. Procedures currently available which could be employed in this condition include:

a. "Time and Material"

1. Upon identification and verification of entitlement, the Contractor may be directed to proceed with the work on a "Time and Material" basis.

2. Under current procedures, this would require authorization from the SEPTA Board. The Change Order would be written in terms of a "not to exceed" value. The actual value would be determined based on a review of the "Time and Material" records maintained during the performance of the work.

3. Specific procedures would include the following:

a. The CM will monitor and record the actual labor, material and equipment employed in the performance of the work on a daily basis.

b. This record will be agreed to by both the Resident Engineer and a responsible representative of the Contractor.

(Continued)

Figure 9-1 (*Continued*)

> c. Determination of the final cost of the work will be negotiated with the Contractor based upon a thorough review and analysis of the Contractor's work effort.
>
> b. MBIP construction contracts indicate that SEPTA may unilaterally direct the Contractor to undertake additional work, for which the compensation would be established via a SEPTA audit of the Contractor's financial records. General procedures in this situation include:
>
> > 1. A directive to proceed with the work involved shall be issued by SEPTA in accord with Section XIV paragraph E of the construction contract.
> > 2. The Contractor shall proceed with the work immediately upon receipt of the Change Order. Costs associated with the work shall be identified and recorded by the Contractor as prescribed by the Construction Manager.
> > 3. Upon completion of the work, an audit of the Contractor's cost records will be performed by SEPTA and serve as the basis for calculation of the reasonable cost of the work.
> > 4. Final cost shall be determined by the CM in conjunction with SEPTA's auditor.
> > > 1. Under this alternate, the CM would maintain records in the field of the actual effort involved, similar to the "Time and Material" procedure, as an additional, independent evaluation of the actual effort involved.
>
> 3. Change Order Tracking
>
> > A. Potential Change Orders
> >
> > > 1. Potential Change Orders (PCO) are initiated by either the Resident Engineer or by a senior member of the CM's staff to identify the existence of an issue or circumstance which may result in a modification to contract cost or time.
> > > 2. All information relevant to the PCO shall be assembled and filed. A log of all PCOs shall be maintained. Each PCO will be identified by construction contract, issue involved, estimated exposure in time of performance and cost as well as current status and recommendation on course of action.
> > > 3. The PCO number will be used to identify the issue prior to its incorporation into the contract as a formal Change Order.
> > > 4. PCO's will be reviewed at monthly management meetings to evaluate their current status and to insure proper action is being taken.
> >
> > B. Change Orders
> >
> > > 1. Upon approval by SEPTA's Board, contract Change Orders will be numbered and logged sequentially for each MBIP contract.
>
> 4. Sources of Change
>
> > A. An awareness of the common sources for contract modifications is essential. Prompt recognition allows the CM to take any available action to mitigate any adverse effects and to begin to collect all relevant data to document the change.
> >
> > > 1. Request for Information
> > >
> > > > a. At the beginning of the construction, the Contractor will be provided with and requested to use standard forms on which to submit requests.
> > > > b. Upon receipt of a request the Resident Engineer will contact the Design Engineer to determine if engineering approval is required. Simultaneously, the Resident Engineer will transmit a copy of the request to the Project Office.
> > > > c. At the Project Office, the request will be entered into the Document Control System for tracking. At the same time, the request will be reviewed by the CM's Office Engineer who will evaluate its impact on other contracts and determine if review is required by organizations other than the Design Engineer.

Figure 9-1 *(Continued)*

> d. Upon receipt of the response from the Design Engineer, or other authority, the Resident Engineer will transmit same to the Contractor, using the remaining section of the RFI form. A copy of this response will also be transmitted to the Project Office for the use of the Office Engineer and entry into the Document Control System.
>
> e. If the response has an effect on another contract, the Office Engineer will initiate a Field Order.
>
> f. If, in the opinion of either the Resident Engineer or the Office Engineer, the response may generate a Change Order, a Potential Change Order will be initiated.
>
> 2. Request for Change
>
>   a. A Request for Change (RFC) can be initiated by SEPTA (which includes the Designer, the CM, or as a result of an interface with the City or other Agency) as well as the Contractor. The procedures for processing an RFC vary depending upon the origin.
>
>   b. Preprinted forms to be used for originating an RFC will be distributed to all participants at the beginning of construction.
>
>   c. If SEPTA originates the RFC, information provided will include:
>   1. Description of the Change
>   2. Brief justification
>   3. Affected contract drawings and specifications
>
>   d. The contractor will acknowledge incorporation of the change within 10 days. If no response to the contrary is received within that time, a confirmation letter will be sent to the Contractor, stating that this revision has been incorporated at no change to either cost or time.
>
>   e. If the Contractor originates the RFC, it will be transmitted to the Resident Engineer and must include the following:
>   1. Description of the change
>   2. Justification for the request
>   3. Affected contract drawings and specifications
>   4. Anticipated effect on project cost and time (if any)
>
>   f. The Resident Engineer shall review the RFC.
>   1. If no change in cost or time is involved the RFC may be handled in a manner similar to the RFI.
>   2. If a change in cost or time is involved the RFC will be handled as detailed in Section 1.B.2.
>
> 3. Value Engineering Change
>
>   a. Value Engineering Proposals
>   1. A Value Engineering incentive provision is included in each construction contract awarded as part of the MBIP. This provision is designed to provide a mechanism by which the Contractor can offer proposals to the CM and SEPTA for reducing the final cost of the contract in a manner that does not impair the essential functions, characteristics or quality of the final product.
>   2. This provision provides as an incentive to the Contractor, the opportunity to share in the cost savings realized by SEPTA for those recommendations that are accepted in accord with the provisions of the construction contract. The Contractor will be entitled to 50% of the value of the cost reduction of an accepted proposal, accounting for the contractor's allowable development and implementation costs.
>
>   b. Submission and Approval
>   1. A Value Engineering Change Notice (VECN) is a document prepared by the Contractor to initiate the Value Engineering process. The VECN will serve

(Continued)

Figure 9-1 (*Continued*)

to notify the CM and SEPTA that the Contractor has identified an area of potential Value Engineering savings. The VECN will contain a brief description of the proposed change and an estimate of total savings involved.

2. The CM will respond to the Contractor within 5 days with approval to proceed with a Value Engineering Change Proposal (VECP) if the proposed change merits further evaluation.

3. Four copies of the VECP will be submitted to the CM.

4. The CM will utilize a two stage review process for all VECPs received:

   a. All VECPs will be reviewed and given an initial evaluation within 24 hours of receipt. Those with true merit will be evaluated in detail. The contractor will be promptly notified of all VECPs that are not selected for further review.

   b. Those VECPs selected for detail evaluation will be reviewed by both the CM and the Design Engineer. All VECPs will be processed expeditiously. The Contractor will be advised of the status of any VECP within 45 days of submission. If additional review is required, the Contractor will be advised of that fact with an explanation of the delay and an estimated date by when a decision will be reached.

5. Those VECPs receiving a favorable Phase II evaluation will be directed to the SEPTA Senior Project Manager. He will have final authority for approving value engineering proposals. Prior to reaching a decision, the Senior Project Manager and the CM will conduct a thorough review of the proposal. The CM will advise the Contractor of SEPTA's decision.

6. If SEPTA accepts all or part of a VECP, the Senior Project Manager will initiate a contract Change Order in accord with procedures contained elsewhere in this manual.

7. If the VECP is not accepted by SEPTA the CM will provide written notification of that fact to the Contractor including the reason or reasons for SEPTA's rejection of the proposal.

c. Contents

   For proper consideration, a VECP must contain, as a minimum, the following:

   1. A detailed description of the difference between the existing contract requirement(s) and the proposal.

   2. A list of contract requirements that must be changed if the VECP is accepted.

   3. A comparative cost analysis of the proposed changes versus the existing requirements. The cost reduction will take into account allowable Contractor development and implementation costs. An estimate of any direct costs that SEPTA may incur in the implementation of the VECP will also be included.

   4. An estimate of indirect costs which SEPTA may incur as a result of implementing the VECP.

   5. A statement regarding any time limitations on the completion of the VECP process which might affect the realization of maximum cost savings. Any effect which the VECP may have on delivery schedules or contract completion time must also be noted.

d. Sharing of Cost Savings

   1. The Contractor's share of the savings is calculated by subtracting SEPTA's costs from the contract savings and multiplying the result by 50%.

4. Field Order

   a. Field Orders are directives to be issued to the Contractor by the Resident Engineer to initiate changes in the work typically resulting from:

      1. Coordination impacts on other contracts

      2. Design Revisions

Figure 9-1 (*Continued*)

3. Scope of Work Changes

    b. Field Orders will be issued to the Contractor by the Resident Engineer. The Contractor will be instructed to notify the Resident Engineer in writing within 10 days if the Field Order will result in a change to project cost or time of performance.

    c. Copies of all Field Orders will be forwarded to the Project Office for logging and tracking.

    d. Field Orders that are expected to result in a change to project cost or time of performance will result in the Initiation of a Potential Change Order.

  5. Proposal Request

    a. Requests for proposal may be generated by SEPTA when the addition or revision to the work clearly entitles the Contractor to additional compensation.

    b. All requests for proposal will be processed by the CM, and directed to the Contractor's address of record.

    c. All requests for proposal will contain a date by which a response is requested.

    d. Proposal requests will be tracked and followed up if no response is received by the due date.

    e. Upon receipt of a Contractor's proposal, it will be reviewed by the CM and forwarded to the Senior Project Manager—MBIP, with an appropriate recommendation.

5. Claims Management

  A. Introduction

    1. Claims and disputes are an integral part of construction project management. A claim is a disagreement between the Owner and the Contractor concerning the contract work and the cost or time associated with it. Claims and changes are basically the same thing. A claim is nothing more than an unresolved disagreement about a Potential Change Order.

    2. Many claims can trace their origins to an unrealistic Changes provision in the contract documents or poor administration of contract provisions. In some cases administrative procedures are not streamlined enough to allow execution of a change order within the time required by field conditions. Directives to perform extra work not included in the contract must be carefully and explicitly written.

    3. No set of contract documents is perfect. All parties must realize that revisions and unanticipated situations will be encountered. Elimination of the unrealistic attitude that "There won't be any change orders on my job!!" is the most important step in avoiding construction claims.

  B. Mitigation

    1. The most effective claims mitigation measures are those applied to the construction documents prior to the bidding of a project. Effective methods to mitigate the effect of claims include:

      a. Thorough review and coordination of the contract plans and technical specifications.

      b. The inclusion of an effective Changes clause in the construction contract. Included should be the means of initiating changes in the work prior to the execution of a format change order.

      c. Realistic scheduling provisions which make the schedule a meaningful tool with which to plan and control the project.

      d. Require the Contractor to submit daily reports of its field activities to the Construction Manager.

      e. Require the inclusion of unit prices in the bid for items of work where revisions in quantity or scope of work are likely.

(*Continued*)

Figure 9-1 (*Continued*)

2. Construction phase measures which can minimize the impact of claims include the following:
   a. Reasonable and intelligent field management. Interference with the Contractor's work plan and/or sequence of operations will result in a claim and must be avoided.
   b. Avoid unnecessary revisions after the start of construction. The cumulative effect of an unreasonably large number of changes will probably be reflected in a delay-disruption and loss of productivity claim.
   c. Prompt and fair processing of legitimate Change Orders. Changes in the work are an inevitable part of most construction contracts. Failure to deal with these changes in a reasonable manner will result in Contractor claims.
   d. Thorough review and evaluation of Contractor schedules. Before approving or accepting a Contractor's schedule, it must be reviewed for practicality, feasibility, and realistic logic and activity durations.
   e. Timely and proper performance of SEPTA obligations. These include timely granting of access, timely responses to submittals, delivery of Owner furnished equipment and timely response to Contractor's notices. The timeliness and proper performance of complementary Contractor obligations should also be monitored for its contribution to any claims situation.
   f. Prompt and effective response to all contractor communications. Notifications of claim situations need prompt response. Notifications that are factually incorrect or based on an erroneous contract interpretation should be rebutted immediately.
   g. Anticipation and early identification of claim situations. Sensitivity to potential claim situations allows the Owner or CM to get an early start on collecting data to determine the merit of a potential claim and to explore possible alternatives. Unusual activity on sites or in correspondence, notifications, requests for information, etc., may provide a preliminary indication that a claim will be made.

3. Evaluation and Resolution
   Upon receipt of a formal claim, or notification of intent to file a claim, a structured claim revue process should be initiated. The following steps are typical of those required to analyze and resolve a construction claim. These individual steps may be modified as required to suit specific situations and are typical of procedures to be utilized on the MBIP.
   a. Familiarization with the Claim
      1. Review and understand what issues are involved. Determine the basis of the claim:
         a. Changed conditions
         b. Delays
         c. Disruptions
         d. Additional Work
      2. Review of contract documents.
         a. Determine what specific portions of the work are involved.
         b. Do the documents support or refute the Contractor's position?
         c. Are they reasonable?
      3. Evaluate overall Contractor entitlement to any compensation for the issues cited in the claim.
      4. Establish a management plan for resolution of the claim and execution of subsequent work.
   b. Assemble all relevant data.
      1. Identify and collect the specific data required to document original plans, as well as actual performance and problems encountered.
      2. If the project is still underway, determine if useful information can be obtained by supplementary monitoring of the work effort in progress.

Figure 9-1   *(Continued)*

> 3. Organize the collected documents chronologically by specific issue or event, and according to the party responsible for the problem.
> c. Evaluation.
> > 1. Determine how the Contractor planned to execute the work. This may take the form of a planned schedule or a measure of craft productivity or other specific performance indicator.
> > 2. Determine the reasonability of the Contractor's claimed or planned level of performance. Was this level of performance attainable?
> > 3. Identify and evaluate deviations from the Contractor's plan and the reasons and responsibility for same.
> > 4. Determine cost and time effects of deviations from the original plan. Summarize cumulative effects by responsibility.
> > 5. Identify and evaluate any secondary or impact damages resulting from the issues involved.
> d. Summarization and presentation of findings.
> > 1. Present summary of findings and recommendation for action to the SEPTA Senior Project Manager—MBIP.
> e. The SEPTA Senior Project Manager will review the results of the analysis and determine the approach to resolve the claim from this point.
> > 1. The CM may provide negotiation or litigation support as required.

CONTRACT CHANGE PROCEDURE

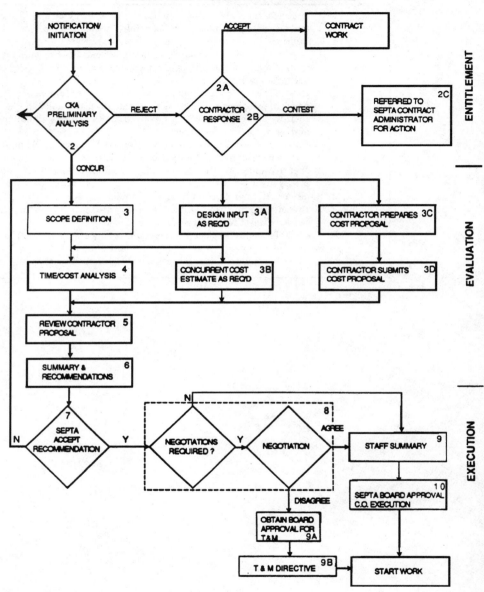

Figure 9-2   Flowchart of contract change procedures.

MAINLINE BRIDGE IMPROVEMENT PROJECT
PROJECT WEEKLY REPORT

MBIP

PROJECT TITLE & NO.	EC-1
CONTRACTOR	Carr & Duff

WEEK ENDING	WEEKLY REPORT No.	

RESIDENT ENGINEER		

CONSTRUCTION SUMMARY

1. CONSTRUCTION ACCOMPLISHMENTS

2. MATERIAL/EQUIPMENT DELIVERIES

3. DELAYS, DISPUTES

4. OPEN ISSUES

5. TESTING

6. CHANGE ORDER WORK

Figure 9-3 Project weekly report.

MAINLINE BRIDGE IMPROVEMENT PROJECT
DAILY INSPECTION REPORT

MBIP

PROJECT TITLE & NO.	**EC-1**		
CONTRACTOR	**Carr & Duff**		
DATE		DAY NO.	WORK HOURS
INSPECTOR (PRINT)			

CONDITIONS   PAGE ____ OF ____

	DAY	S	M	T	W	TH	F	S
WEATHER	Brite Sun	Clear		Overcast	Rain		Snow	
TEMP.	To 32°	32° to 60°	60° to 70°	70° to 85°	85° up			
WIND	Still	Moderate	High	Report Number				
HUMIDITY	Dry	Medium	Humid					

REFERENCE NUMBER:
- 1. Delays and Difficulties
- 2. Changes
- 3. Accomplishments
- 4. Orders or Instructions
- 5. Unsatisfactory Work
- 6. Mat'l/Equip
- 7. Disputes
- 8. Tests
- 9. Safety
- 10. Visitors

REF. No.		SCHEDULE REFERENCE

To the best of my knowledge the work is in conformance with the approved plans and specifications

Signature: _____   Date: _____
    Inspector
Signature: _____   Date: _____
    Resident Engineer

Figure 9-4 Daily inspection report.

MAINLINE BRIDGE IMPROVEMENT PROJECT
DAILY INSPECTION REPORT (Continuation sheet)

MBIP

PROJECT No: EC-1		PAGE:_____OF:_____
Date:		

REF.		SCHEDULE REFERENCE
CONTRACTOR WORKFORCE:		
SEPTA FORCE ACCOUNT:		

Signature: _____   Date: _____
                    Inspector

Signature: _____   Date: _____
                  Resident Engineer

Figure 9-4 *(Continued)*

MAINLINE BRIDGE IMPROVEMENT PROJECT
FIELD ORDER/PROPOSAL REQUEST

MBIP

LETTER NO:_____   DATE: _____

TO:_____

CONTRACT NAME   CONTRACT NO.   FIELD ORDER NO._____

_____   _____   RE: CN NO._____

Check appropriate box:

☐ Please submit an itemized quotation for changes in the Contract Sum and/or Time
   incidental to proposed modifications to the Contract Documents described herein.
   Do not proceed with the work. Your proposal is required by:_____

☐ You are hereby directed to execute promptly this field order which interprets
   the Contract Documents or orders minor changes in the Work without change in
   Contract Sum or Contract Time.

   If you consider that a change in Contract Sum or Contract Time is required, please
   submit your itemized proposal to the Resident Engineer immediately and before
   proceeding with this work. If your proposal is found to be satisfactory and in
   proper order, this Field Order will in that event be superceded by a change order.

DESCRIPTION:

ATTACHMENTS:

By _____
        Resident Engineer   **Date**

Figure 9-5   Field order/proposal request.

MAINLINE BRIDGE IMPROVEMENT PROJECT

☐ **REQUEST FOR INFORMATION**

☐ **REQUEST FOR CHANGE**

MBIP

CONTRACTOR   Carr & Duff, Inc.

CONTRACT NO.   EC-1 _____ SUBJECT _____ DATE _____

CATEGORY:

RFI NO. _____

RFC NO. _____

CM USE ONLY

☐ INFORMATION NOT SHOWN ON
    CONTRACT DOCUMENTS

☐ INTERPRETATION OF CONTRACT
    REQUIREMENTS

☐ COORDINATION PROBLEM

☐ CONFLICT IN CONTRACT REQUIREMENTS

☐ CHANGE IN CONDITIONS

☐ OTHER _____

REFERENCE:

CONTRACT DRAWING NO. _____

SHOP DRAWING NO. _____

SPECIFICATION NO. _____

REQUEST

_____

_____

_____

_____

_____

_____

_____

_____

REPLY REC'D BY _____   REQUEST BY _____
              (Date)   (Name)

REPLY

_____

_____

_____

_____

CC: RESPONSE BY _____

CN NO. (assigned)_____   DATE _____

Figure 9-6 Request for information/request for change.

MAINLINE BRIDGE IMPROVEMENT PROJECT

PRELIMINARY ENTITLEMENT EVALUATION

MBIP

TO: DPM CONTROLS
FROM: _____
RE: CONTRACT No._____   Date: _____

DESCRIPTION

_____
_____
_____
_____

EVALUATION

_____
_____
_____
_____
_____

ENTITLEMENT TO CONTRACT MODIFICATION: ☐ YES   ☐ NO

CLASSIFICATION:

☐ INFORMATION NOT SHOWN ON
    CONTRACT DOCUMENTS

☐ INTERPRETATION OF CONTRACT
    REQUIREMENTS

☐ COORDINATION PROBLEM

☐ CONFLICT IN CONTRACT REQUIREMENTS

☐ CHANGE IN CONDITIONS

☐ OTHER_____

REFERENCE:

CONTRACT DRAWING NO _____
SHOP DRAWING NO._____
SPECIFICATION NO. _____

ACTION REQUIRED:

☐ COST ESTIMATE

☐ SCHEDULE EVALUATION

☐ OTHER _____

SIGNED_____
                RESIDENT ENGINEER

CC: DPM CONSTRUCTION

PCO NO. (ASSIGNED)_____

Figure 9-7   Preliminary entitlement evaluation.

MAINLINE BRIDGE IMPROVEMENT PROJECT

CHANGE ORDER EVALUATION

MBIP

CONTRACTOR, _____ CHANGE NOTICE NO. _____

CONTRACT NO. _____ RFC/RFI NO. _____ DATE _____

FOR REVIEW BY:

☐ ESTIMATING   ☐ OFFICE ENGINEER

☐ SCHEDULING   ☐ OTHER _____

☐ CONSTRUCTION MANAGER

Your review and impact assessment is requested for the attached Draft Change. Please complete this sheet and return to the Deputy PM - Controls. All order comments must be received by _____ . Includes copies of all worksheets and backup.

SCHEDULE IMPACT

COST IMPACT

IMPACT ASSESSMENT

☐ NO OBJECTION   ☐ OBJECTION (reason required)

REVIEWED BY:

NAME	SIGNATURE	POSITION	DATE
NAME	SIGNATURE	POSITION	DATE

Figure 9-8 Change order evaluation.

MAINLINE BRIDGE IMPROVEMENT PROJECT

CONTRACT CHANGE RECOMMENDATION

MBIP

TO: DAN LYNCH, P.E.
FROM: O'BRIEN-KREITZBERG & ASSOC., INC.
RE: CONTRACT No. _____ PCO NO._____ Date: _____

DESCRIPTION

_____

_____

_____

_____

_____

CONTRACTOR POSITION:

OKA POSITION:

RECOMMENDATION:

PROJECT MANAGER: _____ DATE: _____

CC: SEPTA CONTRACT OFFICER

Figure 9-9 Contract change recommendation.

MAINLINE BRIDGE IMPROVEMENT PROJECT

RECORD OF CHANGE ORDER NEGOTIATION

MBIP

CONTRACTOR:	CONTRACT NO.	DATE:
	PCO NO.	CO NO .
	RFI NO.	FI NO.

CONTRACTOR REPRESENTATIVES	OWNER REPRESENTATIVES
NAME AND TITLE	NAME AND TITLE
NAME AND TITLE	NAME AND TITLE
NAME AND TITLE	NAME AND TITLE

DESCRIPTION OF NEGOTIATION

ESTIMATE	NEGOTIATIONS WERE CONCLUDED WITH CONTRACTOR ON	
PROPOSAL	REVISED CONTRACT COMPLETION	
FINAL PRICE ACTUALLY AGREED TO	ADJUSTMENT IN TIME (ATTACH JUSTIFICATION)	
REVIEWING OFFICIAL SIGNATURE	TITLE	DATE

Figure 9-10 Record of change order negotiation.

MAINLINE BRIDGE IMPROVEMENT PROJECT

MEETING ATTENDANCE

MBIP

MEETING TYPE:			NO.
DATE	TIME	LOCATION	

ATTENDEES

NAME	AFFILIATION	PHONE

Figure 9-11   Meeting attendance.

MAINLINE BRIDGE IMPROVEMENT PROJECT
REQUEST FOR
SERVICE SHUTDOWN

MBIP

CONTRACTOR_____ CONTRACT _____

REQUESTED BY_____ DATE_____

LOCATION:_____

_____

_____

NATURE OF WORK TO BE PERFORMED: _____

_____

_____

_____

_____

OUTAGE(S) REQUIRED:

☐ TRACK   ☐ TRACTION POWER   ☐ TRAIN SERVICE

☐ OTHER (DESCRIBE) _____

DURATION OF OUTAGE(S) REQUIRED

FROM _____ AM/PM ON _____ TO _____ AM/PM ON _____
                              DATE   DATE

CONCURRENCE:_____   _____
                    RESIDENT ENGINEER   DATE

- -

SEPTA ☐ APPROVES   THIS OUTAGE.
      ☐ DENIES

RESTRICTIONS _____

_____

_____

_____   _____
SEPTA PROJECT MANAGER   DATE

Figure 9-12 Request for service shutdown.

MAINLINE BRIDGE IMPROVEMENT PROJECT
TELEPHONE QUOTATION

DATE:_____

PROJECT_____ TIME:_____

FIRM QUOTING:_____ PHONE (   )_____

ADDRESS_____ BY_____

ITEM QUOTED_____ RECEIVED BY_____

WORK INCLUDED	AMOUNT OF QUOTATION

DELIVERY TIME   **TOTAL BID**

DOES QUOTATION INCLUDE THE FOLLOWING			If ☐ NO is checked, determine the following
STATE & LOCAL SALES TAXES	☐ YES	☐ NO	MATERIAL VALUE
DELIVERY TO THE JOB SITE	☐ YES	☐ NO	WEIGHT
COMPLETE INSTALLATION	☐ YES	☐ NO	QUANTITY
COMPLETE SECTION AS PER PLANS & SPECIFICATIONS	☐ YES	☐ NO	DESCRIBE BELOW

EXCLUSIONS AND QUALIFICATIONS

AGENDA ACKNOWLEDGEMENT   TOTAL ADJUSTMENTS

Figure 9-13 Telephone quotation.

MAINLINE BRIDGE IMPROVEMENT PROJECT

TIME AND MATERIAL REPORT

MBIP

CONTRACT NO. _____ DATE: _____

CONTRACTOR _____ REPORT NO. _____ PAGE _____ OF _____

The following work was performed this date requiring use of the Labor Force, Materials, and Equipment listed hereon:

DESCRIPTION OF WORK PERFORMED: _____

_____ Certified Correct by _____ Date _____

_____ Contractor's Representative

LABOR				EQUIPMENT			
Name	Trade	Hours	Make	Model	Description	Hours
		OT				
		ST				
		OT				
		ST				
		OT				
		ST				
		OT				
		ST				
		OT				
		ST				
		OT				
		ST				
		OT				
		ST				
		OT				
		ST				

MATERIALS, SPECIAL FORCES AND SERVICES

Quantity	Unit	Description

Verified by: _____ _____
                Inspector   Date

Approved : _____ _____
           Resident Engineer   Date

IMPORTANT

This form must be submitted not later than the day following the date the work was performed.

Figure 9-14 Time and material report.

MAINLINE BRIDGE IMPROVEMENT PROJECT

FIELD INSTRUCTION

MBIP

NAME	FI NO.
ADDRESS	CONTRACT

CONTRACTOR:_____

SUBJECT:_____

INFORMATION OR ACTION:_____

_____
_____
_____
_____
_____
_____
_____
_____
_____
_____
_____
_____
_____
_____
_____
_____
_____
_____
_____
_____
_____

NOTE: PLEASE ADVISE CM IF THIS INSTRUCTION RESULTS IN ANY COST/TIME IMPACT PRIOR TO START OF WORK.

___ CONFIRM IN WRITING   ___ FILE   ___ INSTRUCTIONS   ___ URGENT ACTION

RESIDENT ENG'R:_____   DATE:_____

ATTACHMENTS:

Figure 9-15   Field instruction.

MAINLINE BRIDGE IMPROVEMENT PROJECT

NON COMPLIANCE REPORT

MBIP

CONTRACTOR _____	DATE _____
CONTRACT NO. _____	NCR NO. _____

DESCRIPTION OF NONCOMPLIANCE (equipment, materials, etc.)

DISPOSITION (Check one):

☐ ACCEPT   ☐ REJECT   ☐ RETEST   ☐ REPLACE   ☐ REWORK

DISPOSITION INSTRUCTIONS:

INSPECTOR DATE	RESIDENT ENGINEER DATE

CORRECTIVE ACTION TAKEN:

CORRECTIVE ACTION COMPLETE AND ACCEPTED:

RESIDENT ENGINEER DATE	INSPECTOR DATE

Figure 9-16   Noncompliance report.

MAINLINE BRIDGE IMPROVEMENT PROJECT
SALVAGED MATERIAL RECEIPT

MBIP

CONTRACT NO. _____   DATE: _____

CONTRACTOR: _____

ITEM	SIZE/ DESIGNATION	QUANTITY	REMARKS

☐ THE ABOVE MATERIAL WAS RECEIVED AT: _____
_____ SATISFYING APPLICABLE CONTRACT REQUIREMENTS.

☐ THE ABOVE MATERIAL IS TRANSMITTED TO: _____
_____ VIA (CONTRACTOR/SEPTA) TO BE UNLOADED BY
(CONTRACTOR/SEPTA) SATISFYING APPLICABLE CONTRACT REQUIREMENTS.

CONTRACTOR: _____ CONST. MGR: _____ SEPTA: _____

Figure 9-17 Salvaged material receipt.

Constructive Change Orders

The concept of constructive change orders originated in the federal procurement system, specifically in the agency appeals boards. In *Sweet on Construction Industry Contracts,* Justin Sweet states:[*]

> To keep a dispute in the agency appeals board despite the absence of a written change order, the boards of contract appeal developed the constructive change, a fiction under which the dispute is treated as if a change order had been issued.

The concept by the boards of appeal led to formalization of the concept in federal Standard Form (Procurement) 23-A:

> (b) Any other written order or an oral order (which terms as used in this paragraph (b) shall include direction, instruction, interpretation or determination) from the Contracting Officer, which causes any such change, shall be treated as a change order under this clause, provided that the Contractor gives the Contracting Officer written notice stating the date, circumstances and source of the order and that the Contractor regards the order as a change order.

Change Order Scenarios

As the constructive change concept became established in the federal establishment, it also became accepted in other jurisdictions. The first element of a constructive change order is that a change has been imposed on the contractor. The nature of such changes is far-reaching, as the following scenarios indicate:

1. The contractor plans to perform the work a certain way, but the architect insists on a different approach.

---

[*]Justin Sweet, *Sweet on Construction Industry Contracts* (New York: John Wiley & Sons, 1987), para. 12.13.

2. The contractor submits a material that has been certified as meeting the specifications. The architect directs the use of a different grade of material that also meets the specification but is more expensive.

3. Other changes cause delay, but the owner refuses to grant a time extension.

4. The contractor is required by the architect to provide services that the contractor states are not part of the work scope.

5. The contractor follows procedures as described by the specifications and does not get the desired results. The architect insists that the contractor shall do what it takes to get it right.

6. The contractor carries out standard testing prior to turning over part of a facility. The architect requires an additional level of testing before accepting the unit.

7. The contractor is required to service and test a complex piece of machinery. Using information from the manufacturer, the contractor cannot achieve the required results. The contractor pays for consultation by a manufacturer's representative, who identifies that special modifications had been made to this machinery. This complicates the process, making it more expensive. When asked, the owner acknowledges that customized changes had been made to the machinery.

8. The contractor's marble subcontractor has completed the model bathroom for a high-scale hotel and requests approval. The architect inspects the site and model and agrees that the installation is technically correct but says it is not aesthetically acceptable. The architect describes the parameters that would be acceptable.

9. The contractor advises the owner that excessive noise and vibrations from the owner's adjacent plant (a situation not made known to the contractor at bidding) is interfering with the contractor's ability to perform the work. The owner advises that it is acceptable to the owner if the contractor wants to work on the quiet second or third shifts.

10. The owner emphatically points out that the project must be completed before the winter season. The contractor points out that the contract completion date is not until spring. The owner states that completion before is vital to ensure spring completion. The owner threatens default if the contractor does not complete before winter.

Types of changes

The scenarios given above involve the following types of changes:

1. *Unwarranted interference.* Conflicting approaches to the work means and methods to be used.

2. *Added scope.* Better grade of material required.

3. *Constructive acceleration.*   Changes cause delay, but the owner refuses time extension.

4. *Added scope.*   Contractor required to provide services beyond scope.

5. *Defective specifications.*   Specification procedures do not produce desired results.

6. *Excessive inspection.*   Additional inspection required.

7. *Superior knowledge.*   Service and test of complex machinery difficult because of undisclosed modifications.

8. *Added scope.*   Architect wants better marble pattern.

9. *Owner interference.*   Owner environment is distracting.

10. *Constructive acceleration.*   Owner requires completion date before contract date.

Order elements

The second element of a constructive change order is that an order (or directive) has been given. That is, that the owner (or owner's representative) "by his words or his deeds, must require the contractor to perform work which is not a necessary part of his contract."* In our scenarios, the order elements would be as follows:

1. *Means and methods by contractor.*   Architect insists on alternate approach.

2. *Better grade of material required.*   Architect (in approval process) directs scope increase.

3. Changes cause delay, but owner refuses time extension.   *Owner denial.*

4. Contractor required to provide services beyond scope.   *Architect direction.*

5. Specification procedures do not produce desired results.   *Architect direction.*

6. Additional inspection required.   *Architect direction.*

7. Service and test of complex machinery difficult.   *Owner omission.*

8. Architect wants better marble pattern.   *Architect direction.*

9. Owner environment is distracting—owner offers more expensive solution. *Owner direction.*

10. Owner requires completion before contract date. *Owner direction.*

   For the order element to be valid (i.e., bind the owner), it must be directed by an authorized representative. If the contractor has actual or constructive knowledge that the representative's authority is limited, the contractor cannot successfully claim constructive change based upon direction by that representative.

---

*Industrial Research Associates Inc., DCAB No. WB-5, 68-1.

USPS Course Material

The following comments on constructive changes are extracted from USPS administrative course material:

Constructive Changes

Actions which are construed as changes even though they do not fit the clause's requirements for change orders are called constructive changes. (The term "constructive" comes from construe or interpret.) Such changes can occur in all kinds of contracting.

Any conduct of a USPS representative requiring a contractor to perform work not required by the contract can, in the right circumstances, be deemed a constructive change. Constructive changes can be classified as follows:

1. Interpretations of CORs or inspectors. The PSBCA and the courts have frequently found that individuals who do not have official authority to change a contract can, by their actions or by their interpretation of specifications, actually alter the contract. There is often a fine line between what is technical monitoring and what is direction of a change. In such cases, the contractor may be allowed administrative relief for the additional cost of the work if a change is found. Administration of the clause is difficult, of course, because it is hard for the contracting officer to know in every case what is going on and what actions are being taken by field personnel.

2. Error in interpreting the specifications. Administration of the clause can be particularly difficult when the contractor makes continual requests for interpretation of the specifications or guidance on work methods. The interpretations or decisions made by USPS can lead to contractor claims.

If a contracting officer with authority to make changes makes an error of interpretation, there is ample legal precedent for the error being considered a change. For example, if the contract gives the contractor freedom of choice between various methods of doing the work and the contracting officer insists after award that the contractor must use method A instead of a less expensive method B, a constructive change results. It is likely to be decided that the contracting officer has indeed made a change and it would not be equitable to penalize the contractor because the change was not in writing.

It should be noted that the constructive change order doctrine also applies in situations where a performance specification is used and the contracting officer or another USPS representative insists on a different standard of performance—that is, where USPS's interpretation of the specifications requests the contractor to do more than the contractor expected.

3. Failure to issue a change to correct a defective specification. Many board and court cases have found that failure to issue a change order in the case of a defective specification is basis for declaring that a constructive change had been made. The reasoning has been that, since the contracting officer should have issued a change order but did not, the board or court will do what should have been done so as to allow the contractor some recovery for the additional work and wasted effort necessitated by the specification problems.

The contracting officer should attempt to resolve difficulties with specifications as soon as possible. The Postal Service will likely end up paying for additional work and effort the contractor spends in trying to deal with the defective specifications.

4. Changes outside the scope. A contracting officer may issue a written change order for work that is later found by the board or courts to be outside the general scope of the contract. For example, although the Changes clause allows for unilateral orders changing the place of delivery, the U.S. Court of Claims has found that a substantial difference in point of delivery under a transportation contract was actually outside the scope. In that particular case, the place of delivery was the subject of the contract. Also, in supply contracts, the right of the Postal Service to make changes in the specification is limited to cases in which the goods are to be manufactured for USPS. A change ordered in a standard item would be beyond the scope of the contract.

5. Acceleration. Constructive changes occur when the Postal Service adds performance requirements but fails to extend the completion date in proportion to the additional work to be performed. Since the Postal Service is ordering more work in the same delivery period, the extra cost to the contractor is the difference in cost between performance "as offered" and performance "as accelerated."

The most common constructive acceleration situation involves an order by the contracting officer to meet the original schedule despite the occurrence of excusable delays. This has the same effect as requiring the contractor to perform the work in a shorter time than provided in the contract.

The contracting officer or authorized representative should take special care to grant allowable time extensions in a timely manner.

Failure to cooperate. The Contracting Officer and all other Postal Service representatives have an implied duty to cooperate with the Contractor. This duty includes making sure that the Postal Service keeps all promises it made under the contract (such as furnishing Postal Service property), as well as refraining from actions that would hinder the Contractor's progress. For example, Postal Service actions resulting in a Contractor having to alter the planned sequence of performance are likely to be considered constructive changes. If the actions of a USPS representative are deemed to have failed to keep USPS's side of the bargain or to have hindered the Contractor, the Board or the court will grant the Contractor equitable relief.

Except with respect to defaults of subcontractors, the Contractor will not be in default by reason of any failure in performing the contract in accordance with its terms (including any failure by the Contractor to make progress in the prosecution of the work that endangers performance) if the failure arises out of causes beyond the control and without the fault or negligence of the Contractor. Such causes may include, but are not restricted to, acts of God or of the public enemy, acts of the Government in its sovereign capacity or of the Postal Service in its contractual capacity, and unusually severe weather, but in every case the failure to perform must be beyond the control and without the fault or negligence of the Contractor.

When an excusable delay has occurred, the Contracting Officer is responsible for modifying the contract to recognize the excusable delay and increase the time for performance. If the Postal Service still needs the produce or needs the work done by the original delivery day, the Contracting Officer may require continued performance without modifying the performance period—in effect, accelerating the Contractor's performance. In that case, as noted in above, there is a good chance that a constructive change has occurred. If the excusable delay requires the Contractor to go to additional effort in order to meet the original schedule, the Postal Service will normally have to pay for the acceleration.

Suspensions and Delays Clause

Clause B-16, Suspensions and Delays, addresses situations in which the performance of all or any part of the work is delayed, suspended, or interrupted by an order or act of the Contracting Officer administering the contract, or by the Contracting Officer's failure to take actions required by the contract within the time specified in the contract. The clause provides that in such a situation, the Contractor must receive an equitable adjustment of the price (excluding profit) for any increase caused in the cost of performance, and the delivery or performance dates must be adjusted as well. The contract must be modified accordingly.

The clause states that these adjustments will not be made, however, if performance would have been delayed or interrupted by another cause or if any other item or condition in the contract provides for or excludes adjustments. Further, claims will not be allowed for any cost incurred more than 20 days before the Contractor has notified the Contracting Officer or the Contracting Officer's Representative.

The Court of Claims stated the standard for a constructive change [Calfon Construction, Inc. v. U.S., 18 Ct. Cl. 426 (1989)]:

> To recover under the contract's changes clause on the basis of a constructive change—for work beyond that required by the contract, but without a formal change order—the contractor must show that the requirements of the contract have been met.

Whether an action is or is not a constructive change can be a close call [as shown in the following examples]:

Constructive Change

In Al Johnson Construction Co. v. U.S. (20 Ct. Cl. 184; 1990), Johnson was a contractor on a Mississippi River control structure project. Installation of the dewatering system was a performance specification part of the contract. The Army Corps of Engineers told Johnson that the dewatering plan was inadequate. The Corps did not make a direct order to increase the number of wells, nevertheless, the court did find that the Corps actions constituted a constructive change.

Not a Constructive Change

In RPM Construction Co. v. U.S. (ASBCA No. 36,965, 90-3; 1990) the contractor was directed to install additional groundwater monitoring wells under a contract to install an underground fuel tank. The ASBCA held that additional wells were not a constructive change because the additional work was directed by the state environmental agency rather than the federal government.

Equitable Adjustment

The constructive change concept keeps non-agreed, non-written issues in the change order process from equitable resolution. It does not cover non-causation issues such as: force majeure; and/or financial loss due to price rises, low bids, etc.

11

Change Orders and Claims

If a job has lost money, experience suggests that the pieces (i.e., change orders) often—even usually—do not nearly add up to the whole loss. The successful identification of constructive change orders may help to close the gap, but a gap often does remain.

To close the gap, the project documentation needs to be reviewed to determine which (if any) of the following scenarios applies:

Cardinal change due to change orders

Delays due to change orders

Acceleration
- Directed
- Constructive

Overtime costs
- Premium costs
- Inefficiency

Loss of productivity due to change orders
- Disruption, ripple effect
- Crowding

Measured mile

Consequential cost

Cardinal Change due to Change Orders[*]

A *cardinal change* in a contract is a change (or group of changes) required by the owner that results in a significant breach of contract. This breach, in turn, opens the door to substantial recovery of losses. The cardinal change is so significant

---

[*]Kenneth C. Gibbs, Jack W. Fleming, and James L. Ferro, Chap. 10, "The Cardinal Change," in *Construction Change Order Claims*, ed. Robert F. Cushman and Stephen D. Butler (New York: Wiley Law Publications, John Wiley & Sons, Inc., 1994).

that the contractor is not limited by the provisions of the contract in recovering costs and/or damages from the owner.

Sometimes a cardinal change results from one event, such as collapse of a building in progress (Edward R. Marden v. United States, 442 F. 2d 364, 194 Ct. Cl.; 1971) or a change directing the performance of work materially different from that in the contract (Allied Materials & Equipment Co. v. United States, 569 F. 2d 562, 215 Ct. Cl. 406; 1978).

In other cases, the cardinal change is not a single change but an evolution of changes that finally overwhelm the contractor's ability to manage the contract. For instance, in Charles G. Williams Construction, Inc. v. United States (ASBCA No. 33766; 1989), the owner issued 26 change orders to correct defective drawings and specifications. On each of the change orders, the contractor received 15 percent for overhead. The board ruled that this figure was not sufficient to compensate for the cumulative impact of the deficient design. The contractor was awarded additional compensation for the cumulative impact costs.

Origins

Early cases, which started the path to the cardinal change concept, sought recovery from lost profit when the owners, for their own purposes, deleted major portions of a project. In McMaster v. State of New York (108 N.Y. 542, 15 N.E. 417; 1888), the contractor was to supply and cut sandstone for the exterior walls of 10 wards (5 male, 5 female) on an administration building and several outbuildings at the Buffalo State Asylum for the Insane. After all but the five female wards were completed, managers for the state ordered McMaster to remove all stone from the grounds within 10 days. The five female wards were never built. The court awarded lost profits due to the state's breach of contract in omitting the five female wards from construction.

In 1915, in Kieburtz v. Seattle (84 Wash. 196, 146 P. 400; 1911) the court found that the contractor was entitled to damages caused by changes and lost profits from deleted portions of the original contract work. The Washington State Supreme Court found:

> Where a municipality lets work of a public nature to a contractor to be performed according to specific plans and specifications at a stated price for the completed work, and afterwards radically or materially changes the plan of the work so as to increase the cost of performance or orders and directs the contractor to perform work or finish material not within the contemplation of the original contract, the municipality becomes liable to the contractor for the increased cost of the work, or for the extra cost of the labor and material.

In County of Greenlee v. Webster (25 Ariz. 183, 215 P. 161; 1923), the Arizona Supreme Court found:

> Many alterations in the plans and specifications, resulting in the alteration of the quantity, location and extent of the work as originally estimated and described in the contract. . .were unreasonable and had the effect of changing the general character of the contract. . . . [T]he changes in the alignment of the road were material

and substantial changes in the contract, which . . .greatly increased the rock excavation, . . . compelled the use of pack animals instead of wagons to move excavated material, increased the cost of blasting on account of passing through a settlement of houses, also because of several railroad crossings, and necessitated the removal of material three times in connection with construction of said crossings.

The court found that the proper basis for recovery was the reasonable cost of labor and materials. This essentially quantum meruit approach continues to underlie cardinal change today.

In the federal courts, a 1937 case, General Contracting & Construction Co. v. United States (84 Ct. Cl. 570), used the cardinal change approach. The contract was to construct several buildings and additions to others at the Veteran's Hospital in Somerset, New Jersey. After the contract was signed, but before work began, the contracting officer notified the contractor to omit one nurses building. This was followed by a change order that deleted the building and reduced the price by $99,000. The court found that the contracting officer's authority "did not vest him with the authority to eliminate entirely from the Contract Building 17." It held that this was a cardinal change and awarded payments made in regard to the elimination of the building as well as anticipated profits and overhead.

In another case involving a VA hospital, Wunderlich Contracting Co. v. United States (351 F. 2d 956, 173 Ct. Cl. 180; 1965), the plaintiff claimed a cardinal change. In 1950, Wunderlich was to construct a 14-building hospital complex. The plans and specifications were inadequate and had many ambiguities. Extensive changes and corrections were required by the government. These required Wunderlich to prepare 470 estimates. Most of the estimates were based on actual costs and were incorporated into 35 change orders.

The court acknowledged that 35 change orders was extensive and the "plaintiff's performance has been lengthier and costlier than anticipated. . . .but in the long run they constructed essentially the same project as described in the contract." The court went on to say that the changes were not "so extensive" that they constituted a cardinal change. Odds are that if tried today on the same set of facts, Wunderlich would be ruled a cardinal change.

Further rejections of claims of cardinal change were seen in the 1950s and 1960s for similar reasons:

> F. H. McGraw & Co. v. United States (130 F. Supp. 394, 131 Ct. Cl. 501; 1955) changes to plans and specifications were "extensive and complicated, . . . would materially effect [sic] and curtail the work of the mechanical trades and thus the finishing trades. . . . such changes were not sufficient to constitute a Cardinal Change.

The project added two 10-story wings to an existing hospital building, a separate 3-story building, and other additions and changes.

In Aragona Construction Co. v. United States (165 Ct. Cl. 382; 1964), there was a major substitution of materials. The court found no cardinal change, saying the work done was "essentially the same work as the parties bargained for when the contract was awarded." The hospital was in the same location, looked the same, and had the same facilities and number of floors and rooms.

In another mid-1960s case, J. D. Hedin Construction Co. v. United States (347 F. 2d 235, 171 Ct. Cl. 70; 1965), the court found that a series of changes to the contract for construction of a VA hospital, taken cumulatively, did not change the fundamental work product. The court did find damages for delay due to deficient plans and specifications.

Comments on cardinal change

The cardinal change can be a home run for the contractor. The court or board has great discretion in whether or not to declare a cardinal change. It has been said that the cardinal change concept was invented by the courts to give them a means to establish equity in worthy cases. Claimants should be mindful of the fact that the cardinal change is not a right but, rather, must be earned with a worthy claim.

A strong claim for cardinal change can, and should, have subportions for constructive change, delay, disruption, and so on as appropriate. Thus, if the court or board does not find cardinal change, there are fallback positions that can salvage a substantial portion of the claim.

Delays due to Change Orders

Delay claims may be brought under the contract changes clause. In the John Doe example in Chap. 6, the two change orders with time impact (CO #1, duct bank, and CO #2, sinkhole) could claim cost for delay either under the changes clause or the differing site conditions clause. The cost of time can be calculated as described in Chap. 7.

There is another form of delay that can result from change orders. If the cumulative impact of change orders causes a total delay that pushes weather-sensitive work into an unfavorable weather period (such as either a rainy season or a winter season), then the time lost on critical activities in that unfavorable period would become another change order. Weather delays are usually compensated in time but not money. In this instance, the delay days would be compensable. If the owner does not accept the rationale that change orders forced the work into an unfavorable weather period, the claim for days would be made under the concept of constructive change order and/or constructive acceleration.

Acceleration

Constructive acceleration, as discussed previously, occurs when a time extension is merited but not granted. This is the usual acceleration in construction. Direct acceleration occurs when the owner directs the contractor to accelerate the project. The direction is usually specific as to the amount of acceleration required and as to the method.

The CPM baseline can be useful in identifying the optimum way (or ways) to accelerate the project. Figure 11-1 is a time-scaled warehouse portion of the John Doe Project (original baseline).

To accelerate, activities on the critical path must either be shortened or resequenced. If the need to accelerate were recognized before work started, the critical work on piles, pile caps, grade beams, and slabs (Activities 13-17, 17-18, and 18-29, respectively) could be done in parallel with the site preparation package. This would pick up 34 workdays without shortening any activities.

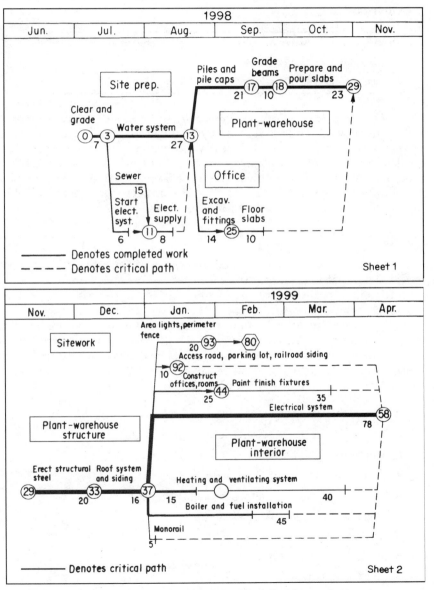

Figure 11-1   Summary diagram drawn to time scale (plotted to early times).

After the plant-warehouse is closed in (Event 37), the critical path is through the electrical system (78 days). The next (almost) critical path is masonry (Events 37-39-42) and drywall (42-44) (total of 25) plus ductwork (Events 44-46-52) and insulate (52-58) (total of 40). If the electrical work is put on overtime, up to 13 days [i.e., 78 − (25 + 40)] could be accelerated without affecting other activities.

From Fig. 11-2, the electrical overtime can be planned activity-by-activity; the following acceleration is 40 percent based on a 60-hour week (6 days @ 10 hours per day):

Description	Activity	Base time	Accelerated time	Days saved
Power conduit	38-43	20	12	8
Branch conduit	43-49	15	9	6

At this point, HVAC becomes critical by one day. Five days could be gained if the restraint (Activity 44-46) could be deleted. If the restraint is deleted (meaning that ductwork installation can proceed before drywall is complete), the 60-hour electrical work week would continue into Activity 49-50 (pull wire). This acceleration would gain six more electrical days. Although the total number of days would now be 20, the project would gain only 18 days because the finish work had only 18 days of float.

Overtime Costs

When overtime is used to achieve acceleration (directed or constructive), the obvious cost is premium time for craft labor. A less obvious cost is loss of labor efficiency (productivity) due to fatigue if the overtime is programmed over a long period of time.

Numerous studies have indicated the quantitative impact of the number of hours per week versus efficiency. These relationships, typically plotted over a number of weeks, indicate that efficiency drops steadily as scheduled overtime is used over a long period of time. The studies have had generally similar results and have been conducted by the following:

- U. S. Department of Labor
- National Electrical Contractors Association
- Mechanical Contractors Association
- Associated General Contractors
- American Subcontractors Association
- Associated Specialty Contractors
- National Contractors Association
- Business Roundtable
- Procter & Gamble

Studies on this topic date back to the early scientific management work done by Taylor and Gilbreth in the early 1900s. In 1972, Henry W. Parker and

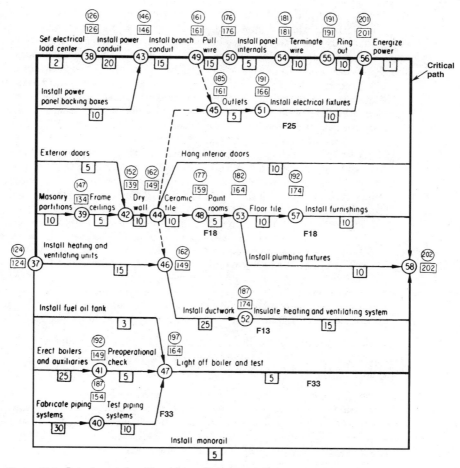

Figure 11-2 Interior work with early and late event times.

Clarkson H. Oglesby published their book *Methods Improvement for Construction Managers* (McGraw-Hill). This work documents many of the productivity studies performed at Stanford by Parker and Oglesby. Their work, as well as the work of John Fondahl and, later, John Borcherding, has been recognized by many as leading the construction industry.

Parker and Oglesby make the following statements in regard to scheduled overtime:[*]

> When labor is in short supply or when for some reason, an owner or contractor wishes to rush completion of a contract, there is a natural inclination to go to overtime on a scheduled or regular basis. This scheduled overtime, by definition, involves regular work for more than 40 hours per week. It is distinguished from

---

[*]Henry W. Parker and Clarkson H. Oglesby, *Methods Improvement for Construction Managers* (New York: McGraw-Hill, 1972), pp. 139–140.

the intermittent overtime required to finish a concrete pour, to do emergency work of short duration, or that may be incidental to many projects.

A study of contractors in the Detroit area on the extra costs of regular overtime brought out the problems of fatigue, absenteeism, mental attitude, personnel turnover, and human error as contributors to the decrements in productivity. . .

A recent survey of the effects of scheduled overtime in industrial construction in labor-intensive work (i.e., not equipment paced) indicates that it has severe adverse effects such as disrupting local wage scales, magnifying apparent labor shortages, and reducing labor productivity. It substantially increases costs but does not materially advance job schedules.

Scheduled overtime operations usually bring a sharp initial drop in productivity as workers repace their daily output to a tolerable level. This is followed by a fairly substantial recovery by the end of the first week. The recovered level may hold fairly steady for the following two to three weeks; it then assumes a steady decline for the next two or three weeks. There is then an even greater drop in productivity which levels at a low point after 9 to 12 weeks of sustained operation. After 6 to 12 weeks, total productivity is less than that for a 40-hour week although labor may be earning twice as much for the time on the job (a 60-hour week at double time for overtime earns a workman twice his 40-hour pay) . . .

It is of interest that, while the three evaluations shown. . .are for different kinds of construction labor, all show essentially the same results. Furthermore, these recent findings are strikingly similar to the surveys mentioned earlier that were published in 1918 and 1947 and were based on studies going back as far as 1900.

Professors Oglesby and Parker provide the basic reason behind the loss of efficiency with extended overtime in a discussion of "energy limitation" and "physical fatigue." Extracts from that discussion follow:[*]

Using the metric measure of energy, the kilocalorie, a man has the capacity of energy conversion of 5 kilocalories per minute and a storage or reserve capacity for work of 25 kilocalories. The sustenance of life's processes (basal metabolism) requires approximately 1 kilocalorie per minute, leaving 4 kilocalories available for work or play. It follows that any physical task that requires an energy output of 4 kilocalories or less per minute can be maintained for long periods. However, when a man is required to expend more than the 4 kilocalories per minute, energy from the reserve of 25 kilocalories is used. When the reserve is depleted, rest is required. Replenishment of the 25 kilocalories takes place at 3.5 kilocalories per minute, slightly less than the 4 kilocalories excess-energy conversion rate. It follows that a man could expend 29 kilocalories in one minute, or 65 kilocalories in ten minutes' work before requiring rest; and in either case he would require a little over seven minutes' rest. . .

Task-setters in industry have set production standards for years without being aware of energy consumption of some of the ordinary tasks. For example, ordinary walking (such as a postman delivering mail) requires 4 kilocalories per minute and thus if continued for 8 hours is the maximum physical exertion that should be required.

. . . Fatigue can be a short-term condition caused by temporary overexertion or a longer-term condition caused by a general exertion above the level that the body can tolerate and still maintain energy equilibrium. A condition of fatigue results in a marked decrement in attention or vigilance, which is reflected directly in accident rates. Studies indicate that accident rates tend to rise when production rates drop, both effects resulting presumably from an increase in fatigue level.

---

[*]Parker and Oglesby, pp. 133–136.

The Business Roundtable (200 Park Avenue, New York, N.Y. 10166) issued a report entitled "Scheduled Overtime Effect on Construction Projects" in November 1980. On the cover page, the heading "Scheduled Overtime Decreases Productivity" is prominent. The following information is extracted from that report:

Executive Summary

This paper reviews an analysis of the impact of scheduled overtime operation on construction projects and the inflationary effects of such operations. The data and findings cited in the previous Business Roundtable report (1974) on the subject have been found still valid and support the following conclusions:

- Placing field construction operations on a project on a scheduled overtime basis disrupts the economy of the affected area, magnifies any apparent labor shortage, reduces labor productivity, and creates excessive inflation of construction labor costs without material benefits to the completion schedule.

- Where a work schedule of 60 or more hours per week is continued longer than about two months, the cumulative effect of decreased productivity will cause a delay in the completion date beyond that which could have been realized with the same crew size on a 40-hour week.

Effects of Overtime on Construction Workers

When a project in an area is placed on a scheduled overtime basis, the movement of workers from other projects in the area to the overtime job creates an "auction" atmosphere. Other jobs go to overtime to hold their labor, and a bidding process is established. The local labor supply is fairly constant, and the additional productivity capacity of transient workers is offset by the reduced productivity of all workers on an overtime schedule. Usually, a major portion of the increase in numbers of workers in the affected areas is a result of permit workers in crafts who are less proficient or poorly qualified.

Disruptions created by unwilling or poorly qualified craft workers, longer working hours each week, increased absenteeism, and reduced effectiveness due to fatigue reduce the productive output of labor materially. On extended overtime, the reduced productivity of workers for a week's work is equal to or greater than the number of overtime hours worked.

Effects on Costs

The premium cost for overtime hours, plus the loss in productivity for the total hours worked, results in an unreasonable inflation of the unit labor cost.

Overtime vs. Productivity

Within narrow limits, workmen expend energy at an accepted pace established by long periods of adaptation. When the hours of work per day or per week are changed, there is an adjustment period. Studies reveal that scheduled overtime operations result in a sharp drop in productivity initially, followed by a fairly substantial recovery by the end of the first week. The recovery level of productivity may then hold fairly steady for a period of two to three weeks but show a steady decline for the following two to three weeks. After five to six weeks of operations, there is a further drop in productivity which levels out at a lower point after nine to twelve weeks of sustained overtime operation. It should be understood that this condition results from normal reactions and does not reflect the effect of other adverse factors such as labor, climate, and poor management.

Survey Results: National Constructors Association

A survey in the late 1960's by members of the National Constructors Association for the scheduled overtime Task Force covered 60 percent of their total membership and showed that 23 percent of their contracts worked on a scheduled overtime basis. They also reported that 20 percent of their dollar volume of construction was on an overtime basis. This indicated that 20 percent of a $2.8 billion labor cost was expended on an overtime schedule representing $560,000,000 of labor payroll.

The survey showed that 65 percent of the overtime schedules were established for the purpose of attracting labor; the remaining 34 percent were to maintain or accelerate construction schedules.

The number of hours per week varied from job to job, but 50 hours represented a conservative average. At 50 hours per week, the inflationary effect on construction labor cost was 60 percent of the cost on a normal 40-hour week. This indicated that the same volume of industrial construction could have been accomplished for $340,000,000 of labor cost if there had been no overtime involved. The added $220,000,000 represents inflation of construction labor cost for only this segment of construction.

As noted in the Roundtable study, the principal reason for the use of scheduled overtime is to attract a particular craft skill or to recruit workers, engineers, and management for remote site work. From that viewpoint, overtime can be justified—and the cost easily established.

Figure 11-3 reflects the results of the Roundtable study. The graph shows deterioration of productivity as scheduled overtime continues over time. The curves reflect decreasing productivity over time when scheduled overtime is in place. The curves are for 40 hours (base), 45 hours, 50 hours, and 60 hours. The reduced productivity is as follows:

	Hours worked	Equivalent hours of work	Hours paid (include premium) at $1\frac{1}{2}$	Net hours (productivity or efficiency)
@ 4 weeks	45	$(0.93 \times 45) = 41.9$	47.5	88.2%
	50	$(0.90 \times 50) = 45.0$	55.0	81.8%
	60	$(0.84 \times 60) = 50.4$	70.0	72.4%
@ 6 weeks	45	$(0.82 \times 45) = 36.9$	47.5	77.7%
	50	$(0.71 \times 50) = 35.5$	55.0	64.5%
	60	$(0.64 \times 60) = 38.4$	70.0	54.9%

For a reasonably short period of time, overtime can produce additional hours of work greater than the traditional 40-hour week. This is achieved at a cost of both the premium time and efficiency. However, if overtime is worked for extended periods, the effect is to pay the higher cost and net the equivalent of just less than the 40-hour week.

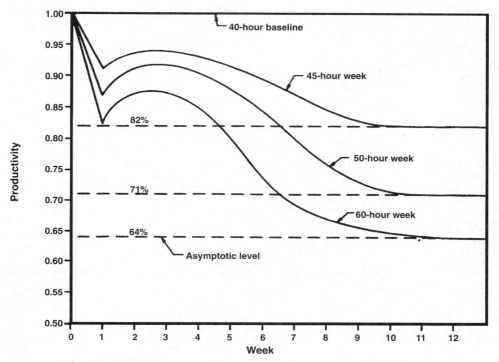

Figure 11-3   Overtime fatigue factors (production based on 40-hour standard week).

Extended overtime was common among the nuclear power plant workers in the 1970s and 1980s. The proof of inefficiency due to extended overhead was obvious in the high labor cost overruns in most of the plants completed in the 1980s. On one nuclear project, management, recognizing the problem of overtime fatigue, had two complete craft labor forces. The first worked four 10-hour days; then the second force took over for four 10-hour days. This was repeated over and over, weekends included. Through union agreements, all overtime was time and a half. Each shift got at least 8 premium hours (paid the equivalent of 44 regular hours). If the "rolling four-tens" included both Saturday and Sunday, 24 hours were at premium (the equivalent of 52 regular hours).

In July 1979, the Office of the Chief of Engineers, U.S. Army Corps of Engineers (COE), issued its *Modification Impact Evaluation Guide* (EP 415-1-3). This guide was issued to advise contracting officers of the potential impacts that contract modifications (i.e., change orders) could have on the remaining (or unchanged) work. Overtime was one of the impacts covered by the guide, which included a graph of efficiency versus time at different levels of overtime (Fig. 11-4).

Comparing the COE efficiency at the end of four weeks (Fig. 11-4) with that shown in the Business Roundtable report (Fig. 11-3) gives similar results:

	COE	Business Roundtable
45-hour week	97.5%	93%
50-hour week	96%	90%
60-hour week	78%	84%

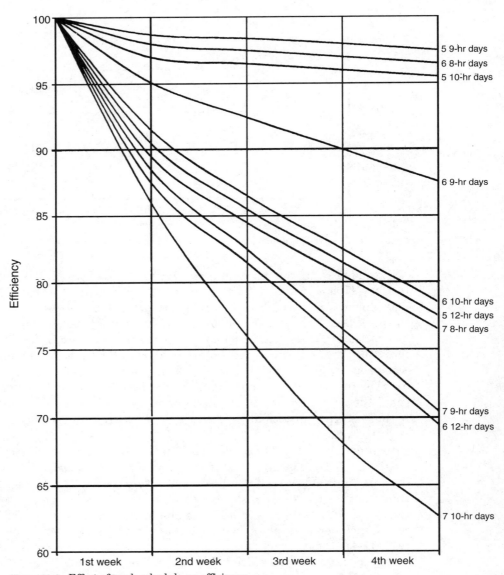

Figure 11-4   Effect of work schedule on efficiency.

Figure 11-5 is an overtime versus efficiency set of curves from "Overtime and Productivity in Electrical Construction" by the National Electrical Contractors Association (NECA; 1989). It is also similar to the COE and Business Roundtable curves:

		COE	Business Roundtable	NECA
Efficiency @ week 4	50-hour week	96%	90%	89%
	60-hour week	78%	84%	83%
Efficiency @ week 12	50-hour week	—	71%	72%
	60-hour week	—	64%	63%

Loss of Productivity due to Change Orders[*]

In the 1960s, a Mechanical Contractors Association (MCA) local committee brought forward the concern/concept that change often caused inefficiencies. The initial thrust was to include the cost of these problems/inefficiencies in the base change order. The concept (now known as "forward pricing") was mostly ignored by owners. Nevertheless, the MCA has continued to publish the factors, and NECA has also published similar impact factors.

Table 11-1 is a list of impact factors from Appendix B of the MCA *Labor Estimating Manual* (1986). In the manual, Appendix B includes a disclaimer that the factors are intended to be a reference only, pointing out that the factors:

- Could be too high or too low for specific individual cases
- Should be modified to suit your own work experience
- May vary from contractor to contractor, crew to crew, and job to job

The MCA manual is silent on the interrelationship between factors—that is, whether they compound or are additive. There is no single answer. The factors are guidelines that can be used to rationalize productivity impacts. These impacts are not limited to change orders as the causation.

Some of the factors lend themselves to measurement more than others. These include:

- Stacking of trades (same place at same time)
- Crew size inefficiency (usually means crowding or overmanning)
- Concurrent operations (may use equipment in less than optimum mode)

---

*William Schwartzkoff, *Calculating Lost Labor Productivity in Construction Claims* (New York: Wiley Law Publications, John Wiley & Sons, Inc., 1995).

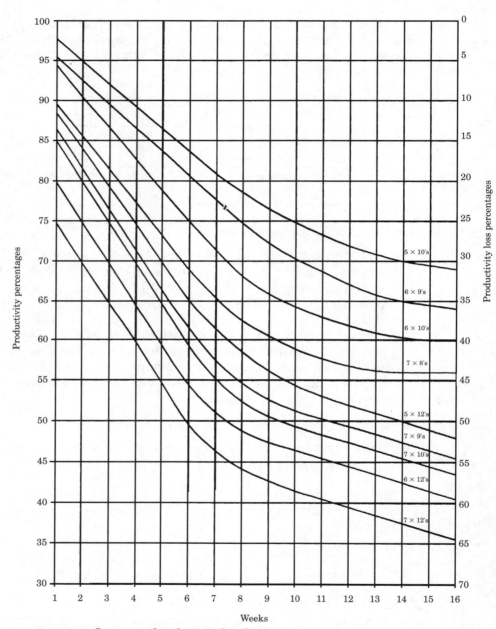

Figure 11-5   Summary of productivity loss data—overtime.

- Overtime (over extended period of time)

Some of the factors can cause or result in other factors, often termed "the ripple effect." Examples include:

- Stacking of trades could cause loss of morale.
- Crowding or overmanning could cause dilution of supervision.
- Overtime could cause fatigue.

The MCA and NECA impact factors are contractor-developed and can, thus, be considered self-serving. The Corps of Engineers' *Modification Impact Evaluation Guide*, on the other hand, was developed by a major owner-oriented agency. It offers the following on impacts on productivity due to modifications (change orders):

> Reduced labor productivity implies a loss from some established normal or antici-pated level of productivity. Although construction does not lend itself to definitive measurement of labor productivity, there are methods a contractor can use to quantify anticipated labor costs when preparing a bid. The most common tech-nique draws heavily on data derived from the contractor's past experiences, including any indicated trends, present labor pay rates, and anticipated labor rate increases during the life of the project.
> The actual labor productivity of a project affects the cost of labor for modifica-tions. On projects where actual labor productivity is running at or better than the

TABLE 11-1   Impact Factors

	Minor	Average	Severe
Stacking of trades	10%	20%	30%
Morale and attitude	5%	15%	30%
Reassigment of manpower	5%	10%	15%
Crew size inefficiency	10%	20%	30%
Concurrent operations	5%	15%	25%
Dilution of supervision	10%	15%	25%
Learning curve	5%	15%	30%
Errors and omissions	1%	3%	6%
Beneficial occupancy	15%	25%	40%
Joint occupancy	5%	12%	20%
Site access	5%	12%	30%
Logistics	10%	25%	50%
Fatigue	8%	10%	12%
Ripple	10%	15%	20%
Overtime	10%	15%	20%
Season and weather change	10%	20%	30%

contractor's anticipated level, data developed from analysis is appropriate for prepricing direct and impact costs of modifications. However, when the contractor's actual labor costs are higher (productivity lower) than those anticipated by the bid, actual experience data should be considered. Depending on the degree to which contractor mismanagement has contributed to the higher labor costs, the estimator may find it expedient to use a combination of actual and anticipated productivity projections in arriving at a reasonable labor cost figure for the modification. This does not imply that a modification should be priced to reimburse the contractor for excess labor costs incurred because of inept management; however, it is possible to incur labor costs higher than those anticipated. The estimator must therefore take this into account when forward-pricing direct or impact-related modification labor costs.

Prepricing of impact costs arising from labor is the most difficult aspect of the estimating process for two reasons. First, the estimator must verify that these determinations are reasonable and well founded. Second, when negotiating it is necessary to convince the contractor that the determinations are reasonable. Most contractors (and many personnel within the Corps of Engineers) would prefer to leave the settlement of impact cost/time until after the modification work is performed. However, such a procedure is not recommended. The preferred approach is to anticipate the costs before the fact, and to include them in the cost estimate. [Figures 11-6 and 11-7] illustrate the effects of various situations on construction manpower efficiency. These figures are included as a source of general information and some estimators may find them helpful in supplementing other data generated in the development of modification cost estimates. However, the validity of the graphs has not been sufficiently tested to warrant their use in preference to established methodologies.

The lowest reasonable price for modifications is estimated by basing the direct and impact costs of labor upon the productivity level established. (Allowances for

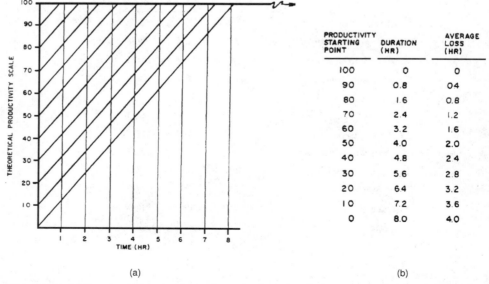

PRODUCTIVITY STARTING POINT	DURATION (HR)	AVERAGE LOSS (HR)
100	0	0
90	0.8	0.4
80	1.6	0.8
70	2.4	1.2
60	3.2	1.6
50	4.0	2.0
40	4.8	2.4
30	5.6	2.8
20	6.4	3.2
10	7.2	3.6
0	8.0	4.0

(a)   (b)

Figure 11-6   (*a*) Construction operations orientation/learning chart. 100 represents the productivity rate required to maintain scheduled progress. (*b*) Productivity losses derived from (*a*).

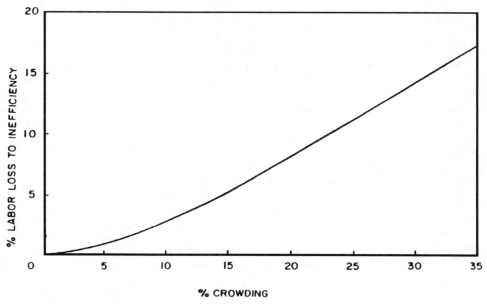

Figure 11-7   Effect of crowding on labor efficiency.

labor impact costs compensate the contractor for losses in productivity.) Typical causes of labor productivity loss on the unchanged work resulting from modifications are as follows.

(1) Disruption. The contractor's progress schedule represents the planned sequence of activities leading to final completion of the project. Workers who know what they are doing, what they will be doing next, and how their activities relate to the successful completion of the project develop a "job rhythm." Labor productivity is at its optimum when there is good job rhythm. When job rhythm is interrupted (i.e., when a contract modification necessitates a revision of the progress schedule), it affects workers on both the directly changed and/or unchanged work and may result in a loss of productivity.

(a) Disruption occurs when workers are prematurely moved from one assigned task to another. Regardless of the competency of the workers involved, some loss in productivity is inevitable during a period of orientation to a new assignment. This loss is repeated if workers are later returned to their original job assignment. Learning curves which graph the relationship between production rate and repeated performance of the same task have been developed for various industrial tasks. The basic principle of all learning curve studies is that efficiency increases as an individual or team repeats an operation over and over; assembly lines are excellent demonstrations of this principle. However, although construction work involves the repetition of similar or related tasks, these tasks are seldom identical. Skilled construction workers are trained to perform a wide variety of tasks related to their specific trade. Therefore, in construction it is more appropriate to consider the time required to become oriented to the task rather than acquiring the skill necessary to perform it. One of the attributes of the construction worker is the ability to perform the duties of their trade in a variety of environments. How long it will take the worker to adjust to a new task and environment depends on how closely related the task is to prior experience or how typical it is to work usually performed by that craft. [Figure 11-6a] assumes that the worker will always be assigned to perform work within the scope of his trade, and that the average worker will require a maximum of one shift (8 hours) to reach full productivity. Full productivity (100 on the Theoretical Productivity Scale) represents optimum productivity for a given project. [Figure 11-6b] is a tabulation of productivity losses derived from [Fig. 11-6a].

(b) The time required for a worker (or crew) to reach full productivity in a new assignment is not constant. It will vary with skill, experience, and the difference between the old and new task. In using the chart or its tabulation, the estimator must decide what point on the Theoretical Productivity Scale represents a composite of these factors. For example, an ironworker is moved from placing reinforcing bars to the structural steel erection crew. The ironworker is qualified by past training to work on structural steel, but the vast majority of his experience has been with rebars, and the two tasks are significantly different. In view of this, a starting point of "0" is appropriate. The estimator can determine from the chart that a "0" starting point indicates the ironworker will need 8 hours to reach full productivity, with a resulting productivity loss of 4 hours. The Government's liability is then 4 hours times the hourly rate times markups. As a second example, assume the same ironworker is moved from placing reinforcing bars for Building A to placing reinforcing bars for Building B. The buildings are similar but not identical. A starting point of "90" is appropriate. The duration of only 0.8 hours is required to reach full productivity, and the productivity loss is 0.4 hours. The Government's liability would then be 0.4 hours times the hourly rate times markups.

(c) The contractor normally absorbs many orientation/learning cycles as labor forces are moved from task to task in the performance of the work. Only those additional manpower moves, caused solely by a contract modification, represent labor disruption costs for which the contractor is entitled extra payment.

(2) Crowding. If a contractor's progress schedule is altered so that more activities must be accomplished concurrently, impact costs caused by crowding can result. Crowding occurs when more workers are placed in a given area than can function effectively. Crowding causes lowered productivity; it can be considered a form of acceleration because it requires the contractor either to accomplish a fixed amount of work within a shorter time frame, or to accomplish more work within a fixed time frame. Granting additional time for completion of the project can eliminate crowding. When the final completion date cannot be slipped, increased stacking of activities must be analyzed and quantified.

(a) Activity stacking does not necessarily result in crowding—when concurrent activities are performed in areas where working room is sufficient, crowding is not a factor. But, if the modification forces the contractor to schedule more activities concurrently in a limited working space, crowding does result. Both increased activity stacking and limited (congested) working space must be present for crowding to become an item of impact costs.

(b) Crowding can be quantified by using techniques similar to those used for acceleration. [Figure 11-7] illustrates the curve developed to represent increases in labor costs from crowding. Before applying this curve, the estimator must determine whether crowding will occur and to what degree. For example, the assumption that the contractor's scheduling of the activities in question is the most efficient sequencing of the work must be verified. Perhaps more workers can work effectively in the applicable work space than the contractor has scheduled; if they cannot, perhaps the crowding is not severe enough to justify using the full percentage of loss indicated by the graph. (The graph should be interpreted as representing the upper limit of productivity loss.) In this case, the estimator's judgment of the specific circumstances may indicate that some lower increase factor is appropriate.

(c) For example, assume that the estimator decides that severe crowding will occur in the following situation: The contractor's schedule indicates three activities concurrently in progress in a limited area of the project. Each of these activities employs five workers, placing a total of 15 workers in the area. One of these activities has a duration of 10 days; the other two have 20-day durations. The modification has required that a fourth activity be scheduled concurrently in the same limited area. This additional activity requires three workers; it has a normal duration of 5 days. There are now 18 workers in an area that can only efficiently accommodate 15. The percent of crowding is 3/15 or 20 percent. On the graph [Fig. 11-7], 20 percent crowding intersects the curve opposite 8 percent loss of efficiency. To find the duration of crowding, the estimator multiplies the normal duration of the added activity by 100 percent plus the percent loss of efficiency. For this example, 5 days times 1.08 equals 5.4 days. Therefore, because of the inefficiency introduced by crowding, the added activity will require 5.4 days to complete. Likewise, on the three affected activities, the first 5 days of normal activity will now require 5.4 days. All four activities will experience loss of productivity resulting from an inefficiency factor equivalent to 0.4 of a single day's labor cost. This is calculated as follows:

Average hourly rate $\times$ hours worked per day $\times$ number of workers $\times$ 0.4 = \$ loss

or

$12.00 \times 8 \times 18 \times 0.4 = \$691$ plus normal labor markups

$3/18 \times \$691$ = direct crowding cost, and should be included in the
Direct Cost section of the modification estimate

$15/18 \times \$691$ = crowding on unchanged activities, and should be placed
in the Impact on Unchanged Work section of the
modification estimate

(3) Acceleration. Acceleration occurs when a modification requires the contractor to accomplish a greater amount of work during the same time period even though he may be entitled to an extension of time to accomplish the changed work. This is sometimes referred to as "buying back time." Acceleration should be distinguished from expediting. Expediting occurs whenever the modification would require the contractor to complete the work before the original completion date included in the contract. Per DAR 18-111, expediting is not permissible in the absence of approval by the Assistant Secretary of Defense (Manpower, Reserve Affairs and Logistics). Acceleration may be accomplished in any of the following ways:

(a) Increasing the size of crews. The optimum crew size (for any construction operation) is the minimum number of workers required to perform the task within the allocated time frame. Optimum crew size for a project or activity represents a balance between an acceptable rate of progress and the maximum return from the labor dollars invested. Increasing crew size above optimum can usually produce a higher rate of progress, but at a higher unit cost. As more workers are added to the optimum crew, each new worker will increase crew productivity less than the previously added worker. Carried to the extreme, adding more workers will contribute nothing to overall crew productivity. [Figure 11-8] indicate[s] the effect of crew overloading.

(b) Increasing shift length and/or days worked per week. The standard work week is 8 hours per day, 5 days per week (Monday through Friday). Working more hours per day or more days per week introduces premium pay rates and efficiency losses. Workers tend to pace themselves for longer shifts and more days per week. An individual or a crew working 10 hours a day, 5 days a week, will not produce 25 percent more than they would working 8 hours a day, 5 days a week. Longer shifts will produce some gain in production, but it will be at a higher unit cost than normal hour work. When modifications make it necessary for the contractor to resort to overtime work, some of the labor costs produce no return because of inefficiency. Costs incurred due to loss of efficiency created by overtime work are an impact element because the increase in overtime results from the introduction of the modification. Contractors occasionally find that to attract sufficient manpower and skilled craftsmen to the job, it is necessary to offer overtime work as an incentive. When this is done, the cost must be borne by the contractor; however, if overtime is necessary to accomplish modification work, the Government must recognize its liability for introducing efficiency losses. [Figure 11-4] is the result of study which attempted to graphically demonstrate efficiency losses over a 4-week period for several combinations of work schedules. These data are included merely as information on trends rather than firm rules which might apply to any project. Although [Fig. 11-4] data do not extend beyond the fourth week, it is assumed that the curves would flatten to a constant efficiency level as each work schedule is continued for longer periods of time.

(c) Multiple shifts. The inefficiencies in labor productivity caused by overtime work can be avoided by working two or three 8-hour shifts per day. However, additional shifts introduce other costs. These costs would include additional administrative personnel, supervision, quality control,

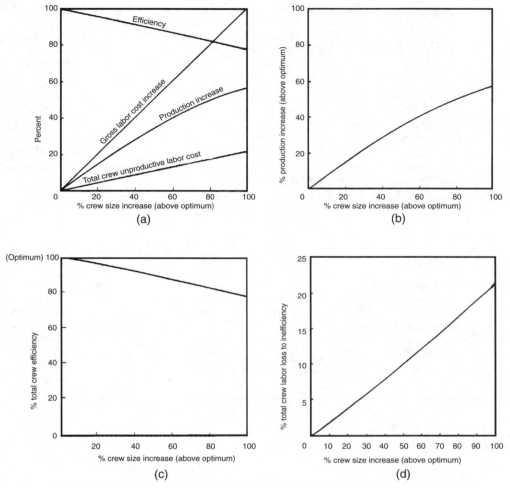

Figure 11-8 (*a*) Composite effects of crew overloading. (*b*) Unproductive labor of crew overloading. (*c*) Efficiency of crew overloading. (*d*) Production gain of crew overloading.

lighting, etc. Modifications that cause the contractor to implement shift work should price the impact cost as appropriate for the activity being accelerated. Environmental conditions such as lighting and cold weather may also influence labor efficiency.

(4) Morale. The responsibility for motivating the work force and providing a psychological environment conducive to optimum productivity rests with the contractor. Morale does exert an influence on productivity, but so many factors interact on morale that their individual effects defy quantification. A project's contract modifications, particularly a large number, have an adverse effect on the morale of the workers. The degree to which this may affect productivity, and consequently the cost of performing the work, would normally be very minor when compared

to the other causes of productivity loss. A contractor would probably find that it would cost more to maintain the records necessary to document productivity losses from lowered morale than justified by the amount recovered. Modification estimates do not consider morale as a factor because whether morale becomes a factor is determined by how effective the contractor is in labor relations responsibilities.

4.5 Quantification. The following example demonstrates how to use [Fig. 11-8] to quantify the impact costs of crew overloading. Assume that the contractor has planned a construction operation with a duration of 15 working days and an optimum crew size of 10. The modification now requires that the contractor accomplish this operation in 10 working days. The rate of production is the unit of work per amount of effort in mandays. The percent increase is new rate minus original rate divided by original rate times 100. Thus,

$$\frac{(1 \text{ job} \div 100 \text{ MD}) - (1 \text{ job} \div 150 \text{ MD})}{1 \text{ job} \div 150 \text{ MD}} \times 100 =$$

$$\frac{.01 - .0067}{.0067} \times 100 = 50 \text{ percent}$$

This represents a 50 percent increase in the crew's rate of production. From [Fig. 11-8a or d], it appears likely that 50 percent production gain can be achieved by increasing the size 80 percent. Other options could be implemented to speed up production: the optimum crew could work longer shifts, more days per week; a second crew could be placed in operation (if allowed by the nature of the work). However, for this example only increasing crew size is considered. The way to quantify the impact cost before the fact is:

	Original Plan	Accelerated Plan
Manpower	10	18
Hourly Rate	$12	$12
Crew Cost/Day (8 hours)	$960	$1,728
Duration (Working Days)	15	10
Crew Cost (Cost/Day × Duration)	$14,400	$17,280
Taxes, Insurance, Fringes (18 percent)	$2,592	$3,110
Total Crew Cost	$16,992	$20,390

Impact Cost (Accelerated − Original) = $3398 ($3400)

or

Impact Cost (Accelerated Plan × Efficiency Loss) =

$20,390 × 16.7 percent [from Fig. 11-8b] = $3,405 ($3,400)

The amount of $3,400 would be placed in the modification estimate, under "Impact on Unchanged Work" and identified by the activity involved. Increased cost of supervision, if necessary, is not included in this crew overloading analysis. Supervision must be costed separately, either as a separate item or as an element of Job Site Overhead, as appropriate.

Measured Mile

In this approach, the productivity problems due to a change order are measured by comparing either the predicted, or actual, labor of a particular trade with similar work of that trade at a more favorable point in time. The similar work is the "measured mile" (i.e., units of work produced per period of time). If similar work does not precede the change order work in this project, the contractor can use production records from a prior project.

The measured mile approach would seem to be a good approach. The problem is identifying a truly similar or analogous project, and if a project is identified, having good production records to support the claim of good production.

The measured mile approach compares the work of one trade with the same trade at another point in the project. It can also compare the work per day on one type of assembly (i.e., pouring elevated floor slabs) with the same type of assembly. Attempts to compare dollars earned per day of one period versus another is not a proper measured mile comparison.

Direct versus Consequential Damages

Actual damages can be either direct damages or consequential damages. *Direct damages* include costs such as additional field management due to extended time in the project, additional home office management and overhead tied up by the job, extended use of equipment, escalation of wages during the extended work period, increase in materials costs, continued items such as security and utilities, and other costs directly related to the longer project implementation period.

Consequential damages result from the change but are not a direct cost due to it. Consequential damages are also not necessarily part of the contract considerations and include claims for losses such as loss of bonding capacity, limitations on workload due to limited working capital, losses due to failure to undertake additional business, and loss of profit or income.

In Roanoke Hospital Association v. Doyle & Russell (Cir. Ct., Roanoke, Va.; review by S.C. of Va. #740212; 1975) the court discussed the distinction between direct and consequential damages:

> Direct damages are those which arise "naturally" or "ordinarily" from a breach of contract; there are damages which, in the ordinary course of human experience, can be expected to result from a breach. Consequential damages are those which arise from the intervention of "special circumstances" not ordinarily predictable. If damages are determined to be direct, they are compensable. If damages are determined to be consequential, they are compensable only if it is determined that the special circumstances were within the "contemplation" of both contracting parties. Where the damages are direct or consequential is a question of law. Where the special cir-

cumstances were within the contemplation of the parties is a question of fact. . . . As a general rule, contemplation must exist at the time the contract was executed.

In that case, the court agreed with the owner and the trial court that extended financing costs are direct damages. Thus, both the cost of interest on the construction loan during the extended delay, as well as the loss of revenue, which could have been anticipated from the facility, are predictable results of the delay on compensable direct damages.

However, the Virginia Supreme Court held that the increases in interest rates that occurred during this particular delay were caused by pressures of the financial market, and for that reason, the increases in the interest rates were in the category of special circumstance, and therefore a consequential damage. In that case, the reviewing court allowed the finding of additional interest costs, but not the increment of interest that was greater than the anticipated rate. The owner's evidence of $690,287 in additional interest costs did not, however, break down into the two areas of "extended interest costs" and "incremental construction interest cost." Therefore, the court held that the jury had no basis upon which to separate the two and remanded the case for a new trial to fix the quantum of direct damages.

12

Documentation

Once approved, a change order is incorporated into the standard pay process documentation. It should maintain its identify for all phases.

Progress Payments

Figure 12-1 is a page from the schedule of values to which the change order information would be added. This schedule lists the cost of each identifiable unit of work. It may be organized by CSI divisions and subdivisions. The change orders may be spread to their appropriate division but must still maintain their identity.

Figure 12-2 is organized in the same order as Fig. 12-1 but shows progress to date in the form of earned value. Figure 12-3 is a monthly request for progress payment submitted by the contractor for approval by the PM/CM or the architect. A special submittal for payment for material or equipment either on hand or in bonded storage, but not yet installed is shown in Fig. 12-4. Figure 12-5 shows an approval routing sheet for a monthly progress payment request.

The basis of progress payments is an estimate of the work actually accomplished. This assessment is determined by actual field observation of weights and quantities, by acceptance of contractor statements, or by a combination of the two. The method and accuracy of work measurements varies with the type of contract. There are many variations of progress payments for construction contracts, but most break down into the following:

- Fixed price, lump sum
- Fixed price per unit quantity
- Time and material

Change orders are negotiated on one of these three bases. When incorporated into the payment process, the change orders are paid in accordance with the basis used to negotiate the change order.

(*Text continues on p. 269*)

SCHEDULE OF VALUES
(Allocation of Total Contract Price)

PROJECT TITLE		PROJECT NO.	SPEC. OR CONTRACT NO.	PAGE	OF
NAME OF CONTRACTOR		CONTRACT TYPE	CONTRACT DATE		
SCHEDULE & ITEM	DESCRIPTION OF COMPONENT PARTS		ALLOCATION*		
			MATERIAL	LABOR	TOTAL
					:
	TOTAL				

* Allocation shall be in percents of total contract price of the items allocated or in dollars and cents.
** This total shall be equal to the total contract price of all items allocated.

Figure 12-1 Schedule of values.

DETAILED BREAKDOWN

PAGE ___ OF ___

			REPORT NO.	DATE
CONTRACT NO.		PROJECT		
ITEM OF WORK (1)	TOTAL VALUE OF WORK (DOLLARS ONLY) (2)	VALUE OF WORK COMPLETED (3)		TOTAL VALUE OF COMPLETED WORK (4)
		TO LAST REPORT (DOLLARS ONLY) (A)	SINCE LAST REPORT (DOLLARS ONLY) (B)	

Figure 12-2 Detailed breakdown.

REQUEST FOR PAYMENT

CONTRACTOR			
LOCATION			
PROJECT TITLE			
CONTRACT NO.	PROJECT NO.		

WORK STATUS:

COMPLETE DATES				PERCENT COMPLETED	
INITIAL CONTRACT	REVISED CONTRACT	ESTIMATED SUBSTANTIAL	ACTUAL SUBSTANTIAL	THROUGH THIS MONTH (SUM OF LINE 4 AND LINE 5 ± LINE 3	NORMAL TO DATE

AVERAGE WORK FORCE	PROGRESS	MATERIALS DELIVERIES
NUMBER EMPLOYED	SATISFACTORY ☐ YES ☐ NO	SATISFACTORY ☐ YES ☐ NO
CONSTRUCTION EQUIPMENT	SHOP DRAWING SUBMISSION	SAMPLE SUBMISSION
SATISFACTORY ☐ YES ☐ NO	SATISFACTORY ☐ YES ☐ NO	SATISFACTORY ☐ YES ☐ NO

REPORT BELOW ANY CIRCUMSTANCES WHICH MAY HAVE ADVERSELY AFFECTED THE PROGRESS SUCH AS STRIKES, WEATHER, DELAYS BY THE OWNER, ETC. INCLUDING EXPLANATION OF ANY "NO" ANSWERS GIVEN IN THE BLOCKS ABOVE.

PROGRESS PAYMENT SUMMARY

1.	INITIAL CONTRACT AMOUNT	
2.	CHANGE ORDERS (Total of Column 2, Form 8-D)	
3.	TOTAL CONTRACT AMOUNT TO DATE (Line 1 plus Line 2)	
4.	VALUE OF WORK COMPLETED TO DATE (Total of Columns 3A and 3B of Form 8-C)	
5.	VALUE OF WORK COMPLETED UNDER CHANGE ORDERS (Total of Column 4, Form 8-D)	
6.	VALUE OF MATERIAL	
	A. MATERIAL ON SITE	
	B. MATERIAL IN STORAGE	
7.	TOTAL VALUE OF MATERIALS (Line 6A plus Line 6B)	
8.	TOTAL VALUE OF COMPLETED WORK AND MATERIALS (Sum of Lines 4, 5, & 7)	
9.	LESS RETAINAGE	
10.	SUB-TOTAL (Line 8 minus Line 9)	
11.	LESS PREVIOUS PAYMENTS	
12.	AMOUNT OF PAYMENT THIS REPORT (Line 10 minus Line 11)	

SIGNATURE	DATE	SIGNATURE	DATE
SIGNATURE	DATE	SIGNATURE	DATE

Figure 12-3 Request for payment.

REQUEST FOR PAYMENT FOR MATERIALS ON HAND

INSTRUCTIONS:

TO CONTRACTORS:

Forward original and one copy to Resident Project Representative. Attach evidence of purchase (and warehouse receipt when required) to the original.

TO RESIDENT PROJECT REPRESENTATIVE:

Retain original in your files with supporting documents for progress payments.

TO (RESIDENT PROJECT REPRESENTATIVE)	DATE
	PROJECT NO.
FROM (CONTRACTOR)	CONTRACT NO.

In accordance with the provisions of the General Conditions of the Contract, request is made for payment for materials on hand for the following materials:

ITEM NUMBER	QUANTITY	UNIT	MATERIAL DESCRIPTION	VALUE	WHERE STORED

AFFIDAVIT

The materials listed above have been purchased exclusively for use on the above referenced project. The material is separate from the other like materials and is physically identified as our property for use only on Contract No. _____. The Owner may enter upon the premises for inspection, checking or auditing, or for any other purpose as you consider necessary. It is expressly understood and agreed that this information and affidavit is furnished to the Owner for the purpose of obtaining payment for the above materials before they are delivered to, or incorporated into, the project described above, and that the storage thereof at the location shown shall not relieve the Contractor of full responsibility for the security and protection of all such materials until acceptance by the Owner of the completed project.

CONTRACTOR: BY_____ TITLE _____ DATE _____

Figure 12-4 Request for payment for materials on hand.

CONTRACT PAYMENT ROUTING SHEET

CONTRACTOR		
CONTRACT NUMBER	INVOICE NUMBER / DATE	
ACTION	**INITIAL**	**DATE**
1. PAYMENT REQUEST RECEIVED BY CONSTRUCTION MANAGER		
2. REVIEW BY ASSISTANT CONSTRUCTION MANAGER COMPLETED		
3. SIGNED BY ASSISTANT CONSTRUCTION MANAGER		
4. SIGNATURE BY CONSTRUCTION MANAGER		
5. RECEIVED BY OWNER		
6. SIGNATURE BY OWNER		
7. RECEIVED BY OWNER'S ACCOUNTS PAYABLE		
COMMENTS		

Figure 12-5 Contract payment routing sheet.

The construction management team plays a vital role in the determination of the actual quantities of work accomplished. Payment is on the basis of units of count, length, area, volume, weight, or for some items, lump sum. The unit of measurement for any particular category is given in the specifications and the unit breakdown list. In the field, the quantities are measured and calculated to a degree of accuracy consistent with the value of the item.

The quantities of some items, such as paving, fencing, and sewer lines, are measured in terms of size, length, or area. Measurement of volume by vehicle loads is not a recommended method of determining quantity because the loading factors vary with the equipment, weather, and operators. Usually, excavation or fill is measured by a field survey or by weight on a truck scale. If it is done by load, a bulking factor of 15 to 20 percent should be deducted. When a change order is approved, an activity (or activities) representing the work in the change order are added to the baseline CPM network.

Figure 12-6 illustrates a sample report of electrical activities from a CPM network. The activities are cost-loaded—that is, each activity has a cost. The cost versus time information can be used to project cumulative costs on both early time and late time bases. Figure 12-7 is an example of two cost curves for a project and uses the figures shown in Table 12-1. If the progress payments plot within the envelope of time formed by the early and late curves, then the project is on schedule. The slope of the payments (i.e., the line plotted between the points) is a good indicator of whether the project will stay on schedule. See Fig. 12-8.

For time and material (T&M) change orders, daily records should be kept for labor (Fig. 12-9), equipment (Fig. 12-10), and materials (Fig. 12-11). The daily sheets list quantities only; they are provided as backup to the monthly invoice for the T&M change order work. The invoice would have a summary sheet (or sheets) for each category with hour costs extended and totaled.

If work is being done under a PCO number, or is being done under protest (with a claim to come later), a disclaimer such as the following should be preprinted on each sheet:

> This T&M sheet relates to work not yet formally approved as a change order. Approval by the resident engineer means only that the work described did occur.

In reviewing the invoice for T&M work, the individual rates should be checked against the contractor's certified payroll. The rates for materials and equipment should be compared against the agreed upon/approved change order.

Field Orders

AIA 201 authorizes the architect to order "minor changes" that do not change project amount or time. AIA 201/CM gives similar authority to the PM/CM. CMAA gives similar authority to the PM/CM.

COST REPORT   /01
SORT KEYS ARE I/J

JOHN DOE BASELINE CPM SCHEDULE
PREPARED BY O'BRIEN-KREITZBERG & ASSOCIATES, INC.

I NODE	J NODE	ACTIVITY DESCRIPTION	REM DUR	CNTR WORK SPEC TYPE CAT. SEC.	TOTAL COST	PERCENT COMPLETE	TO DATE COST
10	11	INSTALL ELECTRICAL MANHOLES	5.0 EL	1-4 0250	$3,800	0	$0
11	12	INSTALL ELECTRICAL DUCTBANK	3.0 EL	1-4 0250	$4,500	0	$0
12	13	PULL IN POWER FEEDER	5.0 EL	1-4 0250	$3,600	0	$0
20	22	UNDERSLAB CONDUIT - P-W	5.0 EL	2-4 1640	$4,500	0	$0
27	28	UNDERSLB CONDUIT OFFICE	3.0 EL	2-4 1640	$3,000	0	$0
37	38	SET ELECTRICAL LOD CENTER - P-W	2.0 EL	3-4 1640	$18,500	0	$0
37	43	POWER PANEL BACKING BOXES P	10.0 EL	3-4 1640	$11,000	0	$0
37	93	AREA LIGHTING	20.0 EL	5-4 0250	$20,000	0	$0
38	43	INSTALL POWER CONDUIT - P-W	20.0 EL	3-4 1640	$12,500	0	$0
43	49	INSTALL BRANCH CONDUIT - P-W	15.0 EL	3-4 1640	$17,500	0	$0
45	51	ROOM OUTLETS P-W	5.0 EL	3-4 0810	$10,000	0	$0
49	50	PULL WIRE P-W	15.0 EL	3-4 1640	$23,000	0	$0
50	54	INSTALL PANEL INTERNALS - P-W	5.0 EL	3-4 1640	$4,500	0	40
51	56	INSTALL ELECTRICAL FIXTURES	10.0 EL	3-4 1650	$19,000	0	$0
54	55	TERMINATE WIRES - P-W	10.0 EL	3-4 1640	$4,500	0	$0
55	56	RINGOUT P-W	10.0 EL	3-4 1640	$1,750	0	$0
56	58	ENERGIZE POWER	1.0 EL	3-4 1640	$1,000	0	$0
61	65	INSTALL ELECTRICAL BACKING BOXES	4.0 EL	4-4 1640	$2,000	0	$0
65	66	INSTALL CONDUIT - OFFICE	10.0 EL	4-4 1640	$6,000	0	$0
66	74	PULL WIRE - OFFICE	10.0 EL	4-4 1640	$6,000	0	$0
74	75	INSTALL PANEL INTERNALS OFFICE	5.0 EL	5-5 1640	$3,000	0	$0
75	79	TERMINATE WIRES - OFFICE	10.0 EL	4-4 1640	$4,000	0	$0
76	79	AIR CONDITIONING ELECTRICAL CONNECTION	4.0 EL	4-4 1640	$1,000	0	$0
79	80	RINGOUT ELECT	5.0 EL	4-4 1640	$1,500	0	$0
		SUBTOTAL	EL		$186,150	0	$0

Figure 12-6 Electrical activities with cost per activity. (*Source: James J. O'Brien, CPM in Construction Management, 4th ed., McGraw-Hill, New York, 1993; reprinted with permission.*)

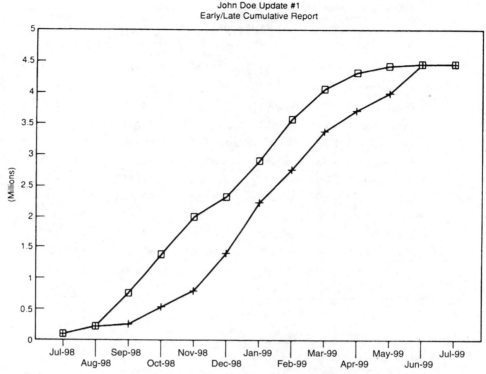

John Doe Update #1
Early/Late Cumulative Report

Figure 12-7 Cost envelope representing baseline schedule. (*Source: James J. O'Brien, CPM in Construction Management, 4th ed., McGraw-Hill, New York, 1993; reprinted with permission.*)

The vehicle to direct such minor changes should be a written field order. If this minor-change approach is abused to actually increase scope, the contractor must proceed with the work (AIA 7.4.1) and must give timely notice of intent to claim cost or time (AIA 7.3.7).

When used correctly, the field order can be used as expeditious documentation in the field. It can be handwritten (usually on multipart paper) or issued on PC or laptop. Examples of field orders include:

- Identification of nonconforming work to the contractor (issued in a nonconforming report, or NCR)
- Clarification of a contract requirement
- Identification of unsafe situations or practices
- Cautioning contractor to be aware of a particular contract requirement
- Direction to hold work in a particular area because of coordination with another contract, utilities, etc.

TABLE 12-1   John Doe Update 1—Cumulative Report

Resr.	Period beginning	Early schedule		Late schedule	
		Usage, $	Cumulative, $	Usage, $	Cumulative, $
Cost	Jul. 1 , 1998	103,568	103,568	103,568	103,568
Cost	Aug. 1, 1998	113,431	217,000	113,432	217,000
Cost	Sep. 1, 1998	530,600	747,600	37,500	254,500
Cost	Oct. 1, 1998	613,950	1,361,550	280,750	535,250
Cost	Nov. 1, 1998	627,917	1,989,467	256,000	791,250
Cost	Dec. 1, 1998	332,083	2,321,550	602,850	1,394,100
Cost	Jan. 1, 1999	573,500	2,895,050	837,067	2,231,167
Cost	Feb. 1, 1999	669,600	3,564,650	534,233	2,765,400
Cost	Mar. 1, 1999	483,200	4,047,850	606,850	3,372,250
Cost	Apr. 1, 1999	269,333	4,317,183	343,233	3,715,483
Cost	May 1, 1999	103,467	4,420,650	277,233	3,992,717
Cost	Jun. 1, 1999	32,000	4,452,650	459,933	4,452,650
Cost	Jul. 1, 1999	0	4,452,649	0	4,452,650

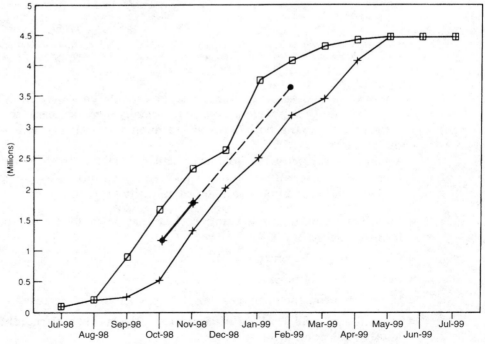

Figure 12-8 Progress payment plot (with projection of expected progress). (*Source: James J. O'Brien, CPM in Construction Management, 4th ed., McGraw-Hill, New York, 1993; reprinted with permission.*)

DAILY LABOR FORM

Sheet _____ of _____

TIME & MATERIAL WORK

Date:_____   Project:_____

Change Order: _____   Contractor:_____

Name	Brass #	Craft	Hours

Submitted: _____   Approved: _____

Figure 12-9 Daily labor form (time and material).

Some field orders should lead directly to identification as a PCO, such as an area hold work order. Others may become the bases for a contractor's claim for a constructive change order, such as when a clarification of work adds scope in the contractor's opinion. The project procedures manual should define the authority of the PM/CM or architect to issue field orders. The manual has to be clearly cross-referenced to the contract documents.

DAILY EQUIPMENT FORM

Sheet _____ of _____

TIME & MATERIAL WORK

Date:_____   Project:_____

Change Order: _____   Contractor:_____

Equipment	Model #	Operator Cost Included		Hours
		Y	N	

Submitted: _____   Approved: _____

Figure 12-10   Daily equipment form (time and material).

DAILY MATERIALS FORM

Sheet _____ of _____

TIME & MATERIAL WORK

Date:_____   Project:_____

Change Order: _____   Contractor:_____

Material Description	Quantity

Submitted: _____   Approved: _____

Figure 12-11   Daily materials form (time and material).

Field orders should be issued in consecutive number, and a log should be maintained by the PM/CM staff. The log should identify and cross-reference appropriate correspondence, RFIs, and other numbered documents.

If AIA 201, AIA 201/CM, or CMAA are not used, the contract documents may contain a clause such as:

> It is understood and agreed that refinement or detailing will be accomplished from time to time with respect to the Plans and Specifications. No adjustment in the Contract Sum or the Completion Date shall be made unless such refinement or detailing results in changes in the scope, quality, function, and/or intent of the Plans and Specifications not reasonably inferable or foreseeable by a Contractor of Contractor's experience and expertise.

The field order can be used to record agreements between the PM/CM and the contractor such as the following:

- The contractor means/methods to achieve a specific result are within the contract intent and/or acceptable to the PM/CM.

- The nature of a clarification, along with the contractor's concurrence that it is not a scope change.

- Confirmation that the contractor is aware of certain contract requirements and specifications of how they will be met.

Where the field order notes defective or nonconforming work, reinspection will be required. Similarly, unsafe condition notices must be revisited.

Field orders can be used to initiate a time and material (i.e., a PCO) when it is imperative that the contractor commence work immediately.

The following appears in the typical field order:

Field	Description
Issued By	The person generating this field order
Specification #	The number applicable to this field order
Instructions	A pop-up pick-list that offers the following choices:
	■ Confirm in writing—Response required
	■ File—Make note, no response required
	■ Instructions—Follow instructions given
	■ Urgent action—Do immediately
Notes	A memo field that allows entry of all directions pertinent to this field order. This will print out on the field order form as the "Information or Action" (body area).
Inspected On	The date this field order work was inspected.
By	The person who did inspection.

These selections will print at the bottom of the field order form. Be sure that your status selection is appropriate to the instructions given—that is, don't give it a status of "no response required" if your instructions are to be confirmed in writing. Once confirmation has been received, however, be sure to change the status flag to OK-Closed.

The Daily Log

From all aspects, it is important to record the daily events on a project. This record should include events that cause or require a change order. Once in place, the daily log should identify daily change order work (versus base contract work) being done. Daily logs should be kept by the PM/CM, the contractor, and the subcontractors. In the wonderful world of computers, this task could be combined; thus far, it never has been. Accordingly, it is in the best interest of each party to keep its own daily log. If kept properly, the facts will be the same, or at least similar. Although the "spin" put on the facts can be anticipated to be self-serving, daily logs, when kept accurately and honestly as a regular business record, are acceptable as evidence. This is a well-established exclusion to the hearsay principle.

The daily logs should also record facts and events that relate to changes such as the following:

- Change order versus base contract work
- Work beyond base contract (i.e., constructive change order)
- Causative connections of change orders to problems (disruption, inefficiency, etc.)
- Unusual events (floods, strikes, fires, accidents, etc.)
- Disruption, ripple effect, required resequencing of work plan due to a change order
- Inefficiencies in the performance of, or due to, a change order

Inspection Report—O'Brien Kreitzberg TRACK 5.0

The inspector's daily inspection report is a legal document that describes all construction activities taking place on the site for a specific day. The report will be used for monthly pay estimates, contract documentation, and claims. The TRACK 5.0 inspection report data-entry screen can be used either to log summary information for search and retrieval purposes or as the inspection report as desired. If used as the inspection report, it must be printed and signed by the inspector.

The top half of the screen records the general information for the day. The contract number, document number, and inspection report number are entered. It will calculate the day as a calendar day and give the day of the week based on the inspection date entered in Insp Date.

Shift, Wind, Weather, and Humidity are pop-up pick-lists with the following values:

Shift: day, swing, or grave

Wind: high, moderate, or still

Weather: overcast, cloudy, rain, or snow

Humidity: high, medium, or low

To use pop-up pick-list fields, leave the field blank and press Enter; the pick-list will pop up. Move the lightbar to the selection by using the arrow keys. Press Enter to select. If you know the number of the selection, you can enter the number directly and press Enter. The pick-list will not pop up, and data entry will be swifter. When entering information, always report the worst weather conditions for that day and, if necessary, add any relevant comments in each task description.

In Fig. 12-12, Sample Report 1 shows how the daily inspection report will print. The top left-hand block gives the general contract information plus the (calendar) day number, date, shift, and the resident and field inspector for this contract. The right-hand block gives the day of the week, the weather, temperature, wind, humidity, and the report number. The third block displays a checklist of concerns.

The next section, "Work Performed," is for individual task information. In the sample, the first line gives the contractor/subcontractor information for that task. The next line shows the equipment, man-hours, and schedule reference. In the following line, "Concern checked," N/A appears. This means that none of the concerns in the list were applicable. This line will automatically print the number of the selection listed in the concerns checklist. When selection 9 is picked, it will print "N/A." If the Checklist of Concerns field has been left blank during data entry, which could happen if PgDn were pressed before the concerns field was reached, the report will print a question mark. To remove the question mark, go to that record and fill in the concerns field with the appropriate selection.

The next line is the start of the description memo field. It will print everything entered in the memo field continuing on the subsequent pages, if necessary.

Figure 12-12 shows how the report handles two different tasks for the same contractor. If there were more tasks or information than would fit on a page, the report would continue to a second page. Total man-hours is the total for each contractor and will always print after all tasks for a particular contractor have printed.

The Work Performed area of Fig. 12-13 illustrates two tasks by different contractors. Each task has a total for man-hours on the division line between the two contractors. In this example, there is only one task per contractor, so the amount of man-hours for each task is the total for that contractor.

In Fig. 12-13, there is very little entered for the description of each task. Figure 12-14 illustrates how the report will handle a task with a lengthy description. Again, there can be as many tasks for as many contractors as needed. As much or as little information about each task can be entered as necessary. When the body area wraps to the second page, it will look like Fig. 12-15. The two header blocks with the contract and weather information will not be carried to the

O'BRIEN-KREITZBERG
TRACK 5.0

DAILY INSPECTION REPORT

PROJECT TITLE & NO. TRN-0001		
METROPOLITAN TRANSIT SYSTEM		

PAGE #1

CONTRACTOR		
GAUSS ELECTRIC CONSTRUCTION CO.		

DATE	DAY NO.	WORK HOURS
04/02/89	85	DAYS

INSPECTOR	
R/I HENRI CLOUSEAU	FIELD-MULVANY

DAY	Wednesday	
WEATHER	Rain	
TEMP.	MIN - 59 MAX. - 62	
WIND	BREEZY	REPORT #
HUMIDITY	HIGH	61

CHECKLIST OF CONCERNS

1. Delays and Difficulties	3. Orders or Instructions	5. Disputes	7. Safety
2. Changes	4. Unsatisfactory Work	6. Tests	8. Visitors

WORK PREFORMED (INCLUDING QUANTITIES)	SCHEDULE REFERENCE

Contractor/Subcontractor   COMSTOCK

Equipment   FLATBED SEMI   Man Hours   4.00   12.58.1

Concern checked   N/A

Loaded 1st load of 19 poles & appurtenances at
Siemens Stockton St. yard. Tractor did
arrive to transport semi to maintenance yard.

Contractor/Subcontractor   COMSTOCK

Equipment   FLATBED SEMI   Man Hours   4.00

Concern checked   N/A

Loaded 2nd load of 17 poles & appurtenances at
Siemens Stockton St. yard.

• •

Total Man Hours   8.00

To the best of my knowledge the work is in conformance with the approved plans and specifications.

Signature _____   Date: _____
Signature _____   Date: _____

Figure 12-12   Sample 1—daily inspection report.

DAILY INSPECTION REPORT

| PROJECT TITLE & NO.   TRN-0001 | | | PAGE #1 | |
| METROPOLITAN TRANSIT SYSTEM | | | | |

			DAY	Wednesday	
CONTRACTOR			WEATHER	Rain	
GAUSS ELECTRIC CONSTRUCTION CO.					

DATE	DAY NO.	WORK HOURS	TEMP.	MIN - 59 MAX. - 62	
04/02/89	85	DAYS	WIND	BREEZY	REPORT #
INSPECTOR					
R/I HENRI CLOUSEAU FIELD-MULVANY			HUMIDITY	HIGH	61

CHECKLIST OF CONCERNS
1. Delays and Difficulties   3. Orders or Instructions   5. Disputes   7. Safety
2. Changes   4. Unsatisfactory Work   6. Tests   8. Visitors

WORK PREFORMED (INCLUDING QUANTITIES)	SCHEDULE REFERENCE

Contractor/Subcontractor   COMSTOCK

Equipment   FLATBED SEMI   Man Hours   4.00   12.58.1

Concern checked   N/A

Loaded 1st load of 19 poles & appurtenances at
Siemens Stockton St. yard. Tractor did
arrive to transport semi to maintenance yard.
• •

Total Man Hours   **4.00**

Contractor/Subcontractor   RUIZ

Equipment   DUMP TRUCK   Man Hours   4.00   12.57.2

Concern checked   3

Completed excavation on substation #15
• •

Total Man Hours   **4.00**

To the best of my knowledge the work is in conformance with the approved plans and specifications.

Signature _____   Date: _____
Signature _____   Date: _____

Figure 12-13   Sample 2—daily inspection report.

O'BRIEN-KREITZBERG
TRACK 5.0

DAILY INSPECTION REPORT

| PROJECT TITLE & NO. TRN-0001 METROPOLITAN TRANSIT SYSTEM | | | PAGE #1 |

| PROJECT TITLE & NO.   TRN-0001 |
| METROPOLITAN TRANSIT SYSTEM |

CONTRACTOR
GAUSS ELECTRIC CONSTRUCTION CO.

| DATE 04/02/89 | DAY NO. 85 | WORK HOURS DAYS |

INSPECTOR
R/I HENRI CLOUSEAU   FIELD-MULVANY

DAY	Wednesday	
WEATHER	Rain	
TEMP.	MIN - 59 MAX. - 62	
WIND	BREEZY	REPORT #
HUMIDITY	HIGH	61

CHECKLIST OF CONCERNS
| 1. Delays and Difficulties | 3. Orders or Instructions | 5. Disputes | 7. Safety |
| 2. Changes | 4. Unsatisfactory Work | 6. Tests | 8. Visitors |

| WORK PREFORMED (INCLUDING QUANTITIES) | SCHEDULE REFERENCE |

Contractor/Subcontractor   NYPL-DR

Equipment   SCRAMBLED GREY CELLS   Man Hours   8.00   99.99.9

Concern checked   4

We like to think we speak logically all the time,
but we are aware that we sometimes use illogical means
to persuade others of our point of view. In the heat of
an impassioned argument, or when we are afraid our
disputant has a stronger case, or when we don't quite
have all the facts we'd like to have, we are prone to
engage in faulty processes of reasoning, using
arguments we hope will appear sound.

Such deceptive arguments are called *fallacies* by
philosophers who, starting with Aristotle, have

To the best of my knowledge the work is in conformance with the approved plans and specifications.

Signature _____   Date: _____
Signature _____   Date: _____

Figure 12-14 Sample 3—daily inspection report.

DAILY INSPECTION REPORT

04/02/89   FIELD INSPECTOR — MULVANY   REP NO: 61(cont.)   PAGE #2

WORK PREFORMED (INCLUDING QUANTITIES)	SCHEDULE REFERENCE

Contractor/Subcontractor   NYPL-DR

Equipment   SCRAMBLED GREY CELLS   Man Hours   8.00   99.99.9

Some arguments have easily recognizable defects. For instance, in the *argument ad hominem* a person's views are criticized because of logically irrelevant personal defect: "You can't take Smith's advice on the stock market: he's a known philanderer." In the genetic fallacy, something is mistakenly reduced to its origins: "We know that emotions are nothing more than physiology; after all, medical research has shown emotions involve the secretion of hormones."

. . . .

Total Man Hours   <u>8.00</u>

To the best of my knowledge the work is in conformance with the approved plans and specifications.

Signature _____   Date: _____
Signature _____   Date: _____

Figure 12-15   Sample 4—daily inspection report.

second or subsequent pages. The top line of the second page, the "Work Performed" header, carries the date of the report, the field inspector's name, the report number, and the page number.

DailyLog

DailyLog is a PC-based software program that documents, expedites, and enforces the project CPM schedule on a daily basis. By integrating daily report writing, potential claim documentation, and contract scheduling updating into one process, DailyLog is a time-saving tool. DailyLog monitors the schedule more efficiently because it selects only those activities scheduled to be currently underway, eliminating sifting through unnecessary data. An exporting feature updates the CPM schedule with daily information and eases the hassle of data-intensive monthly CPM update requirements by eliminating double-entry of data. DailyLog documents delays and their causes on a daily basis. This information is absolutely essential for early settlement of potential claims and for avoiding costly litigation.

The DailyLog report format integrates with the contract schedule, thereby reducing communication gaps that can arise between the owner and the contractor. Because of this, it is a perfect tool for projects utilizing Primavera Project Planner (P3). It enables quick and early problem recognition and can streamline the periodic visits by the home office management to the field office. DailyLog is an excellent tool to link up the planners from the home office and the doers from the field.

DailyLog makes the day-to-day running of the construction project easier by standardizing and computerizing daily diary information, daily checklists, daily reports, and history reports. What it does is extremely simple: it reads the P3/SureTrak schedule data and allows the user to access on screen or on a printed checklist scheduled activities that should be started or in progress that day. These checklists help record daily progress in a standardized format while out on the job site. Figure 12-16 shows this interaction.

The software also allows the user to record details of those activities— whether or not work occurred—including items such as what was accomplished, causes of delays, number of workers for each activity, pieces of equipment, and so on. The program records information about the actual start (AS), the actual finish (AF), the remaining duration (RD), and the percentage complete (PCT). After daily progress is entered, the DailyLog reports are printed and filed, eliminating the need for longhand reports.

DailyLog lets the user record the details of the actual project progress every workday, in a form that neatly folds back into the master project. This allows the documentation of stop-and-go performance of an activity without having to create a complicated CPM model. The information is stored in a standard DBF file to be used later for history reports or for further analysis by the user on dBASE, Lotus, or other software. The data can be printed onto a structured daily report or accumulated for reporting progress over any span of time.

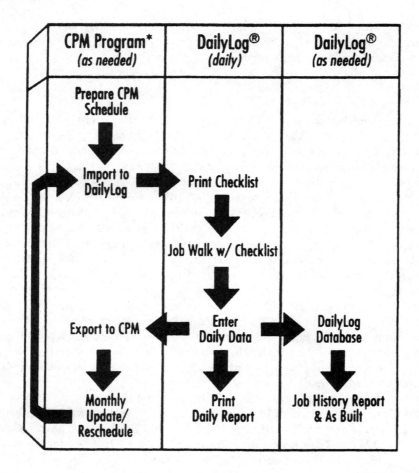

Figure 12-16   DailyLog/CPM program interaction. *Primavera, FinestHour, SureTrak, MS Project, and Timeline.

DailyLog requires the following computer hardware:

- An IBM-compatible computer running on a 386 processor or better
- DOS operating system, Version 3.0 or later
- 1 MB of RAM and at least 20 MB of free hard disk storage
- VGA monitor and adapter
- Graphics printer (IBM graphics– or HP LaserJet–compatible)
- Project schedule that is in Primavera Project Planner (P3), SureTrak, or FinestHour format
- Mouse (recommended)
- Primavera Project Planner (P3), SureTrak, or FinestHour (recommended)

Weather

The Weather pop-up menu displays the predominant weather conditions for the day for which you are entering data. The user can view a list of possible weather condition entries in the same way that the user viewed the list of projects in the Project pop-up menu. The following list gives the possible weather entries:

Clear	Ice
Cloudy	Drizzle
Partly Cloudy	Wind
Rain	Mud
Snow	Missing
Fog	

The entry "Missing" is used if no weather conditions were noted for the day you are recording. Other fields include the following:

Field	Description
High Temp and Low Temp	Allow the user to enter the high and low temperatures for the workday that is being recorded.
Precipitation	Allows the user to enter the amount of rainfall that occurred during the workday that is being recorded.
Report By	The name of the person making the DailyLog entry. An entry in this field is mandatory.
Method	The Method pop-up menu displays a list of three methods the user can use to calculate the remaining duration and percentage complete for activities. The remaining duration (RD) and the percentage complete (PCT) are calculated for exporting data to P3 for project schedule updating. The three methods are as follows: 1. *Manual* allows manual entry of the RD and the PCT independently of each other for each activity. 2. *Worked* calculates the RD by counting the number of days that are actually entered as workdays and subtracting them from the original duration (OD). PCT is then calculated using this same method. 3. *Elapsed* calculates the RD by taking the number of days that have elapsed from the actual start date of the activity to the current date and subtracting them from the original duration. PCT is then calculated using this same method.
Comment	The Comment field allows the user to enter general comments regarding the day's activities. It is a good place to enter a summary of the day's progress, a list of visitors,

Field	Description
	notable incidents, and other information about the project in general. This Comment field uses the DailyLog built-in text editor. Similar to a simple word-processing program, the text automatically wraps to the next line as the user types in comments, and it automatically scrolls down one line when the user types beyond the last line in the field. The arrow keys can be used to move around inside the field, and the backspace and delete keys can be used to delete text.

Printing Checklists and Reports

DailyLog checklists and reports are useful for maintaining hard copies of project information on a daily basis. Checklists may be printed for each day of the project to track daily data and progress for each scheduled activity. Checklists are particularly useful for the project engineer or job foreman to use during a daily job walk.

Using checklists

Because the checklist format is standardized, information is recorded in the same way for each activity. The most common practice is to print a checklist at the start of each day. This checklist is filled in with the current day's information at the job site. As an alternative, the user could print a checklist for the next day's job site walk at the end of the previous workday. (If you print out the checklists the day before, however, be sure to change the Date field on the DailyLog data screen to the date when the checklist will be used.)

The checklist has a title and page number at the top of each page. The first page contains fields for the daily information that is contained on the DailyLog data screen. The project name and the date are filled in for the user. The user can then circle an item for weather conditions and fill in the temperature and precipitation information. Next, the checklist contains resource information for the project. The rest of the checklist is a list of each activity that is scheduled to be started or to be in progress. As on the DailyLog screen, it allows the user to fill in the following items regarding each activity:

- Whether work was performed or a delay was experienced
- Start and finish dates
- Percent complete and remaining duration information
- Descriptions of what was done for each activity work or delay entry
- Resource usage information
- Reference information

Finally, general comments can be entered regarding the work for all activities that were performed on that day.

Using DailyLog reports

DailyLog reports provide a hard copy of the project information recorded in DailyLog for the current day. When these daily reports are printed and filed, the result is a complete, printed record of the project progress. This printed record consolidates all the progress information for the day into an easy-to-read format. It makes an excellent backup of the data entered into the computer on a daily basis.

History reports summarize the project history for all the activities that are listed for the project, as of the current date. History reports are valuable documentation of the project status at periodic or milestone meetings.

Importing and Exporting P3 Projects

Before updating the P3 activities using DailyLog, the user must import the P3 project that he or she wishes to update. This import procedure must be performed only one time per project, unless the user updates the P3 project with changes that affect DailyLog activities. In this case, the user would import the P3 project each time it is updated.

Periodically during the project, and when the project is complete, the user should export the DailyLog data back to the P3 project. Doing this automatically updates the P3 project, allowing it to reflect accurately the events that occurred during the project.

Depending on the length of the project, the user can choose to export DailyLog information to P3 as often as necessary. Many project managers choose a monthly, semimonthly, or quarterly basis for updating P3 projects. The general practice is to save a copy of the original P3 project, as it was originally planned, as the baseline schedule. This baseline schedule should never be overwritten, because it represents the original plan against which all future changes and progress can be compared.

When the DailyLog information is exported to the P3 project, the user can compare the original schedule to the exported and updated actual schedule to construct an accurate, actual versus planned schedule. This can be very useful in tracking the successes and failures of the original project plan.

The DailyLog import and export procedures are written in the P3 batch processing language. This batch system comes with the P3 software package. It is used to extract the necessary information from the P3 project files for importing to and exporting from DailyLog.

Figure 12-17 shows the format for the daily detail checklist (baseline). Figure 12-18 is the daily detail report for the project, showing the connection between the schedule and actual progress. The activity detail report of daily activity by scheduled activity is shown in Fig. 12-19. Figure 12-20 is the as-built status in CPM format. Figure 12-21 is the as-built resource used by activity.

```
┌─────────────────────────────────────────────────────────────────────────┐
│ V160██████          DAILY DETAIL CHECKLIST          Page    1 of    4     │
├─────────────────────────────────────────────────────────────────────────┤
│ Company: DAILYLOG DEMONSTRATION & TUTORIAL                                │
├──────────────────────┬────────────────────────────────────────────────────
│ Project: DEMO        │ Title: JUSTIN CASE RESIDENCE REMODEL (BASE)        │
├──────────────────────┼──────────────────────────┬────────────────────────┤
│ Log Date: FRI 05/14/93 │ Log Time:              │ Logged By:              │
└────────────────────────────────────────────────────────────────────────── 
```

Weather: Clear Cloudy PartlyCloudy Rain Snow Fog Ice Drizzle Wind Mud **Missing**

High Temperature: Low Temperature: Precipitation:

Resource	Title	Units	Amount

Activity	Title	Work	Delay
01302	ARCHITECT/OWNER WATER HEATER SELECTION & BUY	[ ]	[ ]
03201	INSTALL REMEDIAL UNDERPINNING @ MAIN STRUCTURE	[ ]	[ ]
16001	REWIRE MAIN HOUSE & GARAGE	[ ]	[ ]

Figure 12-17   Format for daily detail checklist.

Failure to Keep Records

The lack of good documentation can defeat a claim for a constructive change order. A board or jury may find that the claimant deserves relief; however, the burden of being able to prove the damages is on the claimant. The claimant must prove both liability/causation and related damages. To do this effectively, the claimant must rely on construction documentation kept during the project. This responsibility was addressed in Wunderlich Contracting Co. v. United States (351 F. 2d 956, 173 Ct. Cl. 180; 1965). In this case, the court stated:

A claimant need not prove his damages with absolute certainty or mathematical exactitude. It is sufficient if he furnishes the court with a reasonable basis for computation, even though the result is only approximate. Yet this leniency as to the actual mechanics of computation does not relieve the contractor of his essential burden of establishing the fundamental facts of liability, causation, and resultant injury. It was plaintiff's obligation in the case at bar to prove with reasonable certainty the extent of unreasonable delay which resulted from defendant's actions and to provide a basis for making a reasonably correct approximation of the damages which arose therefrom. Broad generalities and inferences to the effect that defendant must have caused some delay and damage because the contract took 318 days longer to complete than anticipated are not sufficient. . . It is incumbent

V160*****	DAILY DETAIL CHECKLIST	Page 2 of 4

Activity:	01302	Title: ARCHITECT/OWNER WATER HEATER SELECTION & BUY

Start: 05/06/93	Finish: / /	% Complete:	Rem. Duration:

Work Description:

Delay Description:

Delay Reference:

Resource	Amount	Resource	Amount	Resource	Amount

Activity:	03201	Title: INSTALL REMEDIAL UNDERPINNING @ MAIN STRUCTURE

Start: 05/11/93	Finish: / /	% Complete:	Rem. Duration:

Work Description:

Delay Description:

Delay Reference:

Resource	Amount	Resource	Amount	Resource	Amount

Figure 12-17   (*Continued*)

| V160###### | DAILY DETAIL CHECKLIST | Page | 3 of | 4 |

| Activity: | 16001 | Title: REWIRE MAIN HOUSE & GARAGE |

| Start: | 05/11/93 | Finish: | / / | % Complete: | | Rem. Duration: |

Work Description:

Delay Description:

Delay Reference:

Resource	Amount	Resource	Amount	Resource	Amount

Activities Started Out-Of-Sequence:

Figure 12-17   (*Continued*)

V160*****	DAILY DETAIL CHECKLIST	Page 4 of 4

General Comments:

Signed:	Date:

Figure 12-17 *(Continued)*

upon plaintiffs to show the nature and extent of the various delays for which damages are claimed and to connect them to some act of commission or omission on defendant's part.

A lack of records was a major factor in the appeal of Lawrence D. Krause (AGBCA No. 76-1804 at 5989; Nov. 16, 1982). Krause was awarded a contract in the amount of $513,832.90 for the construction of a floodwater-retarding structure (earth dam) in Bexar County, Texas. In a letter of May 1, 1974, Krause made a claim of $62,192.90 for extra work and effort in processing 40,385 cubic yards of additional fill material from extra excavation resulting from a differing site condition.

Bidders were told that in order to generate the material needed to build the embankment, the material would have to be processed through a crushing plant; however, finding enough clay to construct the center section would be difficult. Krause bid $2 per cubic yard, planning to use a scraper and bulldozer rather than front-end loaders, trucks, and front-end dumps. He figured in his bid the cost of processing only the estimated quantities of excavated material in the foundation. He had concluded from the borings that there was sufficient clay material in the borrow area for the earth fill embankment.

About three months into the job, Krause had reached a depth of approximately 15 feet in the proposed foundation excavation. Government borings projected bedrock to be at about 12 to 15 feet below ground surface. Krause was

V160⧸⧸⧸⧸⧸	DAILY DETAIL REPORT	Page 1 of 1

Company: DAILYLOG DEMONSTRATION & TUTORIAL

Project: DEMO	Title: JUSTIN CASE RESIDENCE REMODEL (BASE)

Log Date: FRI 05/14/93	Log Time: 11:41	Logged By: MOY

Weather: Partly-Cloudy

High Temperature: 65	Low Temperature: 45	Precipitation: 0.00

Comments
Underpinning will delay the overall schedule because the work is on the critical path. State inspector was on site early this afternoon to check the safety procedure, and discussed this matter with Jim Bower.

Activity	Title	Work	Delay
01302	ARCHITECT/OWNER WATER HEATER SELECTION & BUY	Yes	No
03201	INSTALL REMEDIAL UNDERPINNING @ MAIN STRUCTURE	Yes	Yes
16001	REWIRE MAIN HOUSE & GARAGE	Yes	Yes

Activity: 01302	Title: ARCHITECT/OWNER WATER HEATER SELECTION & BUY

Start: 05/06/93	Finish: / /	% Complete: 57.1	Rem Duration: 3

Activity: 03201	Title: INSTALL REMEDIAL UNDERPINNING @ MAIN STRUCTURE

Start: 05/11/93	Finish: / /	% Complete: 30.0	Rem Duration: 7

Activity: 16001	Title: REWIRE MAIN HOUSE & GARAGE

Start: 05/11/93	Finish: / /	% Complete: 50.0	Rem Duration: 3

Work Description:
Main kitchen receptacles, lights & hi-voltage appliances. Apprentice cut his finger on broken glass under dishwasher. They have an extra guy today.

Delay Description:
Checked with Wally on the field about completion on this item. He is aware of the minor delay and said he will have half crew here tomorrow (Sat). Said he will eat OT premium.

Delay Reference:
Memo ⧸003   Re: OT premium

Signed:	Date:

Figure 12-18 Daily detail report.

```
┌─────────────────────────────────────────────────────────────────────────────┐
│ V160#####              ACTIVITY DETAIL REPORT           Page    1 of    2     │
│        Date: 05/14/93 Project: DEMO Window: 05/05/93 -> 05/14/93              │
└─────────────────────────────────────────────────────────────────────────────┘
```

Activity: 01001 Title: MOBILIZE ON SITE
          Start   Finish   Od   Ad   Rd   Pct   Tf   Rf
Early:*05/06/93 *05/07/93 Elapsed: 2   2   0   100   0   0
 Late:*05/06/93 *05/07/93 Worked: 1   2   0   100   0   -1
Actual: 05/06/93 05/07/93 Use: 1   2   0   100   0   -1
Activity Codes: AGC   ALL

Dow	Date	Work/Description	Delay/Description/Reference
THU	05/06/93	Work	No Delay
FRI	05/07/93	Work	Extended Duration

Calendar days elapsed:	2	Work:	1	Days worked:	2
Total days documented:	2	No work:	0	Days not worked:	0
Out of sequence start:	No	Delayed start:	0	Delay days:	1
Extended duration:	1	Suspension:	0	Non-delay days:	1

Activity: 01301 Title: NOTICE TO PROCEED FROM OWNER & ARCHITECT
          Start   Finish   Od   Ad   Rd   Pct   Tf   Rf
Early:*05/05/93 *05/05/93 Elapsed: 1   1   0   100   0   0
 Late:*05/05/93 *05/05/93 Worked: 1   1   0   100   0   0
Actual: 05/05/93 05/05/93 Use: 1   1   0   100   0   0
Activity Codes: WBN 01300 A/E

Dow	Date	Work/Description	Delay/Description/Reference
WED	05/05/93	Work	No Delay

Calendar days elapsed:	1	Work:	1	Days worked:	1
Total days documented:	1	No work:	0	Days not worked:	0
Out of sequence start:	No	Delayed start:	0	Delay days:	0
Extended duration:	0	Suspension:	0	Non-delay days:	1

Activity: 01302 Title: ARCHITECT/OWNER WATER HEATER SELECTION & BUY
          Start   Finish   Od   Ad   Rd   Pct   Tf   Rf
Early:*05/06/93 05/12/93 Elapsed: 7   9   0   100   15   14
 Late:*05/06/93 05/27/93 Worked: 7   4   3   57   15   15
Actual: 05/06/93   / /   Use: 7   4   3   57   15   15
Activity Codes: WBN 01300 A/E

Dow	Date	Work/Description	Delay/Description/Reference
THU	05/06/93	Work	No Delay
FRl	05/07/93	No Work	No Delay
SAT	05/08/93	No Record	No Record
SUN	05/09/93	No Record	No Record
MON	05/10/93	Work	No Delay
TUE	05/11/93	No Work	Suspension
WED	05/12/93	Work	No Delay
THU	05/13/93	No Work	No Delay
FRI	05/14/93	Work	No Delay

Calendar days elapsed:	9	Work:	4	Days worked:	4
Total days documented:	7	No work:	2	Days not worked:	3
Out of sequence start:	No	Delayed start:	0	Delay days:	1
Extended duration:	0	Suspension:	1	Non-delay days:	6

Figure 12-19   Activity detail report.

```
┌─────────────────────────────────────────────────────────────────────────────┐
│ V160/////                    ACTIVITY DETAIL REPORT              Page    2 of    2 │
│            Date: 05/14/93 Project: DEMO Window: 05/05/93 -> 05/14/93           │
└─────────────────────────────────────────────────────────────────────────────┘
```

Activity: 03201 Title: INSTALL REMEDIAL UNDERPINNING @ MAIN STRUCTURE
             Start   Finish   Od   Ad   Rd   Pct   Tf   Rf
    Early: *05/11/93   05/20/93 Elapsed: 10   4   6   40   0   1
     Late: *05/11/93   05/20/93 Worked: 10   3   7   30   0   0
   Actual: 05/11/93   /   /   Use: 10   3   7   30   0   0
Activity Codes: FCI 03200 RES

Dow	Date	Work/Description	Delay/Description/Reference
FRI	05/07/93	No Work	Delayed Start
SAT	05/08/93	No Record	No Record
SUN	05/09/93	No Record	No Record
MON	05/10/93	No Work	Delayed Start
TUE	05/11/93	Work	No Delay
WED	05/12/93	Work	No Delay
THU	05/13/93	No Work	No Delay
FRI	05/14/93	Work	Extended Duration

Calendar days elapsed: 8 Work: 2 Days worked: 3
Total days documented: 6 No work: 1 Days not worked: 3
Out of sequence start: No Delayed start: 2 Delay days: 3
Extended duration: 1 Suspension: 0 Non-delay days: 3

Activity: 16001 Title: REWIRE MAIN HOUSE & GARAGE
             Start   Finish   Od   Ad   Rd   Pct   Tf   Rf
    Early: *05/11/93   05/14/93 Elapsed: 4   4   0   100   10   11
     Late: *05/11/93   05/24/93 Worked: 6   3   3   50   6   6
   Actual: 05/11/93   /   /   Use: 6   3   3   50   6   6
Activity Codes: XEL 16000 RES

Dow	Date	Work/Description	Delay/Description/Reference
FRI	05/07/93	No Work	No Delay
SAT	05/08/93	No Record	No Record
SUN	05/09/93	No Record	No Record
MON	05/10/93	No Work	No Delay
TUE	05/11/93	Work	No Delay
WED	05/12/93	Work	No Delay
THU	05/13/93	No Work	No Delay
FRI	05/14/93	Work	Extended Duration
		Main kitchen receptacles, lights	Checked with Wally on the field
		& hi-voltage appliences.	about completion on this item. He
		Apprentice cut his finger on	is aware of the minor delay and
		broken glass under dishwasher.	said he will have half crew here
		They have an extra guy today.	tomorrow (Sat). Said he will eat
			OT premium.
			Reference:
			Memo #003 Re: OT premium

Calendar days elapsed: 8 Work: 2 Days worked: 3
Total days documented: 6 No work: 3 Days not worked: 3
Out of sequence start: No Delayed start: 0 Delay days: 1
Extended duration: 1 Suspension: 0 Non-delay days: 5

Figure 12-19 *(Continued)*

requested to dig a test pit down to about 25 feet below original ground level. He did so, and there was no indication of bedrock. The government stopped work in the foundation area for the next two months, and Krause was unable to use his equipment and crew according to his planned schedule.

A modification was issued in December 1972, directing Krause to remove undesirable material from the entire base of the embankment, an increase in excavation

ACTIVITY SUMMARY REPORT
Date: 05/16/93 Project: DEMO

ACTIVITY	TITLE	ES	EF	OD	AS	AF	ED/AD	MD	PCT	WARNING
01001	MOBILIZE ON SITE	05/06/93	05/07/93	1	05/06/93	05/07/93	2	2	100	XD
01301	NOTICE TO PROCEED FROM OWNER & ARCHITECT	05/05/93	05/05/93	1	05/05/93	05/05/93	1	1	100	
01302	ARCHITECT/OWNER WATER HEATER SELECTION & BUY	05/06/93	05/12/93	7	05/06/93	/	9	4	57	NV
02003	FINE GRADE AROUND @ STRUCTURE PERIMETER	05/28/93	05/31/93	2	/	/	0	0	0	
03201	INSTALL REMEDIAL UNDERPINNING @ MAIN STRUCTURE	05/11/93	05/20/93	10	05/11/93	/	4	3	30	XD DS
03202	INSTALL REMEDIAL UNDERPINNING @ IN-LAW UNIT	05/21/93	05/27/93	5	/	/	0	0	0	
15001	PLUMBING ROUGH-IN FOR KITCHEN EXPANSION	05/17/93	05/18/93	2	/	/	0	0	0	
15002	SEPARATE WATER METERING FOR IN-LAW UNIT	05/19/93	05/19/93	1	/	/	0	0	0	
15003	HOOK-UP NEW WATER HEATER	05/20/93	05/20/93	1	/	/	0	0	0	
16001	REWIRE MAIN HOUSE & GARAGE	05/11/93	05/14/93	6	05/11/93	/	4	3	50	XD
16002	NEW FIXTURES @ MAIN ENTRY HALL	05/17/93	05/17/93	1	/	/	0	0	0	
16003	REWIRE IN-LAW UNIT	05/18/93	05/20/93	3	/	/	0	0	0	
16004	POWER TO NEW WATER HEATER	05/21/93	05/21/93	1	/	/	0	0	0	
99990	PROJECT PUNCHLIST & SELL	06/01/93	06/01/93	1	/	/	0	0	0	

Figure 12-20   As-built status in CPM format.

```
┌─────────────────────────────────────────────────────────────────────────────────┐
│ V1600****                        ACTIVITY RESOURCE DETAIL              Page  1 of  1│
│                   Date: 05/14/93 Project: DEMO Window: 05/06/93 -> 05/14/93         │
└─────────────────────────────────────────────────────────────────────────────────┘
```

Dow Date	Resource Title	Cost Account	D Unit	Amount	Rate	Cost

Activity: 01001 Title: MOBILIZE ON SITE

Dow Date	Resource	Title	Unit	Amount	Rate	Cost
THU 05/06/93 SUP	GENERAL CONTRACT SUPERVISION		MEN	6.00	25.00	150.00
FRI 05/07/93 SUP	GENERAL CONTRACT SUPERVISION		MEN	4.00	25.00	100.00
		Sub-Total:		10.00	25.00	250.00

Activity: 03201 Title: INSTALL REMEDIAL UNDERPINNING @ MAIN STRUCTURE

Dow Date	Resource	Title	Unit	Amount	Rate	Cost
TUE 05/11/93 ELC	ELECTRICIANS		MEN	1.25	20.00	25.00
PLB	PLUMBERS		MEN	4.50	22.50	101.25
SUP	GENERAL CONTRACT SUPERVISION		MEN	0.75	25.00	18.75
WED 05/12/93 ELC	ELECTRICIANS		MEN	1.00	20.00	20.00
PLB	PLUMBERS		MEN	5.00	22.50	112.50
SUP	GENERAL CONTRACT SUPERVISION		MEN	0.50	25.00	12.50
THU 05/13/93						
FRI 05/14/93 ELC	ELECTRICIANS		MEN	0.50	20.00	10.00
PLB	PLUMBERS		MEN	7.00	22.50	157.50
SUP	GENERAL CONTRACT SUPERVISION		MEN	0.25	25.00	6.25
		Sub-Total:		20.75	22.35	463.75

Activity: 16001 Title: REWIRE MAIN HOUSE & GARAGE

Dow Date	Resource	Title	Unit	Amount	Rate	Cost
TUE 05/11/93 ELC	ELECTRICIANS		MEN	0.75	20.00	15.00
FDN	FOUNDATION WORK LABORERS		MEN	5.25	15.00	78.75
WED 05/12/93 ELC	ELECTRICIANS		MEN	1.50	20.00	30.00
FDN	FOUNDATION WORK LABORERS		MEN	6.75	15.00	101.25
THU 05/13/93						
FRI 05/14/93 ELC	ELECTRICIANS		MEN	0.00	20.00	0.00
FDN	FOUNDATION WORK LABORERS		MEN	0.00	15.00	0.00
		Sub-Total:		14.25	15.79	225.00

| | | Total: | | 45.00 | 20.86 | 938.75 |

Figure 12-21   As-built resource used by activity.

of approximately 23,000 cubic yards. Krause requested a price of $4.20 per cubic yard, but the government priced the work at the $2-per-cubic-yard bid price. Krause signed the modification without objection. During the course of this additional excavation, a cave ("hole") approximately 8 feet in diameter at the top and 16 to 18 feet at the bottom was uncovered in the embankment foundation. A modification was issued in March 1973 and accepted by Krause for dental grouting of the cave.

When the additional embankment excavation was completed in April 1973, the actual yardage had increased to 40,385 cubic yards at $2 per cubic yard. Suitable bedrock had been reached at 35 feet. The depth of the excavation in the hole made the use of equipment more difficult, and the hole, which was in the creek channel, filled with water every time it rained. This would not have happened had the appellant reached bedrock by August 18 and been able to start embankment fill at that time and reach ground level by December.

In April 1973, an additional modification was issued to place in the hole an additional 40,000 cubic yards of embankment fill according to plans and specifications at $1 per cubic yard. The quantity and unit price were to be subsequently adjusted after final quantities were determined and an equitable price was established. Time and equipment records were to be maintained for that purpose. The appellant signed the modification because there was to be a later adjustment and because the backfill material was not the excavated material but was from emergency spillway, borrow, and other areas. The material that was used for backfill had been planned for use in the embankment and not in the hole. However, the appellant had decided the material excavated from the hole was too costly and time-consuming to process and therefore abandoned that stockpiled material. The modification made no reference to the excavated material and did not direct Krause to use that material as backfill.

On June 5, 1973, the contracting office issued Modification No. 17, increasing the quantity of additional embankment fill to 40,385 cubic yards and the unit price to $1.10 per cubic yard for a total of $44,423.50. Krause signed the modification over a year after it was issued, on July 31, 1974. He signed it with neither objection nor reservation, although it was more difficult to fill the area involved and the construction of a ramp and other unplanned operations were required. Krause said that he did not consider that the material excavated from the hole was covered by Modification No. 17, only the material that went back into the hole.

In June 1973, the stockpile material was tested and Krause was directed to use the material as backfill. To do so required processing of the material through a crusher. Krause complained that the material had heavy clay content and clogged the screens of the crushing plant. He experienced frequent equipment breakdowns, and the material became wet every time it rained. Equipment deliveries (vibratory roller and rock rack) were delivered late.

The substantial completion date was June 6, 1974, and on June 15, 1974, Krause signed a release of all claims except for the claim of processing the 40,385 cubic yards of material. On August 9, 1974, a claim including a cost breakdown prepared by Krause was filed. It stated that processing of the stockpile occurred between October 1972 and May 1974. The cost breakdown included equipment operating cost sheets and a list of equipment used. Krause's estimate of equipment hours, rates, and so forth was worked up with his foreman based on diaries and other materials. In arriving at these costs (even though he had owned much of his equipment for about nine years and had fully depreciated it over the first five years) to compensate him for his repairs and overhauls, Krause depreciated all of the listed equipment another 20 percent. On average, his records showed more than 1000 hours a year of operation of the equipment, but Krause admitted that he really did not know how many hours a year his equipment was operated.

The board concluded that there was a differing site condition and that a reasonable contractor could have been expected to conclude that limestone, not clay, would be encountered. The problem the board had was in establishing the amount of damage:

> Although Appellant had not kept records of the type needed to document his claim and his foreman's records were too general to show the hours and dates involved in processing the stockpile, Appellant submitted as an attachment to his claim an estimate of his various costs involved in the processing of the excavated material here involved. In addition, appellant presented testimony by his foreman, presumably in support of the amount of his claim. As pointed out, this testimony was of questionable validity. He was obviously not qualified as an expert witness, and he did not substantiate his testimony with any corroborative evidence or documentation. The Board must assess the credibility of a witness on the basis of supporting evidence and other backup material or on his qualifications as an expert. Felton Construction Company, AGBCA No. 406-9, 81-1 BCA [at] 14,932; Policy Research, Inc., ASBCA No. 26144, 82-1 BCA [at] 15,618; Norair Engineering Corp., ENG BCA No. 3981, 80-2 BCA [at] 14,659.
>
> The estimate of equipment hours, rates, and other costs, submitted as an attachment to Appellant's claim letter, is also not supportive of his claim. There were no backup records which would substantiate the hours of use of equipment for processing. The documentation submitted with respect to bond premiums, fuel costs and workman's compensation and insurance costs were imprecise and incapable of being applied only to the operation for which the claim is made.

The board went on to discuss whether Krause had waived his right to any claim with his willingness to sign the modifications, in particular Clauses 16 and 17, which clearly indicated that the price agreed to for the additional work covered all costs associated with such work. There was also a question as to whether the unit prices submitted by Krause and used by the contracting officer were deemed to be all inclusive of costs. The board continued:

> The documentation of extra costs submitted by Appellant is less than convincing, and we would find it difficult to determine an amount which we could conclude was adequately supported thereby. Further, the record does not establish how much of the stockpiled material came from the hole or from other areas of the job, and there is evidence that all materials in the stockpile were mixed to some extent. Nevertheless, it is clear that the Government recognized Appellant had difficulty in processing and using some of the stockpiled material. In this connection, the Board has consistently held:
>
>> Since the purpose of an equitable adjustment is to keep the contractor whole when the contract has been modified by a change, the measure of the adjustment must relate to his altered position. Accordingly, the proper measure of an equitable adjustment where the work has been performed is the difference between the reasonable cost of the work required by the contract and the actual reasonable cost to Appellant of performing the changed work, plus a reasonable amount of overhead and profit.
>
> The burden rests on Appellant to affirmatively prove the amount of money to which he is entitled. Southern Paving Corporation, supra; Bergen Construction, Inc., GSBCA No. 1058,65-1 BCA [at] 4554; Campbell Co., General Contractor, Inc.,

IBCA No. 723, 69-1 BCA [at] 7574; Click Co., Inc., GSBCA No. 3007, 70-1 BCA [at] 8335.

Commingling of base contract costs with change order costs is a documentation problem. After a successful appeal before the ASBCA, the John C. Grimberg Company, Inc., filed for attorney's fees and expenses. From that record:[*]

> The Government also contends that the records supplied by the applicant do not include a meaningful description of services provided and that time indicated on billings was commingled with work on other appeals arising from the contract, making it impossible to verify the amount of attorney's time actually spent on this appeal. Thus, according to the government, the applicant has failed to meet the documentation requirements of Interim Rule 9. Paragraph 9a requires that an application be accompanied by an exhibit fully documenting the fees and other expenses, indicating the date and a description of all services rendered or costs incurred. Paragraph 9b requires limitation of an application to expenses allocable to that portion of the appeal on which appellant prevailed.
>
> We find that the supporting exhibit complies with the requirements of the Rule. Dates and descriptions of services rendered are set forth in sufficient detail in detailed billing reports and statements in the exhibit. Commencing in August 1986, when the then pending four appeals pertaining to this contract were consolidated for hearing, appellant's counsel consolidated them into one billing account. In the fee application counsel apportioned 25 percent of the time spent indicated in an entry to this appeal, unless the description in the billing reports and statements indicated effort solely on another appeal, in which case it was excluded from the application, or specifically pertained to this appeal, in which case it was included in its entirety. We consider this apportionment to be a reasonable allocation of fees and expenses with respect to this appeal.

On a heavy demolition contract in Philadelphia, the contractor claimed that a number of equipment foundations were beyond scope. The city engineer disagreed and instructed his resident not to count trucks carrying the demolished foundations as they left the site. The contractor demanded arbitration and won. When it came to damages, the city had no defense. The contractor got what he asked for. The city's ostrich approach clearly did not work.

Documents Required

Change orders, when approved, become part and parcel of the contract. Accordingly, the documents required to support either the impacts resulting from a change order or to support a claim for a constructive change order are properly kept standard construction documentation. Figure 12-22 shows a partial menu of typical documents that should be available to reflect either the status of a project or, for a completed project, its as-built information. The menu is broken into the following groupings:

---

[*]ASBCA No. 32490, May 9, 1988.

O'Brien-Kreitzberg

Professional Construction Managers

Page 1 of _____

CLIENT _____   DATE _____

PROJECT _____   BY _____

DOCUMENT MENU

RETRIEVAL RECORD

(1)

FINANCIAL DATA

	N/A	YES	NO	ORIG	COPY
Job Cost Report					
Job Cost Detail					
Equipment Cost & Usage Report					
Financial Statement					
Trial Balance					
Labor Productivity Report					
Labor Distribution Report					
Certified Payrolls					
Time Cards					
Bid Estimates					
Labor Budgets					
Material Budgets					
Equipment Budgets					
Unit Prices					

CONTRACT DATA

	N/A	YES	NO	ORIG	COPY
Bid or Proposal Submittals					
Agreements between Owner and Design Professionals					
Agreement between Owner and Construction Manager					
Agreements between Owner and Contractors					
Agreements between Owner and Developer					
Agreements between Construction Mgr and Subcontractors					
Agreements between Contractor(s) and Subcontractors					
Executed Modifications to Agreements or Change Orders					
Technical Specifications					
General Provisions					
Special Provisions					
Supplemental General Provisions					
Supplemental Special Provisions					
Contract Drawings					
Performance Bonds					
Payment Bonds					
Mechanic Liens					
Certificates					
Notice of Substantial Completion					

(1) If a number, see attached sheets

Figure 12-22 Document menu.

O'Brien-Kreitzberg

Professional Construction Managers

DOCUMENT MENU

RETRIEVAL RECORD

N/A	YES	NO	ORIG	COPY

(1) **CONTRACT DATA (Cont'd)**
- Temporary Certificate of Occupancy
- Certificate of Occupancy
- Waiver of Liens
- Warranty and Guaranty data
- Building Permits

PROJECT ORGANIZATION
- List of Owner Key Staff & Function
- List of Construction Manager Key Staff & Function
- List of Design Professional(s), Key Staff & Function
- List of Contractor(s), Key Staff & Function
- List of Developer Key Staff & Function
- List of Subcontractor(s), Key Staff & Function

SCHEDULING DOCUMENTS
- As-planned schedule (Original or Approved by Owner)
- As-built schedule
- Master schedule
- Milestone schedule
- Monthly or Periodic Update schedules with narratives
- Floppy Diskettes for Schedules
- Window schedules (i.e. 2 wk look-back/4 wk look-ahead)

PROJECT REPORTS
- Monthly or periodic progress reports
- Daily, weekly, or monthly construction reports
- Daily, weekly, or monthly inspection reports
- Manpower summary reports
- Geological Site Reports
- Environmental Site Reports

FIELD DOCUMENTS
- Meeting Minutes
- Change Order Log
- Submittal Log
- Outside Correspondence (Letters and Memos)

(1) If a number, see attached sheets

Figure 12-22 *(Continued)*

O'Brien-Kreitzberg

Professional Construction Managers

Page 3 of ____

DOCUMENT MENU

RETRIEVAL RECORD

N/A	YES	NO	ORIG	COPY

(1)   FIELD DOCUMENTS (Cont'd)

Internal Memos					
Quality Control data					
Quality Assurance data					
Material Testing data					
Progress Photographs					
As-built Drawings & Specifications					
Force Account Summaries					
Field Orders					
Requests for Information					
List of Job Files for Owner					
List of Job Files for Design Professionals					
List of Job Files for Construction Manager					
List of Job Files for Contractors					
List of Job Files for Developer					
Monthly Pay Estimates					
Payment Logs					
Safety Records					
Security data					
Purchase Orders					
Subcontractor/Vendor Procurement Records					
Non-conformance notices					
Deficiency notices					
Punch List(s)					
Shop Drawings					
Material Samples					
Notice of Claims					

LEGAL DOCUMENTS

Complaint(s)					
Answer(s)					
Counterclaim(s)					
Interrogatories					
Answers to Interrogatories					
Deposition Transcripts					
Legal Correspondence					
Description of Anticipated Claims					

(1) If a number, see attached sheets

B:DOCMENU1.WK1

Figure 12-22 *(Continued)*

- Financial data
- Contract data
- Project organization
- Scheduling documents
- Project reports
- Field documents
- Legal documents

There are five columns to reflect the retrieval record for the documentation:

- *N/A.* Checked off if the particular record is deemed not applicable
- *Yes.* Checked off if the document is applicable and has been located
- *No.* Checked off if the document has not been found
- *Orig.* Checked off if the document retrieved is an original
- *Copy.* Checked off if the document retrieved is a copy

There are 95 categories listed on the document menu. Each major category has open space at the bottom of its list for additional documentation entries, as appropriate.

The USPS recommends that the contractor should be advised to keep records to document the following types of reimbursable expenses that may be considered in reaching an equitable adjustment for the Postal Service delays or suspensions:[*]

- General office overhead expenses directly attributable to or chargeable against the contract.
- Field overhead, including supervision of the job.
- Idle time of company-owned facilities or equipment. (Idleness of rental equipment is not normally an allowable expense, as the equipment should be returned to the rental agency. However, if the Contractor can show that it is less expensive to continue renting, these costs should be allowed.)
- Increase in material prices. This should be supported by supplier invoices or letters substantiating the price increases.
- Increase in wages. This can normally be verified by consulting local labor union officials and by interviewing Contractor personnel to determine labor standards compliance by the Contractor.
- Loss of efficiency. This is no doubt the most difficult expense to document. The item is best justified and substantiated by actual conferences, and in negotiations the matter can be thoroughly discussed. A written record of the negotiations and the allowances agreed upon should be kept.

---

[*]From USPS Contract Administration Course, May 1993.

- Mobilization and demobilization costs. Like loss of efficiency, these expenses are not easily identifiable. Here again, discussions and negotiations between the Contractor and Postal Service representatives may be the most efficient method of arriving at an equitable adjustment.

- Unusually severe weather conditions that would not have affected performance were it not for the change in time of performance. Conditions of unusually adverse weather may be documented by U.S. Weather Service reports for the affected period. These reports must be compared with those of the original period of performance to document the degree and extent of unusual weather conditions that could have contributed to delays or added work.

- Insurance and bond coverage. This coverage will have to be extended for the period of delay or suspension. Documentation of cost can be obtained readily from the bonding and insurance companies.

As-Built Documentation

In a major claim concerning a U.S. Navy hospital constructed at Bremerton, Washington (Santa Fe Engineers, Inc. v. United States; ASBCA No. 24578, ASBCA No. 25838, 94-2 BCA 26872 at 8040), lack of accurate, detailed, as-built data regarding change orders was a fatal flaw to the presentation of the claim.

This 1994 decision involves 58 claims for delay and inefficiency, as well as its 22 additional claims for inefficiency only, all of which were a substantial portion of numerous claims seeking from the government a total of $7,450,385 (originally $18,971,994), plus a time extension of 408 days.

It was undisputed that Santa Fe substantially completed the contract 408 days after the contract completion date. The parties bilaterally agreed to a 13-day time extension, and the government issued unilateral contract modifications extending the contract completion date by 243 days. Santa Fe sought an additional 152 days to cover the 408 days of alleged delay and compensation for 351 days (admitting that 55 days of delay were concurrent with Santa Fe-responsible delays).

Both parties relied on experts to establish whether there was delay and inefficiency and, if so, which party caused it. They also agreed that the contract required the use of CPM to determine contract delays. It also appears to be undisputed that the CPM schedule had to be revised during the fireproofing of the structural steel, primarily to reflect Santa Fe's resequencing of work activities due to its problems with that fireproofing.

Santa Fe's expert determined the causes of the delays through an analysis of an "as-built" chart plotted from various daily reports of the parties, along with a study of Santa Fe issues files. The government argued this analysis violated the contract requirements that a CPM analysis be used to measure contract delays. It pointed out that the materials used by Warner in preparing the "as-built," as well as those he used in evaluating Santa Fe issues files, were not included in the board's records. The appellant's analysis could therefore not be verified. It claimed that Warner's method was really a total delay method of measuring the delay.

Generally, Santa Fe's expert was unable to identify with specificity when changed work or work for which claims were filed was performed because the data were not available from the project documentation. However, he was able to determine generally when it was done by examining the base contract work related to it.

From the decision:

> In light of the massive effort of appellant's delay expert (findings 147), appel-lant clearly could have reconstructed and inputted the change order informa-tion at the proper times into the CPM schedule had appellant prepared and maintained proper records as to when change order and constructive change work had been performed (finding 167). Appellant's failure to prepare and maintain these records is clearly inexcusable in light of the clear contract requirement that this type of information be provided to maintain the accuracy of the CPM Schedule (finding 16 at pl. 4 & pl. 5). Accordingly, appellant's delay claims cannot be granted.
>
> Even if we were in error as to the proper legal standard of requiring the use of the CPM to prove delay, appellant's delay claims cannot be sustained. Both in the Wunderlich and Bateson cases, the contractor was found to have the burden of proving that the change orders caused delay to the overall project completion date. Appellant relies on its delay expert's opinion to establish the alleged delay. We are unable to accept appellant's delay expert's opinion as being reliable because for the most part we were unable to verify his opinions by looking at the factual data he relied upon.

The contractor, Santa Fe, did not keep clear records describing the perfor-mance of change order work. In particular, its daily records did not differenti-ate between base contract work and change order work. When its expert tried to analyze the time impact of the changes, he could not do so with specific link-age to the project records. Accordingly, the board did not accept his opinions as reliable.

Chapter 12

Change Orders into Claims

The project is over—here come the claims!

The first stage of claim preparation is to establish the owner's liability for damages. Here the focus is on change orders. The contractor must establish a causation connection between the change orders and the damages. The nature of the damages may assist in establishing the causation. Delay is a prime example. When delay is caused by a change order, the causal connection is established. The number of days of delay are readily converted to cost (i.e., general conditions, home office overhead, special facilities, heavy equipment, etc.).

Other impacts, such as overmanning or overtime, are identified from the project records. Similarly, the cost of the impacts can be developed if the project records are detailed and complete.

The John Doe Project Claim

CO #1—Sinkhole caused a delay of 17 calendar days. This was a differing site condition, so the liability is readily established. CO #3—Roof Joist VE caused a 10-calendar-day delay because of the increased erection time. This time extension was by mutual agreement. However, the VE savings did not account for the cost of the additional time. Also, the change order as written claimed half the overhead and profit (OH&P) of the cost as savings. This can be disputed by the contractor, particularly if the contractor had given notice of intent to dispute.

CO #10 delayed the critical delivery of the electrical load center 21 calendar days past its late finish (see Fig. 13-1). CO #11 directed a 48-hour work week for the electrical crew (see Fig. 13-2). Two of the three weeks were regained, leaving a net seven days' delay. The direct cost of the premium time was $31,488 (Fig. 13-3).

In addition to the overtime, the electrical subcontractor worked some activities out-of-sequence (and early) and increased his workforce to 16.

Figure 13-4 is a claim prepared by the Sparks Company in the amount of $139,116. This was presented to the XYZ Construction Company and later included in XYZ's claim.

TIE

January 8, 1999

Change Order #10

Change to the Electric Load Center

On the September 25, 1998 revised baseline, the LF for Fabricate/Delivery of the Electrical Load Center was 15 Jan 99. This change caused a 21-calendar-day delay in the delivery of the Electric Load Center (Activity 221-300).

Figure 13-1 TIE for CO #10.

TIE

January 8, 1999

Change Order #11

Overtime for Electric Work

XYZ electrical subcontractor is directed to put the entire electrical crew on a 48-hour week.

The premium portion of the 8-hours-per-week overtime will be paid on a T&M basis.

Figure 13-2 TIE for CO #11.

TO: XYZ Construction

FM: Sparks Electric, Ltd.

RE: Premium Costs, Feb.–May 1999

The following is a summary of the premium costs:

    12 weeks × 14 craftsmen × 8 hr × $20 = $26,880
    12 weeks × 2 foremen   × 8 hr × $24 =   4,608
    Total $31,488

This figure is supported by certified payrolls on file with you.

Figure 13-3 Cost for electrical premium time.

TO: XYZ Construction

FM: Sparks Electric, Ltd.

RE: Claim for Inefficiency

We are submitting this claim for two areas of inefficiency:

1. *Overtime*

    Sparks was on 48-hour weeks for 12 weeks. From the Business Roundtable curves, this caused a loss of efficiency of 17.5% applied to our Feb.–May 1999 workforce.

    This is 17.5% ($377,856) = $ 66,125

2. *Overcrowding*

    Crowding of our trade resulted in a loss of efficiency of 12% applied to our Feb.–May 1999 workforce.

    This is 12% ($377,856)   = $ 45,343

In addition, we had planned on an optimum 10-person crew. At this level, the foreman would have been a working foreman. Based upon a 16-person crew, we had to have one nonworking foreman. This loss of productivity was:

$(12 \times 1 \times 48 \times \$48)$ = $\underline{\$ 27,648}$

Total $139,116

Figure 13-4   Electrical claim for inefficiency.

Specific Approach

One of the two basic types of claims prepared by contractors is the specific approach. This is presented in elements (i.e., unapproved changes, time impact of all changes, productivity impact of all changes, etc.). The rationale for each element is stated with the associated cost. There are two common problems with specific claims: First, the owner can dispute the purported facts and/or costs, and, second, even at full claim value, the specific claim often falls well short (if not far short) of the contractor's losses.

XYZ Construction Company claim using specific approach

1. Extended overhead

    CO #1—Sinkhole   17 calendar days

    CO #3—Joists   10 calendar days

    CO #10—Electric $\underline{7}$ calendar days

$$\text{Total} \quad \text{34 calendar days}$$

$$\text{Cost per day} = \frac{20\% \ (\$5,000,000)}{296} = \$3378/\text{c.d.}$$

$$\text{Cost} = 34 \text{ c.d.} \times \$3378 = \$114,852$$

$$\text{Profit @ } 10\% = \$11,485$$

2. Markup on CO #11—premium time

$$\text{Markup} = 20\% \ (\$31,488) = \$6298$$

3. Sparks inefficiency claim

$$\text{Sparks claim (\$139,116)} + 20\% = \frac{\$\ \ 27,823}{\$139,116}$$

4. OH&P on VE CO #3

Request that 50% of the OH&P included in the VE savings be returned

$$\frac{\$52,864}{2} = \$26,432$$

5. Inefficiencies due to stacking of trades. XYZ suffered a lost of efficiency of 5% because of stacking of trades during the acceleration period of Feb.–May 1999.

$$\text{This cost 5\% (\$1,750,000)} = \ \ \$87,500$$

$$\text{Profit @ } 10\% = \ \ \ \$8,750$$

$$\text{Subtotal Sparks} \quad \$139,116$$

$$\text{Subtotal XYZ} \quad \underline{\$283,140}$$

$$\text{Total} \quad \$422,256$$

PM/CM answer to XYZ claim

1. *Extended overhead.* The 20% figure includes both overhead and profit:
   Allow $114,852
   Disallow $11,485

2. *Markup on CO #11—premium time.*
   Allow $6298

3. *Sparks inefficiency claim.* The Sparks subcontract is for $576,000. Assuming that labor is 60% of the contract at an average of $41/hour, this equals 8429 craft hours.

From the CPM plan (Fig. 6-1), the electrical work was scheduled to start at activity 38-43 (install power conduit P-W) on December 4, 1998 and conclude at activity 56-58 (energize) on April 8, 1999. This is a period of 104 calendar days. Working days equal ($\frac{5}{7}$ × calendar days) minus holidays. This is $\frac{5}{7}$ (104) − 3, or 71 workdays. The average electrical work crew would be:

$$\frac{8429}{8} \times \frac{1}{71} = 15$$

Sparks says they worked a crew 16 for 12 weeks at 6 days per week. This equals 72 workdays. These facts knock out Sparks claim 2 (crowding) and claim 3 (extra foreman). Claim 1 (overtime inefficiency) would be subject to an audit of planned craft hours versus actual.

4. *OH&P on VE CO #3.* The question of whether or not the contractor keeps the overhead and profit on a VE or credit change orders should have been addressed in the contract. This answer depends upon the contract.

5. *Stacking of trades.* The figures demonstrate that Sparks did not overcrowd its crew. That fact weakens this claim. An audit of actual versus planned labor would be useful in evaluating this claim.

Summary

The claim for $422,256 has been reduced to the following range:

1. Extended overhead	$114,852
2. Markup on CO #11	6,298
3. Sparks inefficiency (overtime)	0 to 66,125
XYZ 20%	0 to 13,225
4. 50% OH&P VE savings	0 to 26,432
5. XYZ stacking	0 to 87,500
	0 to 8,750
Subtotal Sparks	0 to 66,125
Subtotal XYZ	$121,150 to $257,057
Total	$121,150 to $323,182

Total Cost Approach

Most contractors would prefer the total cost approach because it is the simplest and easiest to apply. In a total cost claim, the claimant establishes the total cost of the project. The claim is, then, the total cost less the bid amount (plus any approved change orders). This approach is the most convenient for claimants because cost does not have to be tied to causation. However, for this very reason, courts do not favor this approach. While courts do not like the total cost approach, they do not preclude it. To persist on a total cost claim, the claimant must prevail on the following four points:

1. There is no other reasonable way to estimate the damages.
2. The bid was accurate, with no errors or omissions.
3. The costs claimed are not due to acts of the claimant or others such as force majeure.
4. That costs incurred (and claimed) are reasonable.

Some total cost claims prevail, but the ratio is probably worse than 10 to 1 against Huber, Hunt, Nichols v. Moore (69 Calif. 3d 278, 136 Calif. Rptr. 603; 1977) provides an example of the four guidelines. Huber, Hunt, Nichols (HHN) was the successful bidder on a convention center complex in Fresno, California. After construction, HHN filed a claim for $732,521, saying that the plans and specifications were deficient. It was determined that the approach was total cost. The defense successfully attacked the claim on a piece-by-piece basis.

For example, let's look at Cost Item #1525—Drinking Water. The original budget estimated the cost of drinking water at $1000, but $2293 was expended, resulting in an overrun of $1293. The court stated that in order to arrive at that conclusion, the jury would have to assume at least four elements of proof:

1. The contractor's original estimate of $1000 for drinking water was accurate.
2. The overrun of $1293 in the cost of drinking water was proximately caused by errors and omissions in the architect's plan and specifications.
3. Said errors and omissions in the architect's plans and specifications were proximately caused by the architect's negligence.
4. The overrun of $1293 was not due to other delays caused by change orders, inclement weather, or strikes (of which there were several).

The court agreed that this same fallacy applies to most of the other items of claim damages.

The USPS has the following to say about the total cost approach:[*]

> The total cost approach for determining an equitable adjustment takes the total cost of the work done and subtracts the estimated cost of the work that would have been done were it not for the change. The total cost approach is based on the Contractor's own estimated cost of the work that would have been done. In other words, if the Contractor claims that the final work after the change cost the Contractor $125,000 and the contract award price was $100,000, the total cost approach would recognize $25,000 as being the equitable adjustment for the change.

---

*From USPS Contract Administration Course, May 1993.

This approach is a very risky way of pricing an equitable adjustment because it is based on two weak premises:

- The actual cost incurred is the proper cost for the work (and does not include any costs incurred through the fault of the Contractor).
- The Contractor's estimate is a fair approximation of what it would have cost to actually perform the work. (However, the Contractor may or may not have bid that part of the work realistically; contractors often offer low prices in order to get the contract award—i.e., they "buy in.")

The total cost approach has been used when there is no better method available. In order to overcome the serious objections to this approach, however, the courts have been careful to ascertain the Contractor's actual costs incurred as a result of the change. They have reduced those costs by deducting those due to the fault of the Contractor, thereby eliminating the first major criticism. Next, the courts have attempted to avoid the second major criticism by using an average estimate, derived from the estimates of the Postal Service and other offerors, in order to preclude the possibility of "getting well" on changes after a buy-in.

Before using the total cost method, the Contracting Officer should be sure that all of the following conditions are present:

- No other method is available, and the reliability of supporting evidence is substantiated.
- The nature of the particular losses make it impossible or highly impracticable to determine them with a reasonable degree of accuracy.
- The original offer or estimate was realistic.
- The actual costs are reasonable.
- The Contractor is not responsible for the added expenses.

In other words—and as is often the case in contract administration—the Contracting Officer must review all the facts of the situation and make a reasonable decision based on those facts. The Contracting Officer should negotiate with the Contractor to reach a reasonable price, bearing in mind the basic objective of leaving the Contractor in the same position the Contractor was in before the change.

Total cost claim—John Doe Project

Let's look at how the total cost method can be applied to the John Doe Project.

Contract costs

Direct costs

Labor	$1,505,000
Materials	1,500,000
Subcontracts	1,356,000
	$4,361,000

Indirect costs

General conditions	$1,210,000
Total cost	$5,571,000

Revenue

Base contract	$5,000,000
VE change	(117,160)
Change orders	255,000
Total revenue	$5,137,840
Cost overrun	$433,160

Add:

HOE (Extended time—34 c.d.; see Fig. 13-5)	23,186
Profit (10% of total cost)	557,100
Claim preparation costs	75,000
Total claim	$1,088,446

Backing up these figures is a breakdown of the base bid by CSI division (Fig. 13-6) and a similar breakdown of actual costs (Fig. 13-7). Figure 13-8 compares bid with cost on a CSI division basis.

Modified Total Cost Approach

Recognizing the elements of proof required in the total cost approach, the contractor can "bite the bullet" and modify the total cost to weed out any weak spots. For instance:

- *Bid mistake.* If an item or service cannot be purchased at the bid price, identify it and deduct the difference from the total cost pool.
- *Cost due to weather or strikes.* Identify the cost and deduct it from the total cost pool. One such cost would be "show up" time paid when the crew shows up for work and is sent home because of weather.
- *Accident.* An equipment accident results in costs that are covered by insurance. These costs should be netted out of the total cost pool.
- *Unallowable costs.* Costs that the contractor knows will not pass audit (donations, traffic fines, entertainment, etc.) should be taken out of the total cost pool.

1998 Delay

John Doe Billings 1998	$ 3,050,000
XYZ Total Billings 1998	$25,546,000
XYZ Home Office Overhead 1998	$ 1,021,840

HOE overhead for the John Doe Project in 1998:

$$\text{HOE} = \frac{\$ 3,050,000}{\$ 25,546,000} (\$1,021,840) = \$122,000$$

Calendar days (c.d.) = July 1 through December 31 = 184

HOE/c.d. = $663

HOE (sinkhole) = 17 c.d. × $663
              = $11,271
HOE (joist VE) = 10 c.d. × $663

              = $6,630

1999 Delay

John Doe Billings 1999	$ 2,087,840
XYZ Total Billings 1998	$23,667,000
XYZ Home Office Overhead 1998	$ 1,250,000

HOE overhead for the John Doe Project in 1999:

$$\text{HOE} = \frac{\$ 2,087,840}{\$23,667,000} (\$1,250,000) = \$110,272$$

Calendar days (c.d.) = January 1 through May 26 = 146

HOE/c.d. = $755

HOE (electrical co.) = $ 5,285
Total HOE   = $23,186

Figure 13-5 Eichleay calculation—John Doe Project.

Modified total cost claim—John Doe Project

Now let's apply the modified total cost approach to our example project.

Contract costs

Direct costs

Labor	$1,505,000
Materials	1,500,000
Subcontracts	1,356,000
	$4,361,000

CSI division		Labor	Materials	Subcontracts	Total
2	Site	$ 30,000	$ 20,000	$ 35,000	$ 85,000
3	Concrete	109,000	163,000	—	272,000
4	Masonry	74,000	50,000	—	124,000
5	Metals	270,000	315,000	100,000	685,000
6	Wood/plastics	120,000	80,000	—	200,000
7	Thermal—moisture	298,000	198,000	—	496,000
8	Doors/windows	120,000	80,000	—	200,000
9	Finishes	168,000	112,000	—	280,000
10	Specialties	40,000	40,000	—	80,000
11	Equipment	120,000	300,000	—	420,000
12	Furnishings	60,000	40,000	—	100,000
13	Special construction	—	—	—	—
14	Conveying	10,000	50,000	—	60,000
15	Mechanical			576,000	576,000
16	Electrical			422,000	422,000
					$4,000,000
			General conditions (GC)		1,000,000
	Subtotals	$1,419,000	$1,448,000	$1,133,000	
				Total	$5,000,000

Figure 13-6 Breakdown of base bid by CSI division.

CSI division		Labor	Materials	Subcontracts	Total
2	Site	$ 65,000	$ 35,000	$ 30,000	$130,000
3	Concrete	120,000	180,000	—	300,000
4	Masonry	82,000	60,000	—	142,000
5	Metals	180,000	225,000	120,000	525,000
6	Wood/plastics	160,000	120,000	—	280,000
7	Thermal—moisture	290,000	208,000	—	498,000
8	Doors/windows	150,000	95,000	—	245,000
9	Finishes	208,000	132,000	—	340,000
10	Specialties	60,000	55,000	—	115,000
11	Equipment	120,000	300,000	—	420,000
12	Furnishings	60,000	40,000	—	100,000
13	Special construction	—	—	—	—
14	Conveying	10,000	50,000	—	60,000
15	Mechanical			626,000	626,000
16	Electrical			580,000	580,000
					$4,361,000
			General conditions (GC)		1,210,000
	Subtotals	$1,505,000	$1,500,000	$1,356,000	
				Total	$5,571,000

Figure 13-7 Breakdown of cost by CSI division.

CSI division		Base bid	Costs	Difference
2	Site	$ 85,000	$130,000	$ 45,000
3	Concrete	272,000	330,000	28,000
4	Masonry	124,000	142,000	18,000
5	Metals	685,000	525,000	(160,000)
6	Wood/plastics	200,000	280,000	80,000
7	Thermal—moisture	496,000	498,000	2,000
8	Doors/windows	200,000	245,000	45,000
9	Finishes	280,000	340,000	60,000
10	Specialties	80,000	115,000	35,000
11	Equipment	420,000	420,000	0
12	Furnishings	100,000	100,000	0
13	Special construction	—	—	—
14	Conveying	60,000	60,000	0
15	Mechanical	576,000	626,000	50,000
16	Electrical	422,000	580,000	158,000
		$4,000,000	$4,361,000	$361,000
	General conditions (GC)	1,000,000	1,210,000	210,000
	Total	$5,000,000	$5,571,000	$571,000

Figure 13-8   Comparison of base bid and cost by CSI division.

Indirect costs

General conditions	$1,210,000
Total cost	$5,571,000

Less:

Site work under bid	$15,000
Windows under bid	13,000
Weather show-up time	6,500
Crane accident (covered by insurance)	9,300
Unallowable costs* (per internal audit)	4,300
	$48,100
Modified total cost	$5,522,900

Revenue

Base contract	$5,000,000

---

*Also, claim preparation costs are not added.

VE change	(117,160)
Change orders	<u>255,000</u>
Total revenue	$5,137,840
Cost overrun	385,060
Add:	
HOE (see Fig. 13-5)	23,186
Profit (5% of total cost; applied to modified total cost)	<u>276,145</u>
Modified total claim	$684,391

Jury Verdict Method

Faced with a total damage claim where liability is established but the total damages are questionable, many boards take an alternate approach: the jury verdict. Rather than taking an all-or-nothing damage approach, the board identifies their decision as a jury verdict. The jury verdict approach was used as early as 1960 in Air-a-Plane Corp. v. United States (ASBCA No. 3842, 60-1 BCA 2547; 1960).

In Lawrence D. Krause v. United States (AGBCA No. 76-118-4, 82-2 BCA 16, 129; Nov. 16, 1982) the board found as follows:

> When a contractor has proven entitlement, but cannot define his costs with exact data, courts and appeal boards have been reluctant either to send the dispute back to the Contracting Officer for determination of excess costs or to send the contractor away empty handed when it is at all possible to make a fair and reasonable approximation of the extra costs. Under such circumstances, it is not uncommon to resort to the "jury verdict" method of determining excess costs. Specialty Assembling & Packing Co. v. United States [11 CCF p80,242], 174 Ct. Cl. 153 (1966); River Construction Corp. v. United States (8 CCF p71,905), 159 Ct. Cl. 254 (1962); Western Constructing Corp. v. United States, 144 Ct. Cl. 318 (1958). This Board has employed "jury verdicts" in a number of appeals. Road-Roc, Inc., supra; ICA Southeast, Inc., AGBCA No. 331, 73-1 BCA p9969; Albert J. Demaris, AGBCA No. 437, 75-2 BCA 011,359; Greenwood Construction Co., Inc., AGBCA No. 75-127, 78-1 BCA p12,893; Harold Benson, AGBCA No. 384, 77-1 BCA p12,490. It is not essential that the amount be ascertainable with absolute exactness or mathematical precision. Eastman Kodak Co. v. Southern Photo Co., 273 U.S. 359, 379 (91927); Brand Investment Co. v. United States [2 CCF p826], 102 Ct. Cl. 40 (1944), 58 F. Supp. 749, cert. Denied 324 U.S. 850. It is enough if the testimony and evidence adduced is sufficient to enable the court or board (acting as the jury) to make a fair and reasonable approximation of the amount recoverable. R-D Mounts, Inc., ASBCA Nos. 17668, 17669, 75-1 BCA p11,237. Mann Construction Co., Inc., AGBCA No. 76-111-4, supra; E. Arthur Higgins, AGBCA No. 76-128, 79-2 BCA p14,050.

The record herein does not support the costs alleged by Appellant, but it is doubtful that further proceedings would produce a better foundation for an award (WRB Corporation v. United States [12 CCF p81,781], 183 Ct. Cl. 409; 1968). However, the record does provide a sufficient basis for a fair and reasonable approximation of Appellant's extra costs. Accordingly, we believe a "jury verdict" is appropriate and we will determine an equitable adjustment based on and supported by the record.

Krause had requested $68,953 and 104 days' time to recover liquidated damages. The board went through its own assumptions and calculations and awarded $27,595 and 51 days.

This verdict is certainly in keeping with the following statement:[*]

When a board or court uses the jury verdict approach, the final number chosen is going to be on the low side—the safe side as far as the government dollars are concerned.

In Bechtel National, Inc. (BNI) v. United States (90-1 BCA 22549, NASA BCA No. 1186-7; Dec. 22, 1989) the claim was $4,155,176 on a $13,075,376 fixed price contract. The board found as follows:

We find that Appellant has used a total cost approach. A claim of extra hours worked based on virtually the entire difference between an estimate of hours and the actual hours worked is a typical total cost approach. S.W. Electronics & Manufacturing Corp. v. United States, 228 Ct. Cl. 333, 655 F. 2d 1078 (1981); Nello L. Teer Company, ENG BCA No. 4376, 86-3 p19,326 at 97,768; J.M.T. Machine Company, Inc., ASBCA Nos. 23928, 24298, 24536, 85-1 BCA p17,820, at 89,181. Appellant must, therefore, establish that the Government caused its increased costs with proof that satisfactorily eliminates other causes of inefficiency or delay. Fermont Division, Dynamics Corp. of America v. United States (24 CCF p81,923), 216 Ct. Cl. 448, 578 F.2d 1389 (1978).

The total cost method is not favored by the courts or boards, and to justify its use the contractor has to establish that (1) the nature of the particular losses makes it impossible to determine them with a reasonable degree of accuracy, (2) the contractor's bid or estimate was realistic, (3) the contractor's actual costs were reasonable, and (4) it was not responsible for the added expense. WRB Corp. v. United States, 183 Ct. Cl. 409 (1968); Newell Clothing Co., ASBCA No. 28306, 86-3 BCA p19,093, aff'd, 818 F.2d 876 (Fed. Cir. 1978). Appellant has not satisfied all of these requirements.

And at page 110909 as follows:

The Board will determine the amount of an equitable adjustment by the most reasonable means available when entitlement is found. S.W. Electronics & Manufacturing Corp. v. United States, supra; Servidone Construction Corporation, ENG BCA No. 4736, 88-1 BCA p20,390, at 103,103; Alabama Air-Flo, Inc., ASBCA No. 22638, 78-2 BCA p13,488 at 66,022. We find, however, that we cannot employ the "jury verdict" approach using the Appellant's cost presentation.

---

[*]B. B. Bramble and M. T. Callahan, *Construction Delay Claims* (New York: John Wiley & Sons, 1987), p. 170.

Appellant does not provide a basis in the detailed productivity analysis for us to make a fair and reasonable approximation of the increased costs caused by the NASA RFIs. Appellant's analysis provided the total direct labor manhours claimed by cost category, but was not detailed for the time period or the types of costs alleged to have been incurred. Therefore, the labor manhours in the claim cannot be specifically allocated as the cumulative impact for which we have found Appellant is entitled to recover. To determine the amount of an equitable adjustment some allocation of the causes of increased costs must be made, but BNI's estimates of the large cost increase attributable to Federal Steel were not substantiated by the evidence, and we also cannot segregate the increased costs due to the NASA RFIs from the Federal Steel costs based on Appellant's cost data. Assurance Company v. United States, 813 F.2d 1202 (Fed. Cir. 1978); AGH Industries, Inc., supra.

The Contracting Officer's final decision, however, can properly serve as the basis for a "jury verdict." S.W. Electronics & Manufacturing Corp. v. United States, supra; Schuster Engineering, Inc., ASBCA Nos. 28760, 29306, 30683, 87-3 BCA p20,105, at 101,806, and cases cited therein. Accordingly, BNI would be entitled to recover on its cumulative impact claim at least $275,000 that the Contracting Officer was willing to provide in the final decision (finding 149). We have weighed the record as a whole and concluded that that amount would not fully compensate Appellant for the cumulative impact costs caused by the NASA RFIs. The total cost overrun on the job was $7,954,377, based on a comparison of the total costs incurred and our "should cost" estimate of the cost of the job. Appellant is not responsible for all of this increase, and we have elected to make an independent estimate of the cumulative impact costs caused by the NASA RFIs. See Sovereign Construction Company, Ltd., ASBCA No. 17792, 75-1 BCA p11,251, at 53,606. Respondent is not liable on this claim for other causes of increased costs, i.e. Federal Steel, the tube welding claim, the impact costs related to that claim, Revision F other direct costs, and claim preparation costs, which total $7,510,915. The cost overrun would have to be reduced for these other causes of increased costs, and thus we have allowed as a "jury verdict" $443,462 for Appellant's cumulative impact claim.

The USPS has the following comments on the jury verdict approach:[*]

When costs cannot be isolated and identified, the Postal Service and the Contractor may have to approach an equitable adjustment on the basis of estimate alone. In these cases the Board and the courts have permitted the use of expert opinion to estimate the cost of a change. This approach was first advanced by the Claims Court in 1958. It involves hearing all the evidence, including opinions of qualified experts and specialists on the Contracting Officer's staff and opinions of qualified experts and specialists representing the Contractor, and then basing the equitable adjustment on an overall analysis of the evidence.

The Boards and the courts do not like to use this approach because in most cases they are of the opinion that by the time a case reaches them, there should be actual documentation to review rather than just estimates. Nevertheless, decisions are still based on the jury verdict approach, and the approach may be valid when:

---

[*]From USPS Contract Administration Course, May 1993.

- actual costs cannot be isolated and identified,
- both parties rely on the opinion of qualified experts to estimate the costs incurred, and
- the estimates of the parties are in conflict.

The final amount of price adjustment in a "jury verdict" is a judgment call, based on an evaluation that gives credence to both sides, carefully examines all available data, and weighs the data to arrive at a reasonable decision.

Audits

As noted, final findings often rely on audit results. To successfully audit, the owner must first (through the contract) have the right to perform an audit. Second, the contractor must have proper accounting records.

In Art Metal-U.S.A. Inc. v. United States (GSBCA No. 5245-Rein, GSBCA No. 5998, 83-2 BCA 16881), lack of an adequate accounting system was a key element in determining damages. Art Metal was founded in 1945 in Newark, New Jersey, to make steel kitchen cabinets. In 1952, the company began manufacturing a line of steel office furniture. In 1962, it received from the government the first of several hundred fixed-price contracts for steel office furniture. Between 1962 and 1975, 70 to 80 percent of the company's sales were to the government.

In 1973 and 1974, the government specifications changed from conventional to contemporary office furniture. This required Art Metal to undertake major retooling. This retooling required extensive engineering and retraining of their workforce. Art Metal received no government funding for its added costs.

Findings by the board included the following (at page 101503):

18. Appellant's accounting practice is not to capitalize any of its expenses associated with the engineering and purchase of new tools and equipment. It accumulates those costs in its factory supplies account and charges that expense, and tool room expense as well, as costs of goods sold against its net sales at the end of each of its yearly accounting periods. In doing so, it makes no distinction between general and special purpose equipment. The reason for that accounting practice is to recognize the risks that appellant might never receive another contract for the item those tools and dies were purchased to produce or that the Government might change the design, as in fact happened generally in the period 1972–1975, thereby rendering that tooling and equipment instantly obsolete. Appellant charged the $692,233 costs of new tooling and equipment it had incurred to perform the contract here as changed to cost of goods sold and that against net sales on its profit and loss statement for its fiscal year 1975.

Further (at page 101504):

20. Appellant does not utilize a cost accounting system in its operations. It neither accumulates nor assigns historical costs to products, jobs, departments, or processes. ...Appellant does not analyze the profitability of individual contracts. Neither does it routinely develop indirect cost allocation rates. It develops such allocation rates only for the purpose of submitting bids or claims. Appellant prepares its bids on the advertised

solicitations which it receives by estimating the materal costs it might incur, preparing a bill of materials for the item to be produced, assigning averaged direct labor hours to each item or operation on that bill of materials from data recorded in its production control and statistics system, and developing an allocation rate for all of its indirect costs, which it calculates by collecting its total indirect costs current through the most recent quarter of its fiscal year and dividing those costs by direct labor hours from its payroll records as incurred in its production departments during that same period. It does not adjust its indirect costs to reflect those not allowable under the contract cost principles in part 1-15 of the Federal Procurement Regulations. Appellant routinely factors a larger profit into its bids for definite quantity contracts than it does on its bids for requirements contracts. It does so because it is of the view that definite quantity contracts may well disrupt its production planning. When it anticipates that it may fit orders placed under requirements contracts into its production planning, it accepts a lower margin of profit. Similarly, the profit appellant factors into any of its bids depends on the volume of its business, both actual and anticipated, at that moment. If the volume of business is quite low or expected to decline, appellant will accept diminished profitability in the hopes of garnering business to cover its fixed expenses. On the requirements contract for storage cabinets under which it submitted a $9.9 million claim, appellant bid with a profit on two different sizes of storage cabinets at rates of 5.6 and 2.7 percent of total costs. Obviously, appellant's accounting practice of expensing the costs of new tools and equipment at the end of the fiscal year in which acquired, finding 18, results in enhanced profitability of subsequent production runs of an item of steel office furniture that might occur in subsequent accounting periods.

The government auditors undertook a major review of Art Metal's production control and statistics system, known as short interval scheduling (SIS). The SIS is not part of the financial accounting system. The auditors encountered differing opinions in regard to the accuracy of the SIS and discovered a large quantity of the records had been destroyed. They then decided to bring in Booze Allen Hamilton (BAH) to conduct an independent analysis.

At page 101516, the board states:

37. The Government's auditors reported their audit findings...in a report issued March 23, 1978. They accepted the BAH engineering estimate of additional productive direct labor hours in the press, brake, welding, and assembly departments reasonably required to perform the contract as changed, and rejected appellant's direct labor claim that was premised upon production data posted to operation cards in the SIS system through August 30, 1975. They were of the view that appellant's SIS system was not sufficiently reliable to distribute direct labor hours to particular products, components, or subassemblies; that data entered into the SIS system were not subjected to any sort of control, including that inherent from a division of responsibility between several operating departments or other extrinsic controls, such as verification of postings from SIS interval sheets to SIS summary reports and operation cards, or periodic reconciliation of the SIS system entries to appellant's financial accounting records; and that the SIS system data could easily be manipulated.

At page 101530, the board states:

Here it has always been, and remains, appellant's burden to prove that the additional productive direct labor hours it claims to have incurred are reasonable. Once appellant adduced evidence that those additional productive direct labor hours it claimed were calculated from data recorded and compiled in its SIS system, and that its SIS system was a regularly-maintained system of business records, the Government was charged with the burden of going forward. The Government's burden was to demonstrate, by a fair preponderance of the evidence, that there was no basis for the presumption to be drawn or that the costs as calculated from appellant's business records were unreasonable, unrepresentative, unjustified, or somehow inadequate or incomplete....

We have said that the presumption of reasonableness that attaches to costs calculated from data compiled in the regular course of business activity is supported by the same circumstantial guarantees of reliability and trustworthiness that support the admissibility of business records under the exception to the rule against hearsay. Federal Rule of Evidence 803(6) provides, in pertinent part, that compilations of data regularly and contemporaneously made in the course of routine business activity may be admitted, even through hearsay, unless "the source of information or the method or circumstances of preparation indicate lack of trustworthiness." We have also observed that the presumption of reasonableness of costs calculated from business data compilations may be attacked by a showing that there was no basis for it to be drawn, and we have found that the additional productive direct labor hours appellant is claiming here were calculated from data that are neither trustworthy nor reliable. If it has been demonstrated that the source data are not reliable or trustworthy, are costs calculated from such data still presumed reasonable in the absence of affirmative evidence of objective unreasonability? We think not. The presumption is no more persuasive than the business records exception to the rule against hearsay. Accuracy of business data compilations underlies their trustworthiness, and a tribunal may decline to admit those data compilations whose accuracy may reasonably be questioned...The additional productive direct labor hours appellant claims, even though calculated from data contemporaneously and routinely recorded in the course of appellant's business activity, cannot be presumed to be reasonable because our record demonstrates that the source data from which those hours were calculated are neither reliable nor trustworthy. Appellant offered no other evidence of the additional productive direct labor hours it reasonably incurred to perform the contract as changed, but the Government did just that in the form of the BAH engineering estimates, finding 36, and we have found as fact, as appellant's expert manufacturing engineering witness conceded, that the BAH engineering estimates were accurate and complete estimates of the additional productive direct labor hours reasonably required to perform the contract as changed, finding 40. We decided that the equitable adjustment due appellant is to be determined upon the BAH engineering estimates, not the hours appellant calculated from defective data recorded in its SIS system.

The board's decision continued with other findings in regard to allocation of indirect costs and the cost of special equipment. Art Metal recovered less than half of its claim, due to unorthodox accounting procedures and demonstrated inaccuracies in its production/accounting records.

Chapter 14

Subcontractor Change Orders

Since subcontractors do most of the work on the average construction project, they are the most affected by change orders. Figure 14-1 reflects the traditional relationship between owner, general (prime) contractor, architect, and major subcontractors (electrical and mechanical). Figures 14-2 and 14-3 are variations of this structure. The organizational structure, of course, determines the point-to-point flow or movement of a PCO as it moves toward approval as a change order.

Single Prime Contract

In the traditional single prime contract situation, the owner selects one prime contractor (usually the general contractor), and the prime contractor, in turn, selects many subcontractors and suppliers to perform the work required for the project. In this arrangement, the relationship between the three major participants (the owner, the architect, and the contractor) places the architect in the position of being the representative of the owner and having authority over the contractor during the construction phase. The major role of the architect during this phase is to observe the activity of the contractor and the subcontractors and make certain that the work is in compliance with the architect's design.

Multiple Prime Contract

In the standard multiple prime contract method of project delivery, an owner secures an architect to design the building and prepare the contract documents. However, the owner requires the architect to prepare the documents in such a fashion that the documents can be divided into several prime contract packages. The owner then selects contractors for each of these major portions of the construction project. Typically, these portions include (1) general contracting; (2) plumbing; (3) heating, ventilating, and air conditioning; and (4) electrical. On

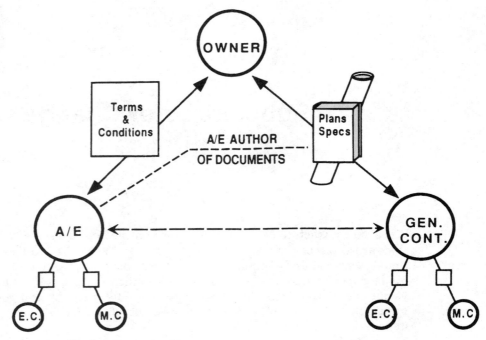

Figure 14-1 Single prime contractor.

more sophisticated projects, the number of packages may be extended to include the elevator/escalator portion, site utilities, telecommunications, structures, and other major project components. Each of these prime contractors, in turn, hires subcontractors and suppliers to perform specific tasks within their disciplines. Figure 14-4 illustrates these relationships.

In some states (including New York and Pennsylvania), state law requires, on public projects, a minimum of four prime contractors in the area of general contracting, heating and ventilating, plumbing, and electrical. These state laws typically do not preclude breaking the job into additional packages.

As in the single prime contract situation, the architect, as representative of the owner, has authority over the contractors during the construction phase. The architect generally monitors the work of each of the prime contractors and subcontractors to guarantee that the work is in compliance with the contract documents. Due to both the increase in the number of prime contractors and the possibility of overlapping areas of responsibility, however, the architect has a much more difficult job of monitoring the work and discerning fault when problems arise. Figure 14-5 reflects a multiple prime contract with the coordination role handled by a PM/CM.

Because of the serious coordination problems that exist in a multiple prime situation, some owners have instituted the usage of a designated coordinator. This is accomplished by indicating, in both the contract and bidding instructions, that one of the prime contractors is designated as the coordinator and that all the other prime contractors must adhere and comply with that requirement. In this arrangement, the owner selects an architect who prepares the multiple prime packages for the owner. The owner, in turn, secures multiple prime contractors, with the designated prime in the role of coordinator. (See Fig. 14-6.) The architect is in the position of authority over the contractors during the construction phase and monitors their work for compliance with the documents. A major problem with the designated coordinator system is that even though the designated prime contractor has the contract authority to coordinate the other prime contractors, there is no means of enforcing the directives.

Figure 14-7 shows a configuration in which the coordination role is assigned to a PM/CM. In some PM/CM assignments, particularly those involving a guaranteed maximum price (GMP), the PM/CM organizationally is the coordinator. In this role, separate prime contracts are awarded to 20 to 30 (more or less)

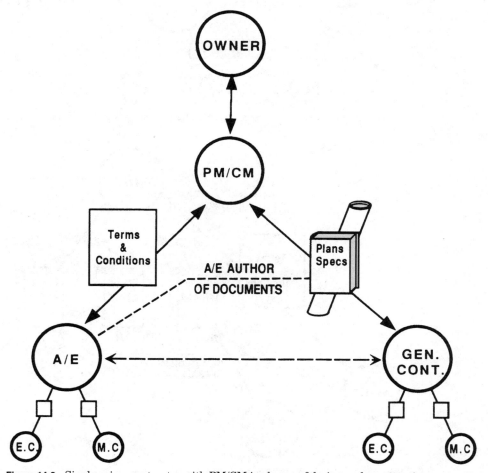

Figure 14-2 Single prime contractor with PM/CM in charge of design and construction.

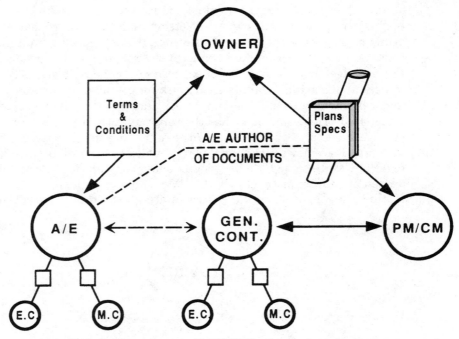

Figure 14-3   Single prime contractor with PM/CM in charge of construction.

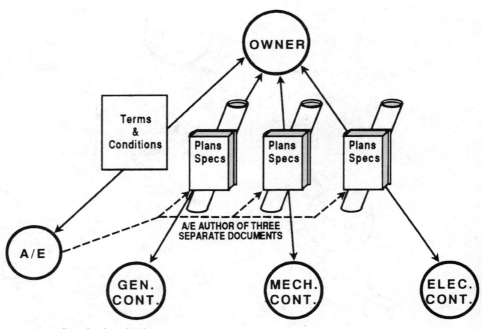

Figure 14-4   Standard multiple prime contracts.

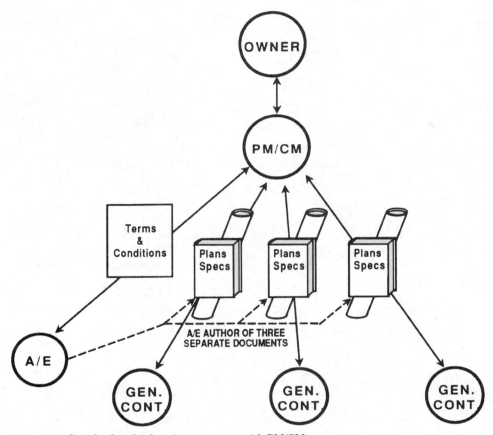

Figure 14-5 Standard multiple prime contracts with PM/CM.

craft contractors. Table 14-1 shows typical prime–subcontractor combinations. This table is an example of the "many prime" approach.

Legal Relationship between Owner and Subcontractors

The subcontractor presents a legal paradox. Even though subcontractors do most of the work on the average construction project, there is no privity of contract between the owner and the subcontractor; they are legally isolated from one another.

In Guerini Stone Co. v. P. J. Carlin Construction Co. (245 U.S. 264; 1915), this isolation was noted:

> [The] prime contract specifically provided that none of its provisions shall be constituted as creating any contractual relationship between the [owner] and any subcontractors . . .and . . . there is no provision in the prime contract which would allow a subcontractor to appear or participate in any way before the [owner] . . . in a dispute which affected . . . the owner . . . Indeed, the [owner], as

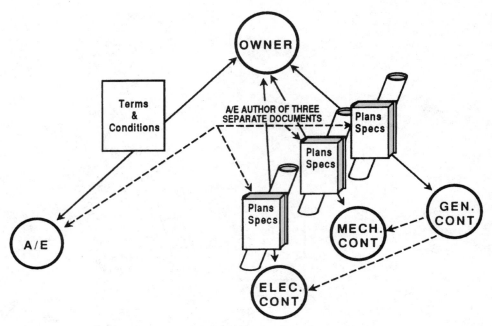

Figure 14-6 Multiple prime contracts with designated coordinator.

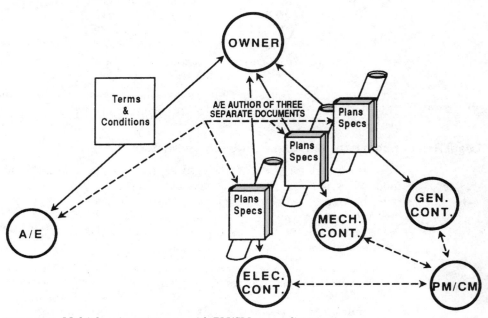

Figure 14-7 Multiple prime contracts with PM/CM as coordinator.

TABLE 14-1   Typical Prime–Subcontractor Combinations

Prime	Subcontractors	
Foundation	Dewatering	Excavation
	Concrete	Piling
General	Concrete	Structural steel
	Rebar	Ornamental iron
	Rough carpentry	Glazing
	Millwork	Roofing
	Masonry	Drywall
	Painting	Tile
	Carpet	Hollow metal
Mechanical	HVAC	Elevators
	Sheet metals	Sprinklers
	Refrigeration	Instrumentation
	Plumbing	
Electrical	Communications	Instrumentation

any other owner, may not wish to deal with a subcontractor at all, and we do not criticize this mode of doing business.

Subcontracts range from the very small to the very large. Traditionally, the subcontractor subcontracted to the general contractor. However, in those states in which there must be separate prime contracts under the law, and more recently in the construction management form of contracting, prime contracts for HVAC, plumbing, and electrical can easily exceed the value of the general contract. These larger prime contractors/subcontractors are experienced in working either way and are experienced in the contractual differences between the two approaches.

Smaller subcontractors, which have traditionally always subcontracted to a prime contractor, may find that the contractual differences in working on a prime contract directly with a construction manager involves substantially different contract administration techniques and requirements. Actually, this type of contract can be, in effect, a contract between the owner and the former subcontractor, since the typical construction management contract places the construction manager in the role of "owner's representative." (See Fig. 14-8.)

The Subcontract

In subcontracting, there is a greater opportunity for the prime contractor and subcontractor to have a meeting of the minds because of the opportunity for negotiation. (In public work, bids are submitted in secret and must be in absolutely proper form to be acceptable; many public corporations follow similar

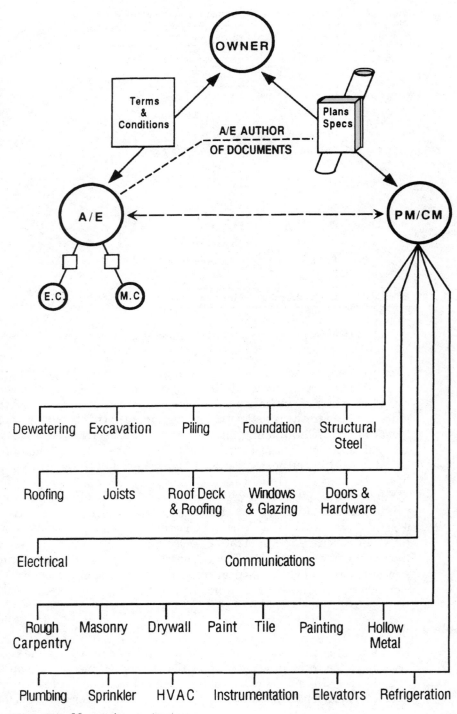

Figure 14-8 Many prime contractors.

practices as a part of their fiduciary responsibility to their public ownership.) Also, because of the more frequent opportunity to select subcontractors instead of merely accepting the lowest bidder, prime contractors often have a long-term relationship with their subcontractors. Usually, a prime contractor will accept prices from a limited number of qualified subcontractors. Thus, reputation and prior working relationships can become factors in the selection process for subcontractors.

Most prime contractors have developed a standard form of contract that they use as their general conditions, adding any specific or special clauses appropriate to the particular subcontract. These general conditions tend to be carefully drawn, but often the individual subcontract is somewhat loose. Also, subcontracting arrangements can be come quite complex.

In International Erectors v. Wilhoit Steel Erectors (400 F. 2d 465, 5th Cir.; 1968) a prime contractor for the construction of an industrial plant subcontracted the fabrication and erection of the structural steel frame. The subcontractor was Southern Engineering, which fabricated the steel, but sub-subcontracted the erection to Wilhoit. Wilhoit, in turn, sub-sub-subcontracted the erection to International Erectors. Southern did not deliver the steel on time, and International, which suffered delay damages, sued Wilhoit. The suit also named Southern as a party.

The court did not find assurances of delivery as part of the contractual obligations that were transferred from party to party. Finding against International, they stated:

> Prudence and perhaps foresight might have insisted that a provision creating such an obligation (i.e., guarantee of delivery) be included in the written contract, but the written [contract] expressly and unequivocally negated any such obligation. This was an arms-length transaction between contractors of considerable experience in such matters, and we cannot rewrite the contract just because one of the parties would in retrospection have written it differently.

The single most important "edge" written into almost every prime contractor subcontract is the contingent payment clause. In essence this clause states, "You get paid when we get paid, and only if we get paid." The contingent payment clause has been the source of subcontractor complaints over many years.

In *Construction Law in Contractors' Language,* McNeill Stokes writes:[*]

> Fortunately, in a number of recent decisions by the courts of Massachusetts, Maryland, North Carolina, New York, Florida, and California, contingent payment clauses have been interpreted to allow recovery by subcontractors even though the general contractor is not paid by the owner. Typically, contingent payment clauses provide that a subcontractor will be paid after the general contractor receives payment from the owner. These recent cases have uniformly construed the contingent payment clauses as not barring the subcontractor's right to recover and that the payment by the owner is not a condition precedent to the subcontractor's right for the payment after a reasonable length of time. In other words, in recent cases courts have construed the contingent payment language as an intermediate timing

---

[*]McNeill Stokes, *Construction Law in Contractors' Language* (New York: McGraw-Hill, 1977), p. 71.

device to defer the subcontractor's payment for a reasonable length of time, but typical contingent payment clauses do not ultimately bar the subcontractor's right to recover from the general contractor.

A recent New York State finding strongly addressed the contingent payment clause. The following comments are by Jeffrey R. Cruz, Esquire, of Postner & Rubin:[*]

> The decision of the New York Court of Appeals, New York's highest court, in West-Fair Electrical Contractors and L.J. Coppola, Inc. v. The Aetna Casualty & Surety Company and Gilbane Building Company, 87 N.Y. 2d 148, 661 N.E. 2d 967, 638 N.S.S. 2d 394 (1995), has evoked equal parts of agony and ecstasy in the construction industry by holding that contingent payment clauses, a.k.a. "pay if paid" clauses, are void and unenforceable . . . .
>
> In West-Fair, the Court held that these contingent payment clauses are void and unenforceable because they violate the New York Lien Law. The Lien Law specifically provides that any contract which waives the right to file or enforce any mechanic's lien "shall be void as against public policy and wholly unenforceable." If a subcontractor's right to be paid could be indefinitely postponed by an owner's insolvency or other inability or failure to pay the general contractor, then the subcontractor's right to enforce its mechanic's lien would be similarly frustrated and would constitute an illegal waiver of lien rights.
>
> The Court did point out that clauses that set a reasonable amount of time for payment to the subcontractor remain enforceable.
>
> The case was very closely watched by the surety community because the Court was also asked to decide whether a contingent payment clause in a subcontract could be a proper defense to a claim against the general contractor's payment bond. The subcontractor argued that the payment bond surety had an independent obligation to pay the claim under the bond, notwithstanding the contingent payment clause in the subcontract. Because the court held the contingent payment clause to be void and unenforceable, the Court did not reach the question of whether the surety's obligation to pay is contingent upon the general contractor's obligation to pay.
>
> The opinion of the Court also contains language which suggests that the subcontractor has a right to recover directly from the owner, on account of the subcontractor's mechanic's lien and "as a matter of contract law."
>
> Consequently, the Lien Law grants the subcontractor an independent right, separate and apart from a general contractor's remedies, to file and enforce a mechanic's lien against a person liable for the debt upon which the lien is founded, such as the owner, and the real estate being improved.
>
> In another part of the opinion, the Court writes:
>
>> As a matter of contract law, the owner and general contractor are liable to plaintiff [the subcontractor] for the work plaintiff has been authorized to perform, and performed, under the subcontract.
>
> These statements by the Court conflict with the well-established rule that a subcontractor cannot recover directly from the owner, under either a contractual or quasi-contractual theory, absent an independent promise to pay by the owner.

---

[*]Postner & Rubin, *Construction and the Law,* Vol. XI, No. 1, Spring 1996. Used with permission.

Postner & Rubin note (partial):

A legislative solution, whose chances do not appear good, would still be years away. General contractors should therefore consider [the following] modifications to their current forms of subcontract.

- A provision in the subcontract which sets a reasonable time for payment to the subcontractor if the owner goes bankrupt or becomes insolvent. The question remains, however, as to what is a "reasonable time."
- A provision in the subcontract which makes the contract subject to Georgia law (a jurisdiction where "pay if paid" has been held enforceable). However, such a clause would most likely fall to the same public policy considerations cited in West-Fair.
- A provision in the subcontract which requires the subcontractor to exhaust its remedies under the Lien Law (i.e., to file a mechanics lien and foreclose on it), as a condition precedent to recovering from the general contractor.
- A provision in the subcontract which obligates the subcontractor to participate in the Owner/General Contractor dispute resolution procedure, and which makes any decisions arising out of the procedure final, binding and conclusive upon the subcontractor.

In general, general contractors have a tough road ahead. Subcontractors whose claims have been held at bay on account of a contingent payment clause in their subcontract will aggressively press their claims in the wake of West-Fair. General contractors need to take a close look at the payment and dispute resolution clauses in their subcontracts if they want to survive in New York.

Incorporation of Prime Contract

It has long been a practice to incorporate the construction documents (plans and specifications) into the subcontract so that the subcontractor is bound by the same scope of work as the prime contractor. This was the case in the contract between Guerini Stone and P. J. Carlin, in which Guerini was hired to do floor and wall concrete work for a post office on which Carlin was the general contractor. Carlin was slow in supplying materials, which delayed Guerini's work. Carlin attempted to avoid liability for a delay damage claim because the prime contract precluded damages for delay. The court refused the defense, holding that "the reference in the subcontract to the drawings and specifications was evidently for the mere purpose of indicating what work was to be done, and in what manner done, by the subcontractor."

However, a more complete incorporation clause is enforceable, such as the following:

Work performed by subcontractor shall be in strict accordance with Contract Documents applicable to the work to be performed and materials, articles and/or equipment to be furnished hereunder. Subcontractors shall be bound by all provisions of these documents and also by applicable provisions of the principal contract to which the Contractor is bound and to the same extent.

With this clause, where applicable, the prime contractor stands in the same position to the subcontractor as the owner does to the prime contractor. There are cases, however, where the incorporation cannot completely transpose the roles of the parties. In Fanderlike-Locke Co. v. United States (285 F. 2d 939, 10th Cir.; 1960), a general contractor working on an air base construction project attempted to invoke the prime contract dispute clause upon the subcontractor. The court rejected the argument stating: "Language of the subcontract references are to the general conditions of the prime contract, not to the general provisions and they relate generally to performance of the subcontract according to the specifications, not to the settlement of disputes, or the subcontractor's right to sue under the Miller Act."

Further, the court noted that the interpretation of total incorporation would purport to have the general contractor entering into contract on the part of the government, which was beyond the general contractor's authority. (This would have been an implied contract giving the subcontractor certain rights to have its disputes settled with government administrative procedures in the same manner that the general contractor's disputes could be settled; however, in an amiable relationship, the prime contractor often pleads the case of its subcontractor to an owner.)

In Johnson Inc. v. Basic Construction (429 F. 2d 764, D.C. Cir.; 1970), general contractor Basic attempted to force subcontractor Johnson to follow the disputes procedure in the prime contract between the owner and Basic. Basic had received a direction to accomplish extra work and had imposed the same obligation upon Johnson. Johnson brought suit, claiming that Basic was obligated to give it a commitment for payment for extra work even though Basic had not received a similar commitment from the owner for the same work. Johnson argued that it was justified in abandoning its work and in being compensated for the work it had performed. Basic argued that it did not have such an obligation because Johnson was bound to the same terms and conditions of the prime contract as was Basic.

The court disagreed and found that the disputes clause was incorporated into the contract only insofar as it was applicable to the work performed, and that the clause, by its terms, did not extend to require an adherence by a subcontractor to an administrative remedy designed to be used only by the parties to the prime contract. This limited application is explained by the fact that the incorporation clause made no mention of binding the subcontractors to the general contract provisions containing the disputes clause to the same extent that the prime contractor was bound.

Accordingly, on the basis of a loosely worded clause, the court refused to force the contractor to relinquish a common-law right to abandon the work. However, if the incorporation clause had been more specific, the subcontractor would have been required to respond to the procedures agreed to by the prime contractor. (In the Johnson Inc. v. Basic Construction case, the incorporation of the prime contract, however, would not have been presumed to have been a waiver of Johnson's statutory right to sue Basic, unless such right was specifically waived in the contract.)

The Subcontractor's Dilemma

The prime contract requires the subcontractor to undertake change order work as well as the basic work scope. To the extent that the subcontractor believes that additional work required is not covered by either the base contract or approved change order, the subcontractor must follow the same claim procedure as the prime contractor. That claim procedure must be against the prime contractor, which usually passes it along to the owner (plus the prime contractor's markup) without change or comment.

The change order process is more cumbersome for the subcontractors because of the added processing time and the requirement to work through the prime contractor. The subcontractors assume (probably correctly) that their claims will always be processed at a lower priority than any claim relating directly to the prime.

Subcontractor claims represent work put in place at a cost in labor and materials. The time required to process and collect a claim can take from months to years and result in a negative impact on the subcontractor's cash flow.

Role of the Prime Contractor

The prime contractor functions as a conduit, passing change order directives from the owner to the subcontractor and returning change order claims from the subcontractor to the owner. The role almost defines itself. If the prime contractor tries to act as a fiduciary filter for the owner and critically reviews change order proposals and/or claims, the result will be a claim against the prime for any funds not granted by the owner. The answer: pass it through to the owner and let the owner be at risk.

On the other hand, if the subcontractor claim is blatantly overpriced, the prime has a problem in loss of credibility in dealing with the owner. (A further problem in federal projects is the requirement that the prime is required to certify under oath that the claim is proper and correct.)

In the case of an obviously overpriced claim, the first approach would be to meet informally with the subcontractor "off the record" to point out the obvious flaws and suggest revision and resubmission of the claim. If the subcontractor refuses to see the light, the prime contractor could forward the claim to the owner (or PM/CM) unapproved for review and comment.

The demeanor between subcontractors and the prime is often informal. Nevertheless, the prime should insist on clear and complete documentation, particularly in the change order process. The format used should mirror those used by the prime.

Constructive Change Orders

Circumstances may give the subcontractor the opportunity to claim that a situation is a de facto change order (i.e., a constructive claim). For instance, in Peter Kiewit Sons' Co. v. Summit Construction Co. (442 F. 2d 242, 18th Cir.; 1969), a subcontractor was required to modify its backfilling procedures and perform them in three phases. The piping had been laid in a manner that was much more difficult than the backfilling in a two-phased operation without any piping, as agreed in the contract and specifications. The material change caused a 300 percent increase in backfill cost and a 50 percent increase in the cost of performing the overall subcontract. The subcontractor was entitled to damages for this substantial breach of contract and change in scope (i.e., a constructive change order).

In Westinghouse Electric Corp. v. Garrett Corp. (437 F. Supp. 1301 D. Md.; 1977), the subcontractor needed source control drawings in order to complete its own shop drawings. The prime contractor failed to produce these source control drawings in a timely manner. This delay had such an impact on the subcontractor's ability to perform its work that the court found the delay represented a constructive change so serious that it was, in fact, a cardinal change.

In S. Leo Harmonay Inc. v. Binks Manufacturing Co. (597 F. Supp. S.D.N.Y.; 1984), the project was the expansion of an automotive assembly plant. The general contractor refused to grant a reasonable time extension, forcing the mechanical subcontractor to go from a single-shift to a triple-shift operation, increasing its workforce from 26 to 98 personnel. The subcontractor was able to demonstrate a resulting loss due to inefficiency. The court found there was constructive acceleration, a constructive change order.

In Blake Construction v. C. J. Coakley (431 A 2d 569 D.C.; 1981) there were substantial delays early in the project. General contractor Blake changed the sequence of work, resulting in the stacking of trades. Coakley's fireproofing work was disrupted. The court found this amounted to "active interference" and that the subcontractor was entitled to damages.

Strategies to Avoid or Mitigate Changes

Change order impact on project cost and/or progress varies with the following:

- The larger the scope (measured by cost), the greater the impact.
- The later in time the change order is implemented, the greater the impact.
- The better the management of the change order process, the less the impact.

Better management must recognize the following:

- The approval process
- The potential macro cost of micromanaging the change order process (i.e., delay)
- Responsibility to proactively process change orders to prevent or mitigate time impact

Mismanagement (or lack of management) in processing change orders is the leading factor in a majority of construction claims and/or litigation.

Change Order Strategies

Zero defects

Other descriptions of this strategy are "freeze scope," "no time extensions—for any reason," and "stonewall the situation." The strategies describe themselves, and they don't work. For instance, an owner might well set out to have no changes during construction. One reason could be a fixed capital resource such as a bond issue. If the approach selected is a proclamation without sufficient planning, one probable result could be a program that reaches too far, with a resulting overdesign. This could, and probably would, lead to a construction price too high to sustain in the face of realities (i.e., change orders).

Commonsense approach

The commonsense strategy recognizes that there probably will be changes and, hence, change orders. The way to minimize change is good preparation in the preconstruction phase:

Program. Many projects go into design without a formal program. A sound program is a good guide for the designer. In situations where the ultimate users are not construction-knowledgeable, such as universities or hospitals, the users should be included in the programming process. To assist their comprehension, suitable use should be made of models, mock-ups, and visits to comparable facilities. This type of user organization has fairly frequent/regular management changes. When this occurs, the program should be revisited with the new user management team.

Design professional. Selection of the design professional (architect, engineer, or architect/engineer) should be pragmatic even if the design professional is Michelangelo reincarnate. Today's owner needs to be concerned with budget and schedule—as well as with quality.

Selection criteria should be made available to potential design professionals. This gives them the opportunity to respond with references and information on how they would meet the requirements of the project. References should be checked, with emphasis on how they met project goals—especially controlling the scope and/or budget.

The PM/CM should be retained early in the process to facilitate regular estimates, design reviews, and constructibility reviews. Value engineering should be encouraged, especially at the end of either preliminary engineering or design development. The using agency and its maintenance department should be part of design, constructibility, and value engineering reviews.

Figures 15-1 to 15-4 are suggestions from a constructibility review. The situation at Griffiss Air Force Base in upstate New York was the need for more power, peaking in the subzero winter cold. The plan was to replicate the existing power plant. However, over time the groundwater level had risen to almost grade at the site for the new plant.

The constructibility review team, including an independent contractor, was very construction-experienced and -oriented. Figures 15-1 to 15-4 include the following ideas:

Figure 15-1: After the new plant is hydro-tested and before it can provide its own heat, there will be a need for temporary heat in the winter season. Establish (1) cost of fuel/energy and (2) standby mechanics.

Figure 15-2: The groundwater is high at the site of the new plant.

Figure 15-3: Raise the structure 4'-8" (or more) to keep the footprint out of the groundwater.

Figure 15-4: Raise the basement slab water level above groundwater. Provide a larger equipment access.

O'BRIEN-KREITZBREG & ASSOC., INC.

Power Plant – Griffiss AFB

RECOMMENDATION
REFERENCE

NO.   1

PROJECT CONSTUCTIBILTY RECOMMENDATION

Include provisions for temporary heat in boiler building &
pollution control building. Establish whether source of heat will
be from existing heating plant, or contractor is to provide.
Establish who pays cost of (1) fuel/energy and (2) standby
mechanics.

COMMENT/ DISCUSSION

There will be a period after the first hydrostatic testing of
mechanical equipment and piping and before the plant is lit off and
able to provide its own heat. The likelihood of this period including
below freezing weather is high, and should be provided for.

DRAWING SPECIFICATION / REFERENCES

Not shown on set reviewed.

COST IMPACT / POTENTIAL

[X] MAJOR   [ ] MEDIUM   [ ] MINOR

CATEGORY

[X] PROVIDE ADEQUATE INFORMATION TO BIDDERS	[X] CLAIM AVOIDANCE
[ ] CONSTR. ACCESS/PROCEDURES	[ ] CONTRACT SCHEDULE/SEQUENC.
[ ] CONSTRUCTIBILITY MATERIALS/ DETAILS/EQUIPMENT	[X] OWNER INTERFACE (O.F.E./ UTILITY SERVICE)
	[ ] OPERATIONAL TESTING, TRAINING, MAINT, & ASSOC. DOCUMENTS

Figure 15-1 Constructibility recommendation #1.

O'BRIEN-KREITZBREG & ASSOC., INC.	RECOMMENDATION REFERENCE
Power Plant – Griffiss AFB	NO. 2

PROJECT CONSTUCTIBILTY RECOMMENDATION

Inform bidders of water condition at Building 37.

COMMENT/ DISCUSSION

Base personnel "recall" water problems in Building 37 construction, which borings substantiate.

Clear notice should be provided.

DRAWING SPECIFICATION / REFERENCES

Dwgs. 2061967-68

COST IMPACT / POTENTIAL

[X] MAJOR   [ ] MEDIUM   [ ] MINOR

CATEGORY

[ ] PROVIDE ADEQUATE INFORMATION TO BIDDERS
[X] CLAIM AVOIDANCE

[ ] CONSTR. ACCESS/PROCEDURES
[ ] CONTRACT SCHEDULE/SEQUENC.

[ ] CONSTRUCTIBILITY MATERIALS/ DETAILS/EQUIPMENT
[ ] OWNER INTERFACE (O.F.E./ UTILITY SERVICE)

[ ] OPERATIONAL TESTING, TRAINING, MAINT, & ASSOC. DOCUMENTS

Figure 15-2 Constructibility recommendation #2.

O'BRIEN-KREITZBREG
& ASSOC., INC.

Power Plant – Griffiss AFB

RECOMMENDATION REFERENCE	
NO.	3

PROJECT CONSTUCTIBILTY RECOMMENDATION

Reduce cost/schedule impact of high water level through raising elevation of boiler house basement floor.

- Raise floor from 456'-0" to 460'-8".
- Accommodate tunnel by raising or holding existing 453-2$^1/_2$ elev.
- Accommodate breeching connection/ash connection to pollution building.

COMMENT/ DISCUSSION

- Would eliminate extensive dewatering.
- Would reduce potential post construction water problems.
- Footing mud slab bottom elevation is 453-4. Water is above 458 – raising by 4'-8" would provide minimum water problem.
- Pollution control building has no basement, is no problem.
- Change equipment doors to tailgate height would ease architectural solutions.
- Simplifying dewatering & eliminating mud slab will shorten time to get out of the ground and make its achievement during 1983 possible.

DRAWING SPECIFICATION / REFERENCES

2062022-24   Boiler House
2062025   Tunnel

COST IMPACT / POTENTIAL

[X] MAJOR   [ ] MEDIUM   [ ] MINOR

CATEGORY

[ ] PROVIDE ADEQUATE INFORMATION TO BIDDERS

[ ] CONSTR. ACCESS/PROCEDURES

[X] CONSTRUCTIBILITY MATERIALS/ DETAILS/EQUIPMENT

[ ] CLAIM AVOIDANCE

[ ] CONTRACT SCHEDULE/SEQUENC.

[ ] OWNER INTERFACE (O.F.E./ UTILITY SERVICE)

[ ] OPERATIONAL TESTING, TRAINING, MAINT, & ASSOC. DOCUMENTS

Figure 15-3 Constructibility recommendation #3.

O'BRIEN-KREITZBREG & ASSOC., INC.

Power Plant – Griffiss AFB

RECOMMENDATION
REFERENCE

NO.   4

PROJECT CONSTUCTIBILTY RECOMMENDATION

Access to basement for installation of large equipment is needed
for: • Emergency Generator (5 x 7 x (7?)
       • Condensate receiver tank 10x10x24
       • Forced draft fans
       • Ash hoppers
Consider raising basement slab elevation.

COMMENT/ DISCUSSION

- Basement is below water level – impracticable to keep wall open unless raised.
- Floor access is 8 x 8 hatch only.
- It will be necessary to either delay erection of one boiler, or leave large opening in 1st floor, which will impact conduit and piping runs.

DRAWING SPECIFICATION / REFERENCES

Dwg. 2061984
     2062057
     2062117

COST IMPACT / POTENTIAL

[X] MAJOR   [ ] MEDIUM   [ ] MINOR

CATEGORY

[ ] PROVIDE ADEQUATE INFORMATION TO BIDDERS

[X] CONSTR. ACCESS/PROCEDURES

[ ] CONSTRUCTIBILITY MATERIALS/ DETAILS/EQUIPMENT

[ ] CLAIM AVOIDANCE

[ ] CONTRACT SCHEDULE/SEQUENC.

[ ] OWNER INTERFACE (O.F.E./ UTILITY SERVICE)

[ ] OPERATIONAL TESTING, TRAINING, MAINT, & ASSOC. DOCUMENTS

Figure 15-4   Constructibility recommendation #4.

Budget. A well-constructed budget with sufficient funding behind it is a necessary foundation to avoid change orders. When appropriate, the budget should include a comprehensive geotechnical study. The geotechnical study should do all that is necessary to identify all underground problems, thus avoiding (or at least minimizing) "unforeseen underground conditions."

To the extent known, special circumstances should be included in the budget. These can include special work hours, special environmental conditions, overtime requirements, and so on. For example, for priority work on prisons at Rikers Island, the New York City specifications instructed bidders to build premium time and shift costs into their price.

At key milestones in the preconstruction time frame, the PM/CM should conduct independent estimates of total program costs. These estimates should include program "soft" costs (design, owner costs, financing costs, PM/CM, etc.).

A major New York City builder (quasi-public) said about estimates: "Never let the architect do the estimate; it's a conflict of interest. Get back on budget before you go on to the next [design] phase."

Good plans and specifications. The better the design documents, the fewer the change orders. Changes in the design scope or concept are better handled in the earlier stages of design. If the design scope can be settled (or even "frozen") at the design development stage, the designer can focus on coordination between the disciplines. Good coordination reduces the probability of change orders due to errors or omissions.

Construction Phase

The PM/CM

The PM/CM in the construction phase should avoid change when it is either optional or unnecessary. More usually, changes will not be avoidable; in these cases, the PM/CM role is early identification and early resolution to mitigate the impact of changes. Denial and/or protracted negotiations waste time and amplify the impact of the change.

Delegation of authority

A number of federal government agencies use the contracting officer approach to managing contracts. The contracting officers have predetermined levels of authority (ranging upward to "unlimited"). They can apply this authority to the unilateral issue of change orders. This directly addresses two of the problems in managing change orders: the processing time and the ability to pay progress payments when the change order work is being performed.

The federal process goes a step farther; it delegates the prerogative of approving change orders. The limit ranges, typically, from $5000 at the resident engineer level to $50,000 to $100,000 at the project manager level. Settling the smaller, annoying changes on an expeditious basis puts a positive spin on project relationships.

Dispute resolution

The use of alternative dispute resolution (ADR) procedures can help settle disputed work (i.e., constructive change orders). On larger projects, the establishment of a dispute resolution board can settle many problem areas. This board is, typically, set up by each prime contractor and the owner selecting a board member. The board members are to be qualified, experienced, and independent (i.e., not working on the project). The selected members then choose a neutral member. The neutral member usually chairs the board. Typically, the board meets quarterly at the site to maintain contact with progress and to address any disputes presented to them.

Partnering

Partnering came out of the construction roundtable work during the early 1980s and was continued by the Construction Industry Institute (CII). It is clear that the early CII work was directed toward private industry where negotiated and/or cost-plus contracts are permitted. The U.S. Army Corps of Engineers became interested in the potential of partnering for publicly bid work. By 1990, the Corps was ready to make partnering policy. In a policy memorandum on ADR, the chief of engineers segued from ADR to partnering, as follows:[*]

3. I believe the concepts of cooperative, interest-based problem solving, which are basic to ADR, are part of a new spirit which we must encourage in responding to the challenges of the 1990s. The Corps of Engineers must be part of a partnership among the people we work with and those we serve. In the spirit of partnership, we must emphasize common interests, cooperative working relations, communication and understanding. This calls for new ways to deal with conflict. I believe that ADR offers management tools for dealing effectively with conflict while avoiding the expense and delay of adversarial proceedings. My Focus 90 program recognizes the partnership concept as one of two areas of emphasis.

4. Partnership demands the will to resolve disputes. Clearly, the best dispute resolution is dispute prevention. Acting to prevent disputes before they occur is key to building cooperative relationships. By taking the time at the start of a project to identify common goals, common interests, lines of communication and a commitment to a cooperative problem solving, we encourage the will to resolve disputes and achieve project goals.

5. I support and encourage everyone in the Corps to be aware of opportunities for productive partnerships and to seek appropriate ways, including ADR, to resolve disputes.

---

*Lt. General H. J. Hatch, Commander's Policy Memorandum #11, Subject: Alternative Dispute Resolution, August 7, 1990.

Partnering is not contractual—in other words, it does not modify or change the contract. Instead, partnering is done within the bounds of the contract.

The owner should indicate an interest in partnering but cannot insist on partnering—to do so would be like forcing a shotgun marriage. When an owner has decided to encourage partnering, the owner can direct the owner's team to participate. The contractor (or contractors) must still be convinced to participate.

Partnering breaks down the adversarial relationship evident on most job sites. Time and effort spent on nonproductive tasks required for litigation is now directed toward problem solving. Contentious items on normal job sites become minor irritants. In addition, the team concept has shown to produce measurable increase in productivity and quality control. The desires of the team members have a greater probability of being met.

Successful partnering requires continued interest from senior management throughout the project. This requires time, effort, and some money from all parties. Senior management must be available to participate actively in follow-on meetings after the initial work session. As in any system, there is always the risk of having partners who will attempt to work the system purely to their advantage. When this occurs, the system begins to unravel and the old "business as usual" adversarial way of conducting business appears. Fortunately, when a partnering agreement deteriorates to the point where it becomes a burden on the participants, the partnering agreement can be dissolved easily since it is not a binding legal agreement.

The rules of the specific project are developed jointly in the form of a partnering agreement. This agreement is not a change order. It works within the terms of the contract. The partnering agreement is usually drawn up during the initial partnering meeting. Figure 15-5 (p. 348) is an addendum on partnering to Corps of Engineers Engineering Circular No. 1110-1-71. The addendum discusses partnering and provides an example, shown in Fig. 15-6 (p. 349), of a partnering agreement.

Partnering offers a challenge to the stakeholders to find mutual advantages in addressing change orders.

When a change order is processed, it should be reviewed from several aspects. It is appropriate for the PM/CM and/or the owner to set up a change order review checklist, so that each change order is reviewed on the same basis.

Change Order Checklist

Figure 15-7 (p. 350) is a change order checklist included in the Corps of Engineers' *Modification Impact Evaluation Guide*. It addresses the following: (1) Justification for the change order, (2) the request for proposal, (3) the government estimate, (4) reviewing the contractor's proposal, (5) audit, (6) negotiations.

Appendix A—Partnering

1. The partnering concept seeks a cooperative environment, not a confrontational one. A win/win outcome for all parties is the ultimate goal. Experience has demonstrated that when win/lose strategies are employed by one or more parties to gain advantage, a lose/lose reality results (i.e., quality degradation and/or unreasonable cost and time growth for the Corps and its customers, and unprofitable ventures for private sector A-E firms and construction contractors). Partnering agreements accomplished by Engineering Divisions must be consistent with, and part of, the total project partnering plans of PMP.

2. The "partnering mode" seeks to identify and communicate the needs, expectations and strengths of all parties (participants). The partnering model recognizes that a synergistic approach to accomplishing the required activity will enhance the opportunity to produce a quality service or product on schedule and within budget, to the mutual satisfaction of all participants. In the cooperative environment of the partnering model, creative solutions to problems can be developed. The partnering model supports the use of alternate dispute resolutions processes, in lieu of litigation.

3. To be successful, however, partnering must first and foremost be a volunteer effort. Second, all participants must exhibit an enthusiastic willingness and commitment to embracing the concept. Third, successful partnering must be focused on the communication of needs, strengths and expectations of each party at appropriately specified milestones during the performance of the required activity. Therefore, a "partnering process" must be mutually developed and followed. Fourth, goals must be established so that the degree of success of the partnering effort can be measured throughout the performance period. The quality of the partnering effort must be considered a factor in the evaluation of the overall success of the delivered service or product. A sample "Partnering Process" is anticipated to be included in a future Quality Management ER.

4. A sample "Design Quality Partnering Agreement" is provided for information. This agreement provides a framework for all parties to obtain a quality service on schedule and within budget. It also provides the basis for the development of the follow-up partnering process document.

5. Neither the partnering agreement, nor the partnering process is contractually binding. The rights and responsibilities of the contractual parties are not changed by "partnering."

Figure 15-5   Appendix A—Partnering.

SAMPLE DESIGN QUALITY PARTNERING AGREEMENT
BETWEEN
THE CORPS CUSTOMER, USACE, AND THE PRIVATE SECTOR ARCHITECT-ENGINEER.

1. Mission Statement. As signatories of this agreement we dedicate ourselves to a professional, enjoyable and productive relationship. We will strive to work as a team to produce quality work, on schedule and within budget.

2. Objectives. We support this mission statement through our voluntary and enthusiastic commitment to subscribe to the following objectives.

   a. To increase the resources available to the Government by effective, friendly, knowledgeable teaming.

   b. To include all work-related participants in our partnership.

   c. To develop an informed, practical understanding of the needs, strengths and expectations of all partners.

   d. To reach a common understanding of the needed requirements including participation in scope development, criteria review and technical guidance evaluation.

   e. To strive for open, honest, clear, and timely communications among all participants.

   f. To respond proactively and swiftly to concerns, deadlines and requests.

   g. To mutually explore and utilize new, innovative and proven technologies and applications to produce technically excellent products which advance the state-of-the-art.

   h. To do "the right thing right, the first time" throughout the performance of the work.

   i. To eliminate the need for litigation by producing a quality service and/or product that is worthy of an "exceptional" rating.

   j. To maintain a steady, uniform work flow; minimizing processing time, finalization of technical requirements, and promptly processing payment invoices.

   k. To recognize that safety and health are primary concerns. Our goal will be to complete all work without injury or death from any controllable cause.

   l. To evaluate the effectiveness of this partnering agreement at predetermined points throughout the performance period.

3. Signatories.

   For (The Army Corps of Engineers' Customer/Partner):

   | (Name) | (Position) | (Date) |

   | (Name) | (Position) | (Date) |

   For (The USACE Command):

   | (Name) | (Chief, Engr Div) | (Date) |

   | (Name) | (DOE (PM)) | (Date) |

   For (The Architect-Engineer):

   | (Name) | (Firm Principal) | (Date) |

   | (Name) | (Firm Proj. Mgr.) | (Date) |

Figure 15-6 Sample design quality partnering agreement.

Modification Impact Checklist

1. Justification for change order
   a. Is change essential for proper function of facility?
   b. Is contract modification preferable to accomplishing change work under separate contract?
   c. Is funding available to accomplish the changed work, including impact costs?
2. Request for proposal
   a. Is scope of work clearly defined?
   b. Is a specific date stipulated for receipt of the proposal?
   c. Is the request accompanied by sufficient instructions on format, detail desired, etc.?
3. Government estimate
   a. Are both direct and indirect costs considered?
   b. Are both direct and indirect delaying factors considered?
   c. Are cost/time for direct work and affect on unchanged work itemized separately?
   d. Is the impact on unchanged work based upon current project status and the approved progress schedule?
   e. If no impact exists, does estimate so indicate?
   f. Is appropriate allowance for revising progress schedule included?
4. Review of contractor's proposal
   a. Is proposal in required format, with items adequately identified and justified?
   b. Does proposal reflect a thorough understanding of the scope of work?
   c. Is impact on the unchanged work, if applicable, identified separately from direct cost/time?
   d. Does the proposal use current job status and the approved progress schedule as the basis for justifying additional cost/time, both direct and impact?
5. Audit
   a. Was the audit agency given advance notice of impending request for audit?
   b. Did the request include desire for verification of specific items in addition to routine audit procedures?
6. Negotiations
   a. Was impact discussed (even if none exists) and this fact included in the record of negotiations?

Figure 15-7   Modification impact checklist.

Description of John Doe Project

This project description is taken from James J. O'Brien, *Preconstruction Estimating: Budget through Bid*, McGraw-Hill, New York, 1994.

WHATCO (John Doe) Project

The Widget Housing and Transmission Company (WHATCO) wants the ability in the Philadelphia area to assemble 120 units (minimum) to 180 units (maximum) per year. Company industrial engineers advise that production of one 50-ton unit requires two months and 1000 sf of work space. Design will be in 1998 and construction in 1998–1999.

The plant is to be supported by a parts warehouse. Major parts will be stored on pallets on a 4-ft-high forklift-loaded storage area. At maximum pace, 30 units will be in production. Industrial engineering advises that each unit must be supported by 1000 sf of storage. This includes flat storage and bin storage.

Plant

100 ft × 300 ft × 24 ft high
    Steel frame
    Roof: 100-ft girders with 20-ft LS joist secondary system
    Precast plank
    20-year 3-ply built-up roof

Siding: Insulated metal panel on girt system

Slab-on-grade 12 in thick; perimeter grade beam 8 in thick × 36 in deep

Railroad siding: 3 roll-up doors 12 ft w × 20 ft ht

Truck loading dock: 3 roll-up doors 12 ft w × 20 ft ht, 3 dock levelers

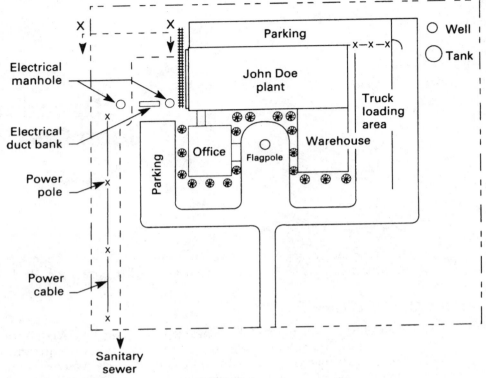

Figure A-1 Site plan, WHATCO (John Doe) Project.

Crane

50-ton gantry crane, craneway 300 lf

Supported by columns at 30-ft space

Column foundations, 20 cast-in-place piles

Steel frame foundations
    Sides, 20 piles
    End, 4 piles
    Pile caps, 6 ft × 4 ft × 2 ft (common with crane piles)

Switchgear

5000 kW

Primary 13.2 kV

Secondary 440 V

Fire protection. The hydrant system measures 700 lf × 6 in diameter

Warehouse

120 ft × 80 ft × 24 ft high

Roof and siding: Same as plant

Structural system:
    80-ft girders with 20-ft LS joist secondary system
    Column foundations
      12 cast-in-place piles
      Pile caps 4 ft × 4 ft × 3 ft
    Grade beams 8 in × 36 in deep

HVAC: Same as plant

Lighting: Same as plant

Power: From plant

Fire protection: Sprinkler

Rack system: By owner

Bin system: By owner

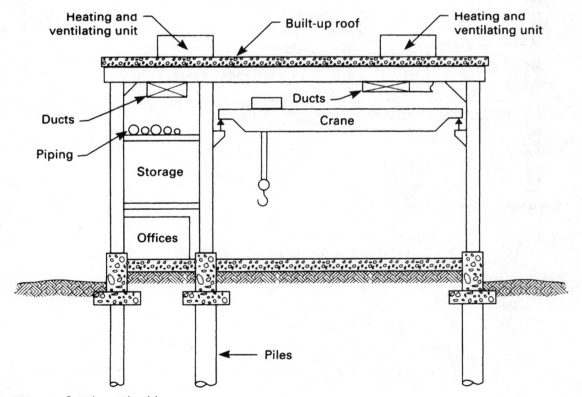

Figure A-2 Interior section AA.

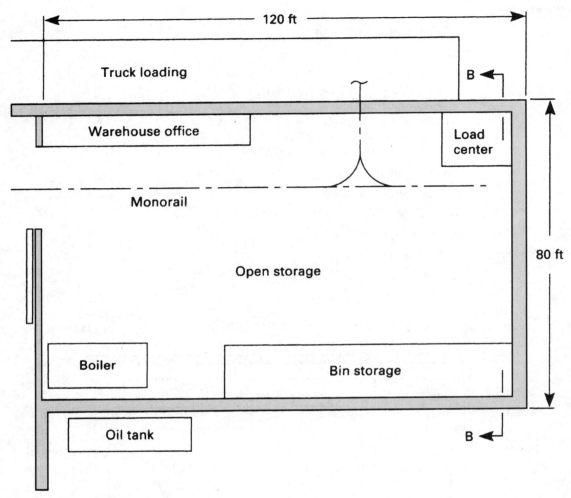

Figure A-3 Warehouse floor plan.

Bathrooms
  Men 500 sf: 4 water closets; 4 urinals (wall-hung)
  Women 300 sf: 3 water closets

Office building

80 ft × 80 ft × 15 ft high

Precast frame and roof

Roofing: 20-year 3-ply built-up roof on lightweight fill and 1-in insulation board

(*Text continues on p. 360.*)

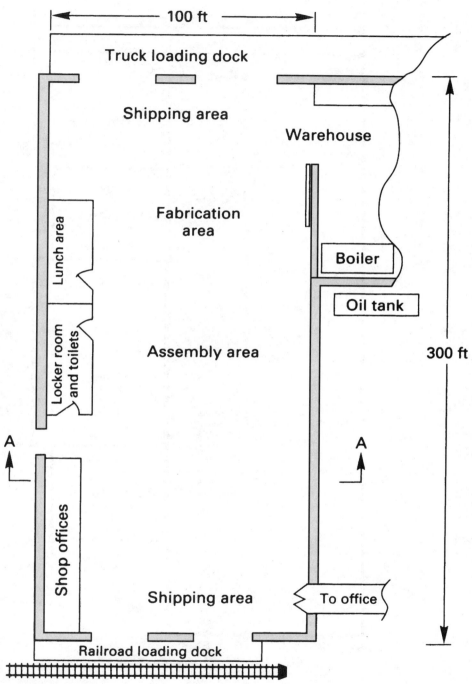

Figure A-4 Plant floor plan.

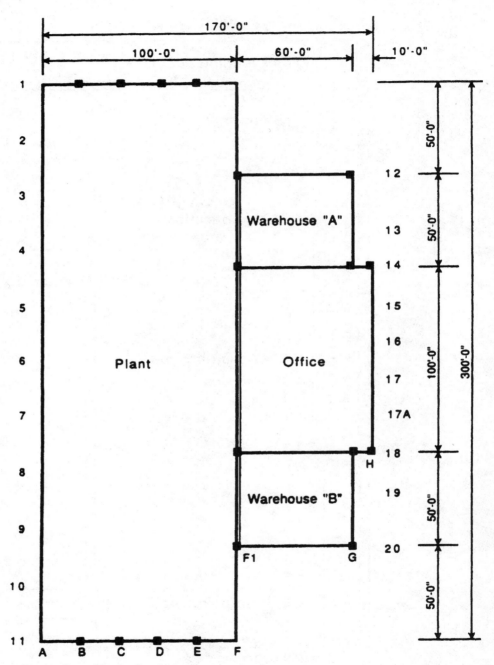

Figure A-5 Spread footing locations.

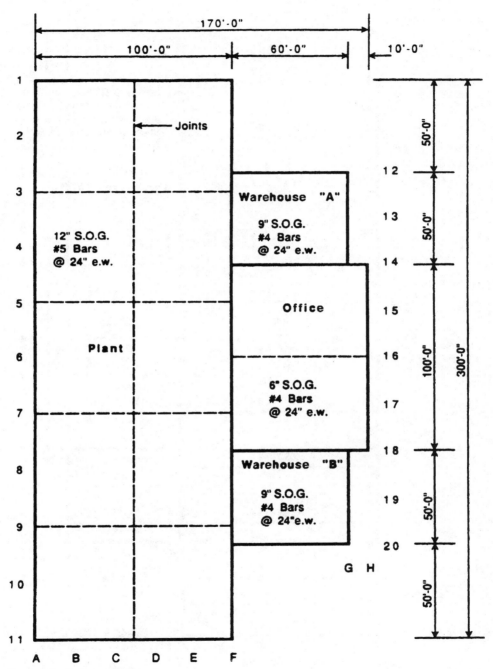

Figure A-6 Slabs-on-grade.

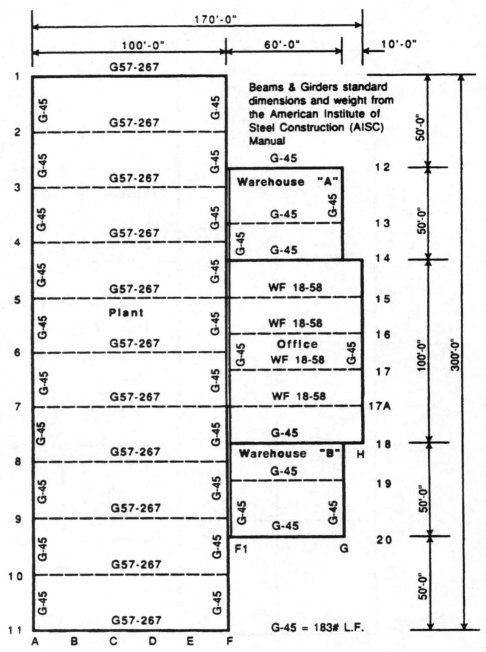

Figure A-7 Beams and girders.

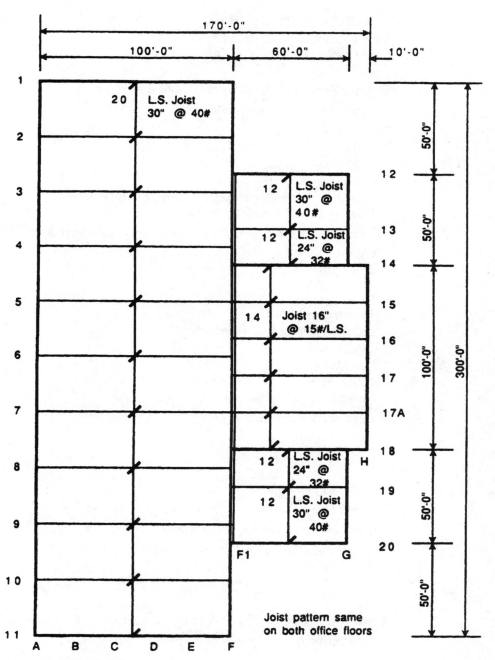

Figure A-8 Joist schedule.

Windows: Andersen thermopane (nonoperable)

Doors
  Entrance, double plate glass
  Exits (2), metal

Interior partitions: ⅜-in dry wall, metal stud

Interior doors: Prehung wood

Flooring
  Offices and halls—carpet
  Bathrooms—tile

Bathrooms:
  Men 225 sf: 2 water closets, 2 urinals (wall-hung)
  Women 225 sf: 3 water closets

Lunchroom: Dishwasher, double sink, stove, microwave, cabinets, 400 sf

HVAC
  750-kBtu heater, air circulation
  20-ton chiller
  Duct distribution
  Ventilation fans
  Exhaust fans, as necessary

Electric
  300-kW service
  100 fc light
  Receptacles at 1/100 sf = 64

Fire protection: Sprinkler

Site

Clear, strip, and grade, 6 acres

Electric 13.2-kV service, 3000-lf pole

Service entrance access, 200-lf duct

Water tower, 250,000 gal at 75-ft head

Other

Codes: BOCA, NFPA, Pennsylvania Department of Labor and Industry

Construction schedule: 1 year

HVAC

Ventilation: Rooftop fans, 4 air changes/h

Heat:
  2 package boilers with hot water (1 million Btu each)
  Radiant heaters (6) at doors

Lighting

Industrial 100 fc

B

Cost Estimate of John Doe Project

This estimate is taken from Chap. 10 of James J. O'Brien, *Preconstruction Estimating: Budget through Bid*. McGraw-Hill, New York, 1994. The unit costs are for 1995, but are still in the right order-of-magnitude.

TABLE B-1 Summary of Quantities and Costs, Divisions 1 to 14

Item	CSI No.	R.S. Means I.D. No.	Unit	No.	Unit price 1995	Cost
		Plant				
Site preparation (clear)	02110	021 140 2000	acre	0.7	4,640	3,248
Excavation, footprint	02220	022 246 4220	cy	3,345	5.36	17,929
Excavation, spread footing	02220	022 250 0100	cy	86	70	6,020
Borrow	02220	022 216 0020	cy	1,672	3.10	5,183
Deliver borrow	02220	022 216 3800	cy	1,672	9.34	15,616
Spread	02220	022 216 2500	cy	1,672	5.22	8,728
Compaction	02220	022 226 5040	cy	1,672	0.88	1,471
Fill	02220	022 262 0170	cy	112	2.64	296
Concrete reinf., footing	03200	032 107 0500	lb	6,056	0.78	4,724
Concrete reinf., SOG, load deck	03200	032 107 0750	lb	37,328	0.60	22,397
Slab-on-grade 12 in	03310	033 130 4900	sf	30,000	3.85	115,500
Spread footers	03310	033 130 3850	cy	86	160	13,760
Loading dock concrete	03310	033 130 3150	sf	4,000	2.07	8,280
Insulating lightweight fill	03500	035 212 0250	sf	30,000	1.14	34,200
Structural steel, heavy	05120	051 255 4900	ton	1.53	2,240	342,720
Structural steel, light	05120	051 255 4600	ton	06.1	2,560	153,956
Base plates	05120	051 255 4300	lb	3,708	1.05	3,893
Joists	05210	052 110 0030	tons	72	1,504	108,288
Metal deck	05310	053 104 0440	sf	31,000	5.12	158,720
Sand blast	05120	051 255 6130	sf	22,387	0.81	18,133
Primer	05120	051 255 6520	sf	22,387	0.54	12,089

TABLE B-1 Summary of Quantities and Costs, Divisions 1 to 14 (*Continued*)

Item	CSI No.	R.S. Means I.D. No.	Unit	No.	Unit price 1995	Cost
		Plant (*continued*)				
Intermediate coat	05120	051 255 6630	sf	22,387	0.78	17,462
Top coat (elevated)	05120	051255 70.30	sf	22,387	1.19	26,641
Insulation roof	07220	072 203 0100	sf	30,000	1.04	31,200
MFG wall panels	04710	074 202 0400	sf	18,000	14.53	259,740
Roofing	07510	075 102 0200	square	300	170	48,000
Flashing	07620	076 201 5800	lf	800	4.44	3,552
Ext. downspouts	07630	076 201 2500	lf	245	9.14	2,239
Ext. gutters	07630	076 210 3400	lf	400	13.31	5,324
Steel doors and frames	08110	081 103 0100	each	5	209	1,045
Steel rollup doors	08365	083 604 0010	each	4	15,188	60,750
Door hardware	08710	087 125 3110	each	5	429	2,145
Exterior paint	09910	099 106 2400	sf	4,516	0.54	2,439
Interior paint, block	09920	099 106 2400	sf	6,000	0.54	3,240
Dock levelers	11161	111 601 4800	each	4	7,296	29,184
Bridge crane	14630	Quote	each	1	NA	418,264
Total						$1,966,276
		Warehouse(s)				
Site preparation (clear)	02110	021 104 2000	acre	0.14	4,640	650
Excavation, footprint	02220	022 246 4220	cy	669	5.36	3586
Excavation, spread footing	02220	022 250 0100	cy	9	70	630
Borrow	02220	022 216 0020	cy	334	3.1	1,035
Deliver borrow	02220	022 216 3800	cy	334	9.34	3,120
Spread	02220	022 216 2500	cy	334	5.22	1,743
Compaction	02220	022 226 5040	cy	334	0.88	294
Fill	02220	022 262 0170	cy	39	2.64	103
Concrete reinf., footings	03200	032 107 0500	lb	1,018	0.78	794
Concrete reinf., SOG, load deck	03200	032 107 0750	lb	5,002	0.60	3,001
Slab-on-grade 9 in	03310	033 130 4840	sf	6,000	2.79	16,740
Spread footers	03310	033 130 3850	cy	9	160	1,440
Loading dock concrete	03310	033 130 3150	sf	400	2.07	828
Insulating lightweight fill	03500	035 212 0250	sf	6,000	1.14	6,840
Structural steel, light	05120	051 255 4600	ton	13.5	2,560	34,560

TABLE B-1 Summary of Quantities and Costs, Divisions 1 to 14 (*Continued*)

Item	CSI No.	R.S. Means I.D. No.	Unit	No.	Unit price 1995	Cost
		Warehouse(s) (*continued*)				
Structural steel, heavy	05120	051 255 4900	ton	20	2,240	44,800
Base plates	05120	051 255 4300	lb	504	1.05	529
Joists	05210	052 110 0030	tons	13	1,504	19,552
Metal deck	05310	053 104 0440	sf	6,400	5.12	32,768
Sandblast	05120	051 255 6130	sf	6,519	0.87	5,672
Primer	05120	051 255 6520	sf	6,519	0.54	3,520
Intermediate coat	05120	051 255 6630	sf	6,519	0.78	5,085
Top coat (elevated)	05120	051 255 7030	sf	6,519	1.19	7,758
Insulation roof	07220	072 203 0100	sf	6,000	1.04	6,240
MFG wall panels	07410	074 202 0400	sf	6,000	14.53	95,898
Roofing	07510	075 102 0200	square	60	160	9,600
Flashing	07620	076 201 5800	lf	440	4.44	1,954
Ext. downspouts	07630	076 201 2500	lf	70	9.14	640
Ext. gutters	07630	076 210 3400	lf	100	13.31	1,331
Steel doors and frames	08110	081 103 0100	each	2	209	418
Steel rollup doors	08365	083 604 0010	each	2	15,188	30,376
Door hardware	08710	087 125 3110	each	2	429	858
Exterior paint	09910	099 106 2400	sf	2,216	0.54	1,197
Interior paint, block	09920	099 106 2400	sf	3,600	0.54	1,944
Dock levelers	11161	111 601 4800	each	4	7,300	29,200
Fork lifts	NA	Quote	each	2	20,000	40,000
Total						$414,704
		Office				
Site preparation (clear)	02110	021 104 2000	acre	0.16	4,640	742
Excavation, footprint	02220	022 246 4220	cy	764	5.36	4,095
Excavation, spread footing	02220	022 250 0100	cy	6	70	420
Excavation, grade beam	02220	022 254 0050	cy	45	5.95	268
Borrow	02220	022 216 0020	cy	382	3.10	1,184
Deliver borrow	02220	022 216 3800	cy	382	9.34	3,568
Spread	02220	022 216 2500	cy	382	5.22	1,994
Compaction	02220	022 226 5040	cy	382	0.88	336
Fill	02220	022 262 0170	cy	22	2.64	58
Concrete reinf., footings	03200	032 107 0500	lb	710	0.78	554
Concrete reinf., SOG, load deck	03200	032 107 0750	lb	5,170	0.60	3,102

TABLE B-1   Summary of Quantities and Costs, Divisions 1 to 14 (*Continued*)

Item	CSI No.	R.S. Means I.D. No.	Unit	No.	Unit price 1995	Cost
		Office (*continued*)				
Concrete reinf., grade beam	03200	032 107 0100	lb	1,248	0.67	836
Concrete reinf., second floor	03200	032 107 0750	lb	10,528	0.60	6,317
Slab-on-grade 6 in	03310	033 130 4820	sf	7,000	1.92	13,440
Spread footings	03310	033 130 3850	cy	6	160	960
Grade beams	03310	033 130 4260	cy	23	486	11,178
Insulating lightweight fill	03500	035 212 0250	sf	7,000	1.14	7,980
Scaffold, 2-story	01525	015 254 4100	csf	36	82	2,952
Coping, precast, 14 in wide	04210	042 116 0150	lf	120	35	4,200
Face brick common, 8 in long, 4 in thick	04210	042 184 0800	sf	2,700	9.98	26,946
Backup wall, masonry block, 8 in thick, reinforced	04220	042 216 1150	sf	2,700	6.11	16,497
Block partitions, 8 in, block, reinforced	04220	042 216 1150	sf	12,670	6.11	77,414
Structural steel, heavy	05120	051 255 4900	ton	64	2,240	143,360
Base plates	05120	051 255 4300	lb	336	1.05	353
Joists	05210	052 110 0030	tons	13	1,504	19,552
Metal deck	05310	053 104 0440	sf	7,000	5.12	35,840
Handrails, aluminum 3-rail anodized	05520	055 203 0220	if	75	42	3,150
Sandblast	05120	051 255 6130	sf	14,175	0.87	12,332
Primer	05120	051 255 6520	sf	14,175	0.54	7,655
Intermediate coat	05120	051 255 6630	sf	14,175	0.78	11,057
Top coat (elevated)	05120	051 255 7030	sf	14,175	1.19	16,868
Dampproof grade beam	07160	071 602 0300	sf	367	0.45	165
Insulate grade beam	07210	072 203 1740	sf	367	1.28	470
Insulation building	07210	072 203 0100	sf	8,850	1.28	11,328
Insulation roof	07220	072 203 0100	sf	7,000	1.04	7,280
Roofing	07510	075 102 0200	square	70	160	11,200
Flashing	07620	076 201 5800	lf	140	4.44	622
Ext. downspouts	07630	076 201 2500	lf	70	9.14	640
Ext. gutters	07630	076 201 3400	lf	100	13.31	1,331
Steel doors (3 ft × 7 ft)	08110	081 103 0100	each	8	209	1,672
Wood doors (3 ft × 6 ft 8 in)	08205	08 082 1640	each	36	314	11,304

TABLE B-1   Summary of Quantities and Costs, Divisions 1 to 14 (*Continued*)

Item	CSI No.	R.S. Means I.D. No.	Unit	No.	Unit price 1995	Cost
		Office (*continued*)				
Entrance doors	08410	084 105 1000	sf	42	21.82	916
Hardware	08710	087 125 0030	each	45	262	11,790
Glass windows	08920	089 204 5100	sf	900	23.17	20,853
Acoustical suspension system	09130	091 304 0050	sf	12,300	0.92	11,316
Dry wall partitions, metal stud, 2 sides	09260	092 608 0350	sf	17,800	1.05	18,690
Metal studs	09260	092 612 2200	sf	8,900	1.19	10,591
Drywall on block (drywall only)	09260	092 608 0150	sf	7,950	0.59	46,905
Ceramic tile	09310	093 102 0010	sf	900	14.85	13,365
Acoustic tile	09510	095 106 0810	sf	12,300	2.46	30,258
Resilient tile	09660	096 601 0100	sf	3,825	1.98	7,574
Carpet	09685	096 852 1100	sy	875	38	33,250
Exterior paint	09910	099 106 2400	sf	42	0.54	23
Interior paint, masonry	09920	099 106 2400	sf	10,640	0.54	5,746
Interior paint, drywall	09920	099 106 2400	sf	51,500	0.54	27,810
Interior paint, doors	09920	099 204 2500	sf	3,444	0.91	3,134
Metal toilet partitions	10160	101 602 0400	each	12	1,146	13,752
Lockers	10505	105 054 0300	each	20	84	1,680
Elevator	14240	142 011 1000	each	1	51,200	51,200
Total						$789,473

TABLE B-2   Division 15, Mechanical, Summary of Quantities and Costs

Item	CSI No.	R.S. Means I.D. No.	Quantity	Unit	Unit price 1995	Total
		Plant, Mechanical, HVAC				
Water pipe, 3-in steel Sch. 40 on hangers	16060	151 701 2090	700	lf	26.25	18,375
Gate valves, 3-in steel, flanged	15100	151 980 0850	4	each	1,725.00	6,900
Check valves, 3-in steel, flanged	15100	151 980 1450	2	each	950.00	1,900
Unit heaters, 140 MBH, vertical flow	15620	155 630 5080	20	each	995.00	19,900

TABLE B-2   Division 15, Mechanical, Summary of Quantities and Costs (*Continued*)

Item	CSI No.	R.S. Means I.D. No.	Quantity	Unit	Unit price 1995	Total
		Plant, Mechanical, HVAC (*continued*)				
Heating pipe (HW), 2-in steel	15060	151 701 2070	1200	lf	16.81	20,172
Gate valves, 2-in FLG	15100	151 980 0830	4	each	1,550.00	6,200
Insulation (2-in pipe)	15260	155 651 9360	1200	lf	7.00	8,400
Air handling units, 18,620 cfm, 15 hp	15850	157 290 4220	10	each	5,530.00	55,300
Total						$137,147
		Warehouse, Mechanical, HVAC				
Water pipe, 2-in steel, Sch. 40	15060	151 701 2070	200	lf	16.81	3,362
2-in gate valves, steel, flanged	15100	151 980 0830	4	each	1,550.00	6,200
Unit heaters, 140 mbh, vertical flow	15620	155 630 5020	8	each	995.00	7,960
HW pipe, 2-in steel	15060	151 701 6070	400	lf	16.81	6,724
2-in gate valves, steel, flanged	15100	151 980 0830	4	each	1,550.00	6,200
Insulation, 2-in pipe	15260	155 651 9360	400	lf	7.00	2,800
Air handling units, 18,620 cfm, 15 hp	15850	157 290 4220	4	each	5,530.00	22,120
		Warehouse, Plumbing				
Water cooler, nonrecess, wall mount, 8 gph, wheelchair type	15450	153 105 2600	2	each	1,375.00	2,750
Total						$58,116
		Office, Plumbing				
Support and carriers						
Drinking fountain	15400	151 170 0500	2	each	73.00	146
Water closet, concealed arm, flat slab	15400	151 170 3200	12	each	175.00	2,100
Urinal, wall	15400	151 170 4200	2	each	295.00	590
Lavatories, wall	15400	151 170 5400	9	each	156.00	1,404
Lavatories, vanity top, porcelain enamel on C.I., 20 in × 18 in	15440	152 136 0600	9	each	330.00	2,970
Kitchen counter top, 32 in × 21 in double bowl	15440	152 152 220	1	each	440.00	440

TABLE B-2   Division 15, Mechanical, Summary of Quantities and Costs (*Continued*)

Item	CSI No.	R.S. Means I.D. No.	Quantity	Unit	Unit price 1995	Total
		Office, Plumbing (*continued*)				
Urinals, wall-hung	15440	152 168 3000	2	each	660.00	1,320
Urinals, rough-in	15440	152 168 330	2	each	330.00	660
Wash fountain, S/S 54 in	15440	152 176 3100	1	each	2,625.00	2,625
Wash fountain, rough-in	15440	152 176 5700	1	each	520.00	520
Hot water heater, gas fired, 100 gal	15450	153 110 2120	1	each	1,250.00	1,250
Water closets, floor mounted	15440	152 180 1000	12	each	706.00	8,472
Water closet, rough-in	15440	152 180 1980	12	each	475.00	5,700
Water closet, seats	15440	152 164 0150	12	each	40.00	480
Water cooler, wall mounted nonrecessed, 8 gph, wheelchair type	15450	153 150 2600	2	each	1,375.00	2,750
Steel piping, 1 in	15060	151 401 2200	250	lf	8.75	2,188
Steel piping, $\frac{1}{2}$ in	15060	151 401 2140	100	lf	7.00	700
Cast iron, 2 in	15060	151 301 2120	40	lf	13.88	555
Cast iron, 8 in	15060	151 301 2220	100	lf	38.75	3,875
WC bends 16 in × 16 in	15060	151 320 0260	12	each	94.00	1,128
P traps, 2 in, pipe size	15060	151 181 3000	9	each	52.50	472
General service PVC, Sch. 40, $1\frac{1}{2}$ in	15060	151 551 0710	150	lf	11.94	1,910
Vent flashing, $1\frac{1}{2}$ in copper	15060	151 195 1430	4	each	37.50	150
Total						$41,001
		Office, Mechanical, HVAC				
Fire extinguisher cabinet, single, S/S door and frame	10522	154 115 1200	8	each	287.00	2,296
Fire extinguisher, 20 lb	10522	154 125 0180	8	each	206.00	1,648
Boiler, hot water, gas fired, 4488 mbh	15550	155 155 3420	1	each	35,400.00	35,400
Fin tube radiation, $\frac{1}{2}$-in copper tube, $4\frac{1}{4}$-in Al fin	15515	155 630 1200	600	lf	60.00	36,000
Steel pipe, 2 in	15060	151 701 2070	300	lf	16.81	5,043
Insulation, 2 in	15260	155 651 9360	300	lf	7.00	2,100
Chimney vent, prefab, 18 in D, metal, gas, double wall, galv.	15575	155 680 0280	30	lf	90.00	2,700

TABLE B-2   Division 15, Mechanical, Summary of Quantities and Costs *(Continued)*

Item	CSI No.	R.S. Means I.D. No.	Quantity	Unit	Unit price 1995	Total
		Office, Mechanical, HVAC *(continued)*				
Chimney vent elbow	15575	155 680 0760	1	each	194.00	194
Chiller, 30-ton, 12,000 cfm, pkg.	12,000	15655157 125	3400	1	8,095.00	8,095
Ductwork, Al	15880	157 250 0155	8,920	lb	10.00	89,200
Air handling units, 18,620 cfm	15850	157 290 4220	4	each	5,531.00	22,124
Registers, 10 in × 10 in	15880	157 470 1080	30	each	46.00	1,380
Air balancing	15880	NA	1	ls	30,000.00	30,000
Wet pipe sprinkler, steel, black, Sch. 40	15330	8.2-111	14,000	sf	3.25	45,500
Total						$281,680

TABLE B-3   Division 16, Electrical, Summary of Quantities and Costs

Item	CSI No.	R.S. Means I.D. No.	Quantity	Unit	Unit price 1995	Total
		Plant, Electrical				
Rigid galv. steel conduit 3 in	16110	160 205 1930	1800	lf	24.63	44,334
Rigid galv. steel conduit, 2 in	16110	160 205 1870	1500	lf	12.88	19,320
Motor connections, 15 hp	16050	160 275 0150	10	each	125.00	1,250
Motor connections, 100 hp	16050	160 275 1590	1	each	181.00	181
Wire, 3-conductor, 600 V, No. 4	16120	161 140 3000	2	clf	1,594.00	3,188
Wire, 3-conductor, 600 V, No. 12	16120	161 140 2200	32	clf	763.00	24,416
Wire, 2-conductor, 600 V, No. 12	16120	161 155 1500	30	clf	650.00	19,500
Terminations, No. 4	16120	161 520 4800	40	each	19.30	772
Terminations, No. 12	16120	161 520 4520	20	each	7.88	1,576
Receptacles duplex, 120 V, grounded, 20 A	16050	162 320 2470	60	each	19.38	1,163
Grounding	16050	NA	1	ls	10,000.00	10,000

TABLE B-3 Division 16, Electrical, Summary of Quantities and Costs (*Continued*)

Item	CSI No.	R.S. Means I.D. No.	Quantity	Unit	Unit Price 1995	Total
		Plant, Electrical (*continued*)				
Outlet boxes, pressed steel, octagon	16050	162 110 0020	60	each	19.88	1,193
Pull boxes and cabinets, 12 in × 12 in × 8 in	16050	162 130 0350	12	each	105.00	1,260
Telephone, transition, wall fitting	16700	161 160 2510	10	each	38.00	380
Thermostat, No. 18, 2-conductor	16120	161 155 0500	6	clf	55.00	330
Fire alarm wire	16120	161 155 1600	6	clf	175.00	1,050
Starters, MCC class 1, type B, combined MCP, FVNR, 25 hp	16300	163 110 0200	10	each	1,125.00	11,250
Starters, MCC class 1, type B, combined MCP, FVNR, 100 hp	16300	163 100 0400	1	each	3,220.00	3,220
Circuit breakers, enclosed, NEMA 1, 3-pole, 15 hp	16300	163 205 2040	10	each	320.00	3,200
Circuit breakers, enclosed, NEMA 1, 3-pole, 100 hp	16300	163 205 2200	1	each	1,725.00	1,725
Lights, metal halide, 1500 W	16500	166 115 7650	36	each	445.00	16,020
Fixture whips, ⅜-in Greenfield, 2 connectors, 6 ft long	16500	166 125 0300	36	each	21.88	788
Exit lights	16500	166 110 0080	8	each	110.00	880
Total						$166,996
		Warehouse, Electrical				
Rigid galv. steel conduit, 2 in	16110	160 205 1870	1000	lf	24.63	2,463
Motor connections, 15 hp	16050	160 275 0150	4	each	125.00	500
Wire, 2-conductor, 600 V, No. 12	16120	161 155 1500	15	clf	650.00	9,750
Termination, No. 12	16120	161 520 4520	120	each	7.88	9,456
Receptacles duplex, 120 V, grounded, 20 A	16050	162 320 2470	30	each	19.38	581
Grounding	16050	NA	1	ls	20,000.00	20,000
Outlet boxes, pressed steel, octagon	16050	162 110 0020	30	each	19.88	596

TABLE B-3   Division 16, Electrical, Summary of Quantities and Costs (*Continued*)

Item	CSI No.	R.S. Means I.D. No.	Quantity	Unit	Unit Price 1995	Total
Warehouse, Electrical (continued)						
Pull boxes and cabinets, 12 in × 12 in × 8 in	16050	162 130 0350	4	each	105.00	420
Telephone, transition, wall fitting	16700	161 160 2510	10	each	38.00	380
Thermostat, No. 18, 2-conductor	16120	161 155 0500	2	clf	55.00	110
Fire alarm wire	16120	161 155 1600	2	clf	175.00	350
Starters, MCC class 1, type B, combined MCP, FVNR, 25 hp	16300	163 110 0200	4	each	1,125.00	4,500
Circuit breakers, enclosed, NEMA 1, 3-pole, 15 hp	16300	163 205 2040	4	each	320.00	1,280
Lights, metal halide, 1500 W	16500	166 115 7650	12	each	445.00	5,340
Fixture whips, $\frac{3}{8}$-in Greenfield, 2 connectors, 6 ft long	16500	166 125 0300	12	each	21.88	263
Exit lights	16500	166 110 0080	6	each	110.00	660
Total						$56,649
Office, Electrical						
Rigid galv. steel conduit, 4 in	16110	160 205 2220	100	lf	175.00	17,500
Rigid galv. steel conduit, 1$\frac{1}{2}$ in	16110	160 205 1850	2000	lf	10.56	21,120
Motor connections, 15 hp	16050	160 275 0150	4	each	125.00	500
Motor connections, 50 hp	16050	160 275 1560	1	each	88.00	88
Wire, 3-conductor, 600 V, No. 4	16120	161 140 3000	2	clf	1,594.00	3,188
Wire, 2-conductor, 600 V, No. 12	16120	161 155 1500	40	clf	650.00	26,000
Terminations, No. 4	16120	161 520 4800	4	each	19.30	77
Terminations, No. 12	16120	161 520 4520	200	each	7.88	1,576
Grounding	16050	NA	1	ls	10,000.00	10,000
Outlet boxes, pressed steel, octagon	16050	162 110 0020	60	each	19.88	1,193
Pull boxes and cabinets, 12 in × 12 in × 8 in	16050	162 130 0350	8	each	105.00	840

TABLE B-3   Division 16, Electrical, Summary of Quantities and Costs (*Continued*)

Item	CSI No.	R.S. Means I.D. No.	Quantity	Unit	Unit Price 1995	Total
		Office, Electrical (*continued*)				
Telephone, transition, wall fitting	16700	161 160 2510	40	each	38.00	1,520
Thermostat, No. 18, 2-conductor	16120	161 155 0500	2	clf	55.00	110
Fire alarm wire	16120	161 155 1600	2	clf	175.00	350
Starters, MCC class 1, type B, combined MCP, FVNR, 25 hp	16300	163 110 0200	4	each	1,125.00	4,500
Starters, MCC class 1, type B, combined MCP, FVNR, 50 hp	16300	163 100 0300	1	each	1,800.00	1,800
Circuit breakers, enclosed, NEMA 1, 3-pole, 200 A	16300	163 205 1260	1	each	9,125.00	9,125
Circuit breakers, enclosed, NEMA 1, 3-pole, 15 hp	16300	163 205 2040	4	each	320.00	1,280
Transformer, 1500 kVA, oil-filled, pad-mounted, 15KV/277-480, secondary, 3-phase	16400	164 160 0600	1	each	37,375.00	37,375
Main service switchgear	16320	Quote	1	each	15,000.00	15,000
Switchboard	16320	Quote	1	each	12,500.00	12,500
Lighting load centers	16320	163 230 0700	4	each	788.00	3,152
Fluorescent CW lamps, Troffer grid, 2 ft × 4 ft × 3 lamps	16500	166 130 0500	150	each	155.00	23,250
Fixture whips	16500	166 125 0300	150	each	21.50	3,225
Surface incandescent	16500	166 130 4920	12	each	94.00	1,128
Receptacles, duplex, 120 V, grounded, 15 A	16050	162 320 2460	60	each	11.44	686
Total						$198,083

C

Schedule Networks for John Doe Project

These schedule networks are taken from James J. O'Brien, *CPM in Construction Management*, 4th ed., McGraw-Hill, New York, 1993.

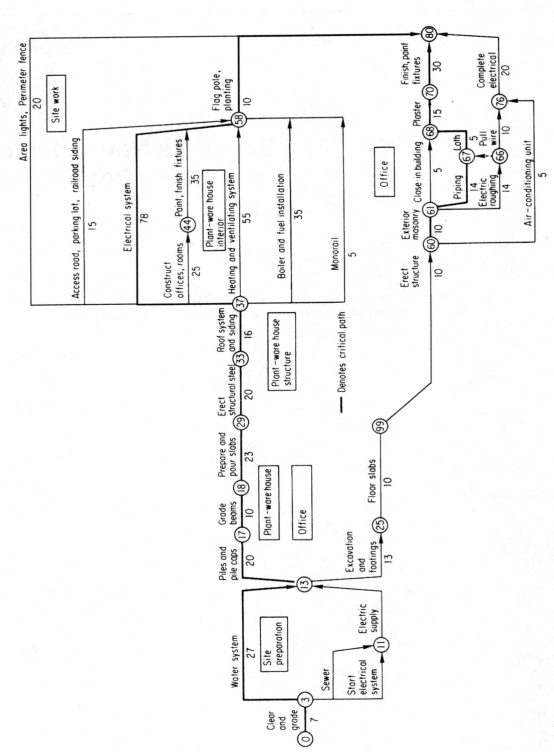

Figure C-1 Summary John Doe project network.

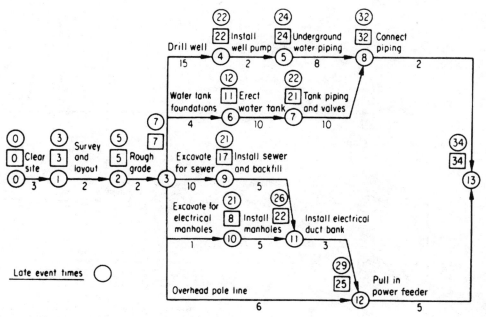

Figure C-2 Site preparation.

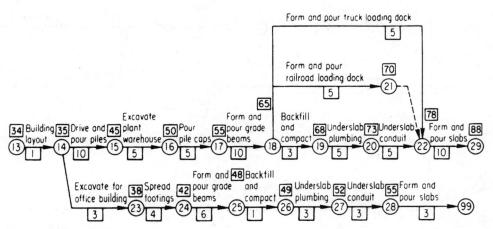

Figure C-3 Foundation network.

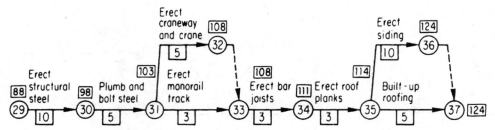

Figure C-4   Close-in of plant and warehouse.

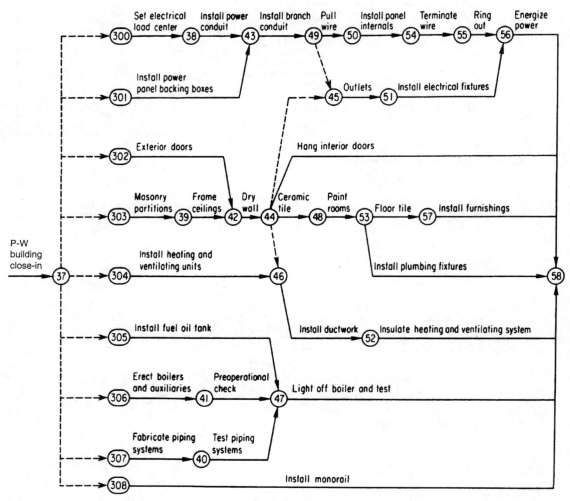

Figure C-5   CPM network: interior work, plant, and warehouse.

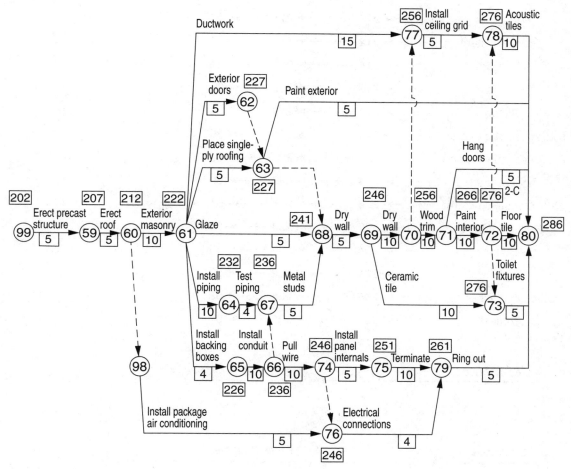

Figure C-6   Office building.

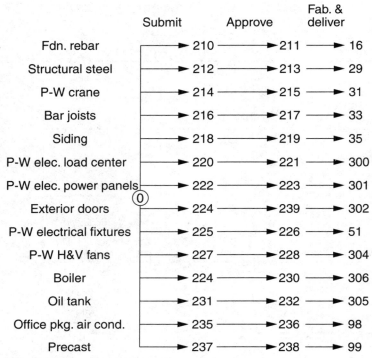

Figure C-7 Procurement network.

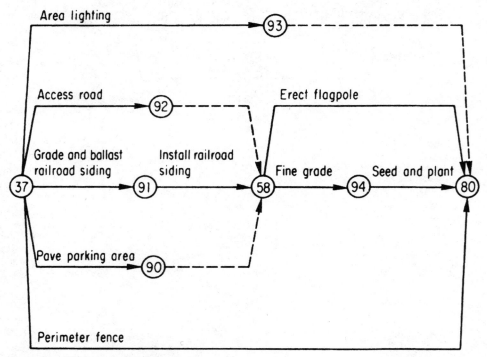

Figure C-8 CPM network: site work.

Index

ABOUT THE AUTHOR

James J. O'Brien, P.E., is Vice Chairman of the Board of O'Brien Kreitzberg, a major construction management company that has handled such projects as the renovation of San Francisco's cable car system and the redevelopment of New York's JFK International Airport. The recipient of various industry awards and honors, he is the author or co-author of numerous books, including McGraw-Hill's *Preconstruction Estimating, CPM in Construction Management, Scheduling Handbook, Value Analysis in Design Construction, Contractor's Management Handbook,* and *Construction Management.*